Unspeakable

Unspeakable

Father-Daughter Incest in American History

LYNN SACCO

The Johns Hopkins University Press
Baltimore

Printed in the United States of America on acid-free paper
9 8 7 6 5 4 3 2 1

The Johns Hopkins University Press
2715 North Charles Street
Baltimore, Maryland 21218-4363
www.press.jhu.edu

Library of Congress Cataloging-in-Publication Data
Sacco, Lynn.
Unspeakable : father-daughter incest in American history / Lynn Sacco.
p. cm.
Includes bibliographical references and index.
ISBN-13: 978-0-8018-9300-1 (hardcover : alk. paper)
ISBN-10: 0-8018-9300-3 (hardcover : alk. paper)
1. Incest—United States—History. 2. Child sexual abuse—United States.
3. Incest victims—United States. I. Title.
HV6570.7.S33 2008
364.15′36—dc22 2008043992

A catalog record for this book is available from the British Library.

Special discounts are available for bulk purchases of this book. For more information, please contact Special Sales at 410-516-6936 or specialsales@press.jhu.edu.

The Johns Hopkins University Press uses environmentally friendly book materials, including recycled text paper that is composed of at least 30 percent post-consumer waste, whenever possible. All of our book papers are acid-free, and our jackets and covers are printed on paper with recycled content.

CONTENTS

Acknowledgments *vii*

Introduction 1

1 Incest in the Nineteenth Century 19

2 Medicine and the Law Weigh In 53

3 Gonorrhea and Incest Break Out 88

4 Protecting Fathers, Blaming Mothers 120

5 Incest Disappears from View 157

6 Incest in the Twentieth Century 182

Epilogue 209

Abbreviations *229*

Notes *231*

Index *343*

ACKNOWLEDGMENTS

A midlife career change is not for the faint of heart. And it would not have been possible without the many people I encountered along the way who offered me their time, friendship, criticism, advice, and encouragement. The Women's Studies program in the Department of American Studies at SUNY-Buffalo ushered me into academia, and Elizabeth Lapovsky Kennedy, Susan Cahn, Carla McKenzie, and Minjoo Oh helped me start the process of moving from attorney to academic. Thank you.

Moving next to the University of Southern California for doctoral work in history was one of the best decisions I ever made. My dissertation adviser, Lois W. Banner, was unstinting in her efforts to support my research, push my thinking farther, help me to understand and become an active member of the profession, convince me to work on my writing, and be the strong ally that is the stuff of graduate student dreams. And did I mention she is a terrific friend? Without her mentorship I would never have finished graduate school, let alone this book. The other members of my dissertation committee, Steven J. Ross and Marsha Kinder, provided criticism and encouragement in perfect measure, and their enthusiasm for film and theory has enriched my life. I am grateful to the History Department for both supporting and challenging me, particularly Elinor Accampo, Philippa Levine, Terry Seip, Joseph Styles, and especially Carole Shammas, and fellow graduate students Roger Brown, Pete La Chapelle, and Sharon Sekhon.

I spent the 2000–2001 academic year as a Dissertation Fellow in the Women's Studies Department at the University of California, Santa Barbara, where the amazing core and affiliated faculty lavished attention on me, especially Jacqueline Bobo, Eileen Boris, Patricia Cline Cohen, Laury Oaks, Erika Rappaport, and Janet Walker. Thanks also to dissertation fellows Ruby Tapia and Sandra Soto and to history grad students Sarah Case and Beverly Schwartzberg, who invited me to join their dissertation writing group. I first met Pat Cohen at a fancy reception during a conference in Montreal, where she demonstrated the meaning of patience by kindly listening to me prattle on about my work. When I received the fellowship at Santa Barbara, Pat worked hard on my behalf, including supporting my application for a UC President's Postdoctoral Fellowship by volunteering to serve as my mentor for two years. You can't dream up people like her.

My good fortune continued when I returned to Los Angeles and Emily Abel invited me into her medical history reading group. With Emily, Janet Farrell Brodie, Sharla Fett, and Alice Wexler I discovered the Aztecs, the woman who walked into the sea, the courage and joy that is possible when career and family collide, all things tuberculosis, and the pleasures of dining near the ocean with women who know how to cook and appreciate food. Their insights reshaped my project from a dissertation to a book—and a much better book than I would have imagined without them. For their intellectual support and modeling of the importance of friendship possible between academic women, thank you.

When I left Chicago, no one would have predicted I would move ten years later to Knoxville. My transition from city girl to southerner will never be complete, but my colleagues in the History Department at the University of Tennessee, especially Janis Appier, Steve Ash, Lorri Glover, and Jeri McIntosh, have made it possible. Janis, Jeri, and Lorri read much of this manuscript. Steve and Lorri read it all, and Lorri read it more times than I can count; she alone is responsible for all errors. I found the Taoist Tai Chi Society while in Buffalo, and connecting with its Knoxville members has helped me, more than anything else, to understand and find a place for myself in this nonurban culture. With Jenny Arthur, Bob Riehl, Peter Rose, and Allen Johnson I have discovered Remote Area Medical, green burial grounds, wood-fired pottery, and every brew pub in town.

Katherine Donohue, head of History and Special Collections, the Darling Biomedical Library, UCLA, offered to introduce me to physicians who had trained in the 1930s through 1950s. Doctors Willard L. Marmelzat, Gertrude Finklestein, and Helen B. Wolff enthusiastically shared with me their insights into the diagnostic process and commented on my readings of the medical literature. Finding records for behavior that was often denied or simply unrecorded was a challenge, and one that I would not have accomplished without the sustained interest of archivists across the county. I greatly appreciated the time, thoughtfulness, and assistance of archivists and librarians, including Carol L. Bowers, Daniel M. Davis, and Rick Ewig, the American Heritage Center, University of Wyoming; Carolyn Duroselle-Melish and Ed Mormon at the New York Academy of Medicine; Jack Eckert and Molly Craig, Rare Books and Special Collections, Countway Library of Medicine, Harvard; Tony Gardner, Special Collections, California State University, Northridge; Anne Engelhart, Diane Hamer, and Ellen Shea, Schlesinger Library, Harvard; David J. Klaassen, Social Welfare History Archive, University of Minnesota; Kevin B. Leonard, University Archives, Northwestern University Library; Adele A. Lerner and Jim Gehrlich, Archives, New York Weill Cornell Center; Jeffrey Mifflin, Special Collections, Massachusetts General Hospital; Susan Mitchem, the Salvation Army Archives; John Parascandola, National Library of Medicine; Hynda Rudd, Los Angeles City Archives; Diane Shannon, Rush–Presbyterian–St. Luke's Medical Center Archives; Gerard Shorb, the Chesney Medical Archives, Johns Hopkins Medical Institutions; Dace Taube, Regional History Collection, USC; Bob Vietrogoski, Long Health Sciences Library, Columbia; Richard J. Behles, Health Sciences and Human Services Library, University of Maryland, Baltimore; Peggy Cabrera, University of Arizona Library Special Collections, Main Library, Tucson; Kenneth R. Cobb, Municipal Archives of the City of New York; James Gerencser, Library and Information Services, Archives and Special Collections Division, Waidner-Spark Library, Dickinson College, Carlisle, Pennsylvania; Stephen Kiesow, Humanities Department, Seattle Public Library, Washington; Patrick T. Lawlor, Rare Book and Manuscript Library, Butler Library, Columbia University, New York; Avril J. Madison, at the University of Washington Libraries' Manuscripts and University Archives, Seattle; Adrienne Millon, Ehrman Medical Library Archives, New York University Medical Center, New York; Robert L. Schuler, Northwest Room and Special Collections, Tacoma Public Library, Washington; staff at the Charles E. Young Research Library Department of Special Collections, UCLA.

Thanks also to the institutions and organizations that provided financial support for this project: the University of Tennessee, Knoxville; the UC President's Postdoctoral program; the Women's Studies Program, UCSB; the USC Lambda Alumni Association; the Coordinating Council for Women in History / Berkshire Conference of Women Historians; the

Schlesinger Library, Harvard; the Woodrow Wilson Foundation and Johnson & Johnson; the Western Association of Women Historians; the American Heritage Center, University of Wyoming; the Center for Feminist Research, USC; and the Haynes Foundation.

Gail Bederman and Regina Kunzel offered astute criticism of papers I gave on some of this material at conferences, as did Emily Abel and Jennifer Freyd, who read parts of the manuscript.

My editor at the Johns Hopkins University Press, Jacqueline Wehmueller, first talked to me about publishing this book with Johns Hopkins an embarrassing number of years ago and did not toss me overboard even when it seemed I would never figure out what I wanted to say and how I wanted to say it. When Rachel Devlin reviewed the final manuscript for the press, she raised questions that were tough to answer—in exactly the way you would hope another scholar would engage with your work. Thank you both. Grace Carino's careful copyediting and excellent suggestions deserve all the credit for getting the final kinks out of my writing; what a blessing, thank you.

Thanks finally to the friends and family who kept me afloat through my midlife adventure and its unexpected turns: Theresa Albini, Challis Gibbs, Joy Goldsmith, Janina Hoffman, Marcia Johnson, LV Jordan, Art Kallow, Bruce Kranzberg, Emily Moore, Cathy Sacco, Chuck Sacco, and Ivan Uldall; my fantastic "girlfriends" who put the fun into Los Angeles: Kate Rocca, Susan Spinks, and Suzanne St. Clair; and special thanks to Sara Epstein, Lucia Melito, and Christy Russell for guiding me through a year that rocked my world but that I wouldn't wish on anyone.

My parents, Charles and Marge Sacco, were proud to have their eldest child be the first in their extended families to attend college. And while the vicissitudes of my career paths no doubt raised their blood pressure, they were unfailingly supportive and encouraging. From my father I learned to be fearless, and from my mother I learned to acknowledge fear and then work through it. What a mess I would have been without them. Thank you.

Unspeakable

Introduction

By all appearances, Marilyn Van Derbur Atler had enjoyed a charmed life. Raised by parents who were highly regarded members of Denver's social elite, Atler was crowned Miss America in 1958, when she was 21 years old. After graduating college Phi Beta Kappa, she married her high school sweetheart and began a successful career as a television celebrity and motivational speaker for Fortune 500 companies. But in June 1991, Atler's celebrity took a disturbing turn. Her photo appeared on the cover of *People* magazine, emblazoned by a headline promising that she would tell all about "Her Shocking 13-Year Ordeal of Sexual Abuse by Her Millionaire Father." Atler's first-person announcement that, since childhood, her father had repeatedly sexually assaulted her seemed extraordinary. Yet her accusation, which included not only the sordid details of her childhood but also an account of debilitating psychological illnesses in adulthood that she claimed were caused by the abuse, proved to be only the tip of an iceberg.[1] Over the next two years so many women offered accounts of their own fathers' predatory assaults that Morley Safer introduced a *60 Minutes* segment on the phenomenon by observing that "there's now almost an epidemic of people who say they're incest survivors."[2]

The accusations raised more questions than they answered, particularly because of their sheer number. Was it the case that, within the privacy of their own homes, American family men behaved like the most repellent of sexual predators? If father-daughter incest was as widespread as the revelations suggested, how had it remained unnoticed for so long? Were the women's claims, as one prominent psychiatrist and an adviser to the False Memory Syndrome Foundation charged, simply the latest manifestation of "victim feminism," a female mass hysteria that owed more to the Salem witch trials and feminist ideologues than to twentieth-century science and reason?[3]

Between 1988 and 1998, these questions spun into a debate over the incidence of father-daughter incest in American homes, and they remain unanswered today.[4] Two deeply ingrained assumptions about father-daughter incest emerged as the bases of contention. The first was that father-daughter incest rarely occurred; the second was that it never occurred in white middle- and upper-class families. These assumptions were not new. From the late nineteenth century on, professional and popular culture fixed father-daughter incest in the American mind as rare behavior that occurred only among socially marginal groups: the poor and working classes, immigrants, and people of color. Marilyn Van Derbur Atler's accusations against her "millionaire father" were sensational because they clashed with the more comfortable notion that incestuous fathers were easily identifiable outcasts, like the supposedly backward and vicious men of the southern Appalachians or Missouri Ozarks, where, a writer for the *New York Times* noted in 1931, people believe that buzzards "seek out persons guilty of incest and vomit over them."[5]

How had beliefs about the racial and economic alterity of male sexuality, especially regarding a topic so "taboo" as incest, come to be so firmly established? Historians have demonstrated the inaccuracy of assuming that certain sexual activities have been uncommon because they were illicit.[6] Like homosexuality, contraception, interracial liaisons, and abortion, father-daughter incest, particularly among the white middle and upper classes, has lurked in the country's past as an open secret. But the twentieth century's reliance on psychoanalytic theory to explain sexual behavior in general and father-daughter incest in particular has obscured the historical roles played by medicine, law, and popular culture in shaping ideas about these topics.

When I looked for evidence of father-daughter incest in medical, legal, and popular culture sources, I discovered that nineteenth-century Americans were familiar with allegations of father-daughter incest. I found more than five hundred reports of father-daughter incest, published in more than nine hundred newspaper articles across the country, mostly between 1817 and 1899. The frequency with which the discovery of father-daughter incest made news suggests that, while it was not common, neither was it rare. Nor did it occur only within socially marginalized families. Until the last quarter of the century, nineteenth-century reports most often identified "incest fiends" as respectable—even prominent—

white men: clergymen, local officials, men with long and deep ties to the community, and men who had earned or inherited wealth. It was only after the Civil War, when white Americans began reacting to unprecedented immigration in the urban North and the postslavery reorganization of society in the South, that published reports about incest in destitute families and families of color suddenly increased.

Newspapers and trial court records also show that public revelations of father-daughter incest were widely discussed—not just among lawyers, court personnel, and the newspapermen who covered the trials but also by churchgoers, women's groups, and men idling in barbershops and saloons. They did not react to the revelation of incest with disbelief. When Americans discovered that a man whom they had respected was, in fact, a "brute in human form," inflamed communities quickly turned against him. If an accused man survived a lynch mob, jurors were often so eager to convict him that they did so without even leaving their seats to deliberate.[7] And record-breaking crowds eager to hear the victim testify pushed their way into courtrooms "crowded to suffocation."[8]

After the turn of the twentieth century, however, the number of newspaper reports of father-daughter incest, particularly among white middle- and upper-class Americans, abruptly declined. I was able to locate only 136 cases of father-daughter incest reported between 1900 and 1940, more than half of which occurred in the first decade of the century. Even when multiple stories about the same case are included in the count, the total number of articles increases to only 225. At the same time, public knowledge of the details of most cases dropped sharply as professionals took control over investigations and trial judges routinely granted defense lawyers' motions to close the courtroom to spectators. What had changed?

Unspeakable argues that a seemingly minor footnote in medical history—an abrupt turn-of-the-century reversal in medical views about the etiology of gonococcal vulvovaginitis, or gonorrhea infection of the vagina in prepubertal girls—is key to understanding what had changed. Doctors' interest in, and sudden rejection of, long-standing medical science concerning venereal disease—which most doctors and hospitals in the nineteenth century considered too shameful to treat, let alone discuss publicly—was a sign of the times. In the 1890s, many Americans feared that seemingly unsolvable social issues would at best upend the existing social hierarchy and at worst overwhelm the country and result in its implosion as a capitalist democracy. These issues included violent and widespread labor unrest, recurrent economic depressions, a flood of immigrants into cities lacking the infrastructure to absorb them, and African American mobility.

White, native-born, self-consciously modern Americans, including doctors who became newly interested in eradicating venereal disease, struggled to keep the dream of America afloat. They advocated wide-ranging social and political reforms that they believed could organize and contain the chaos. This period of reform, the Progressive Era, began in the 1890s and ended with World War I. The hallmarks of this era that are important for this story include a rise in professional careers and authority, a shift from health care being primarily the domain of female kin to that of licensed physicians, an insistence that the male-headed nuclear family be the central organizing element of a well-functioning society, and a nearly religious devotion to the idea that science, education, and reason could be applied to solve any social problem. Each of these factors contributed to elevating highly specialized forms of knowledge, which required education, training, and expertise, over knowledge based on personal or lived experience, or what individuals or one's community might generally accept as "common sense."

And just as many of the professions were gendered (women were thought to make ideal social workers but not medical doctors), women in particular found the social value of their experiential knowledge diminishing with increasing speed. When Progressive Era professionals set about to use their new expertise to remake society, their focus included the details of women's daily life. They instructed women on how to experience pregnancy and childbirth, how to respond to illness or disability (whether their own or that of a family member), and how to keep house—down to how to clean the toilet. Social workers coerced poor and immigrant women into adopting these new habits as part of their program of "Americanization." For women with disposable income, new trends in marketing, advertising, and mass production made it appear easy to transform their lives and improve their standing as women in a newly emerging middle class.[9] But however much they found professional knowledge useful for improving their families' health and standard of living, the process required nonprofessional women to concede some of their own authority.

This struggle over knowledge and authority, however, was not new. Doctors had long displayed their contempt for female knowledge, even a mother's observations about the health and welfare of her children. In the nineteenth century, mothers had often been the first to identify physical signs that their daughters had been sexually assaulted, most often when they discovered an unusual, foul-smelling vaginal discharge that also stained their daughters' bedding and clothing. Many of the girls had not reached puberty and were not, to their mothers' knowledge, sexually active. Understanding that their daughters might have symptoms of venereal disease, which they knew to be sexually transmitted, mothers sought confirmation from doctors, whom they asked to examine their daughters, for signs they had been raped.

Stories of eminent physicians mocking distressed mothers abound in the nineteenth-century medical literature. Doctors laughed at some of the frightened women, whom they dismissed as ignorant and hysterical. Physicians passed severe judgment on other mothers, whom they assumed were conniving liars who had injured their own daughters in a scheme to get revenge or to blackmail a wealthy man by threatening him with an accusation that would lead to criminal prosecution and social condemnation. After the turn of the century, health care professionals and social workers constantly expressed their exasperation with mothers who clung to the "old-fashioned" view that their daughters' vaginal discharges were symptoms of gonorrhea infection and evidence that they had been sexually assaulted. Doctors belittled women's popular knowledge and blamed them for their daughters' condition, pointing to their poor housekeeping and not to a male assailant as the source of their daughters' disease.

But the mothers of infected girls had not simply imagined a connection between sexual assault and vaginal infection. Elite doctors and medical jurists may have enjoyed sufficient luxury to indulge in condescending dismissals, but ordinary practitioners, whose livelihood depended on gaining the trust of the families in the communities in which they worked, often had the sad duty of confirming a mother's worst suspicions: that her daughter had been sexually assaulted and infected with gonorrhea, a venereal disease. And they did so with such frequency that prestigious medical textbooks railed against "ignorant" practitioners. But why did doctors engage in a prolonged struggle, both among themselves and with mothers, over whose knowledge was authoritative in diagnosing gonorrhea infection in girls, and why was the subject of child rape, including incest, fraught?

Western physicians considered vaginal gonorrhea a venereal, or sexually transmitted, disease. In trials at the Old Bailey in the seventeenth century, and in America by the early nineteenth century, vaginal gonorrhea infection in a girl who claimed to have been sexually assaulted was valid physical evidence (as was the presence of semen, pregnancy, or severe genital injuries) that might corroborate her accusation. Because rape is a crime that often has no witnesses, and because children may not be able or legally competent to testify as to what happened, prosecutors often relied on a medical opinion when deciding whether sufficient evidence existed to file charges against a man accused of sexually assaulting a child. In the prebacterial era, doctors had no tools to aid them in making this diagnosis. They relied only on their experience in visually assessing a girl's symptoms and their opinion of whether the girl or her mother had any motive to lie. Because medical knowledge about gonorrhea infection in women, particularly girls, was scant, doctors could not even agree on what symptoms indicated gonorrhea, let alone how to tell if a girl's vaginal discharge might be due to poor hygiene or to a nonsexually transmitted disease like pinworms. Because doctors knew that their ability to make a differential diagnosis was limited, they were often reluctant to offer a medical opinion that, if based on a misdiagnosis, could lead to the imprisonment or even the execution of an innocent man.

By the 1890s, a number of revolutionary breakthroughs in medical and diagnostic science, including acceptance of the germ theory of disease and a bacteriologic test for gonorrhea, offered doctors the ability to diagnose gonorrhea with an unprecedented degree of certainty. The new technology stimulated the interest of Progressive Era doctors and reformers who were keen to identify the source of infection so that they could help to clean up America—morally and medically—by eradicating venereal disease. Doctors and social reformers assumed, both from the phenomenal reach of an increasingly commercialized prostitution industry and from the ideology positing that only extramarital, or immoral, sexual contacts spread venereal disease, that they would be able to conclusively identify prostitution as the locus of infection.[10] And since medicine and popular culture had long associated promiscuity with venereal disease, no one would have been surprised if the new diagnostic tools did confirm a link that everyone already assumed existed.

Doctors and reformers were shocked, however, when they discovered instead that gonorrhea vulvovaginitis was "epidemic" among girls, including the daughters of the white middle and upper classes. Most infected girls had not yet reached puberty; they were between the ages of 5 and 9 and had no obvious history of sexual contact. Doctors were vexed—how had so many girls become infected with a sexually transmitted disease? As newspapers in that decade confirmed respectable white Americans' social biases by increasingly reporting the occurrence of father-daughter incest primarily among African Americans and immigrants, doctors assumed that girls from respectable white families were rarely the victims of sexual assault, especially incest. White Americans believed they had evolved into the highest form of civilization in the history of the world. And they saw the dark-skinned immigrants who both frightened and disgusted them, and the African American men whom many depicted as sexually unrestrained predators who threatened white women and girls, as primitive savages. Thus doctors readily believed that such men retained a natural proclivity to commit acts as heinous as incest.

But as the evidence increasingly pointed to men from their own class and threatened white professionals' ideology about their racial and national superiority, doctors refused even to consider the possibility that a respectable white American man, even one who was infected with gonorrhea at the same time as his daughter, could have spread his infection to her "in the usual manner." Doctors did not hesitate to identify a boarder, uncle, or brother who lived in the same residence as a possible assailant and source of infection. But when a girl's father seemed the most likely source of infection, even when father and daughter admitted sleeping in the same bed, doctors raised the possibility of incest—but only to reject it out of hand. Hard pressed to explain how else American girls had become saturated with a "loathsome" disease, doctors swiftly revised their views about the etiology of gonorrhea, not their assumptions about the sexual behavior of white Americans.

Unwilling to let medical science challenge their social biases, health care professionals and social reformers who wrote about girls' infections relied on their social biases to inform their new medical science. Physicians knew that, because gonorrhea bacteria can survive only very briefly outside a warm, moist environment, objects such as bed linens and toilet seats were unlikely to be sources of infection. But physicians also knew that the epithelial layer of girls' genitals forms a thinner, less resistant barrier to bacteria before puberty, and they used this fact to propose that it actually *might* be possible for young girls to contract gonorrhea through contact with everyday objects. Doctors saw in this slim possibility a way to acknowledge the prevalence of infection without also having to acknowledge incest. In their eagerness to avoid the implications of the evidence before them, doctors even forgot the supposed proclivity of socially marginalized men to sexually assault their daughters. Instead, from the turn of the century until World War II, they urged, in terms they admitted were often illogical, contradictory, and unproven, that it was possible, however improbable, that girls—and girls alone—were susceptible to gonorrhea infection from everyday nonsexual contacts, including their mothers' hands, bed linens, other girls, and toilet seats. In this discourse health care professionals and social reformers erased the connection between men, sexual assault, and infection by casting suspicion instead at "dangerous" objects and infected girls, whom they labeled "menaces to society."[11]

Some nineteenth-century doctors had suggested that girls might be susceptible to gonorrhea infection from unsanitary contacts, such as with a wet sponge that moved directly from the genitals of an infected woman to a girl. But they never suggested that such a narrow chain of events could explain how every girl, or even most girls, became infected. Only after they discovered that infection was widespread among girls from their own class did twentieth-century doctors expand the possible virulence of nonsexual contacts with the gonorrhea bacterium beyond the probable. Health care professionals and social reformers almost unanimously explained the fact of widespread infection among girls by pointing to the supposed ease with which girls became infected from nonsexual contacts. They had no evidence to support such a connection but repeated it so often that by the 1940s medical textbooks removed gonorrhea vulvovaginitis, gonorrhea infection of girls' genitals, from the category of sexually transmitted diseases.

Doctors were not naive to the fact that men sexually assaulted girls, including their own daughters. But doctors and reformers could not believe that incest occurred frequently enough among white middle- and upper-class Americans to account for the incidence of infection. The speculations of anxious health care professionals and social reformers pitched a noisy silence over the implications of the new data on the prevalence and incidence of infection, signaling their resolve to find a source, any source, of infection other than the girls' fathers. The silence is audible in the medical records of social welfare institutions and private doctors; in hundreds of articles in the medical, public health, nursing, and medical social work literature; in the reports of blue ribbon committees organized by charitable organizations; in scores of criminal court records and newspaper reports; and in the private and published writings of reformers of all stripes.

The silence in which early twentieth-century professionals shrouded their discovery of father-daughter incest among respectable white Americans was not the result of a conspiracy. It reflects the worldview of health care professionals and social reformers, and their desperation to sustain it, even if that meant mislabeling or ignoring evidence that they themselves had uncovered. The sexual exploitation of anyone is a repugnant act, but this book examines only discourses about father-daughter incest. I found no similar hesitation on the part of doctors to identify the sexual abuse of boys or to acknowledge that boys assaulted their sisters or that men assaulted their nieces, sisters-in-law, and granddaughters. Nor did anyone suggest that boys became infected with gonorrhea other than by sexual contacts, whether with males or females, family members or servants. When a nurse, apprentice, or servant lived with a white family, doctors often assumed they were the source of a girl's infection. Only when a "respectable" white father seemed the most likely source of his daughter's infection did doctors insist that imagining that such a man had infected his daughter through sexual contact "tried credulity."[12]

The eagerness with which doctors attributed girls' infections to nonsexual contacts stands in stark contrast to their response to the new data on venereal disease infection among the rest of the American population. In the first decades of the twentieth century, doctors joined with a wide array of social reformers to promote "social hygiene," a public health and social reform movement dedicated to eradicating venereal disease from America by, among other tactics, educating Americans about how gonorrhea is transmitted to adults and boys: by sexual contact, and not by casual contacts with toilet seats or doorknobs or by eating certain foods. Yet their intransigence on the etiology of girls' infections ran so deep that even as New Dealers launched social hygiene into federal public health policy in the late 1930s, doctors continued to insist that girls were the glaring exception to the rule.

Doing so stripped the diagnosis of its forensic value, a maneuver that made it impossible in most cases for an infected girl to receive a medical opinion that might both affirm her experience and provide the medico-legal authority she needed to successfully prosecute and protect herself from her assailant. Court records and doctors' accounts dating from the late eighteenth century show that misogynist beliefs about the credibility of women and girls often contributed to doctors' unwillingness to provide a medical opinion that supported a girl's claim that she had been sexually assaulted. Still, only when faced with evidence that strongly suggested that father-daughter incest might be widespread among the white middle and upper classes did Progressive Era doctors uniformly and summarily dismiss centuries of medical knowledge about the etiology of gonorrhea in girls in favor of wild speculation.

In the 1970s and 1980s, second-wave feminism began a far-ranging critique of assumptions about the safety of women and children within the nuclear family.[13] Recasting domestic violence from personal psychopathology to a social issue, or from a private to a political issue, was one of the feminist movement's major accomplishments. Still, the idea that father-daughter incest occurred in seemingly ordinary families, or even, as in Atler's case, families that appeared to be dreamily perfect, cast so alarming a shadow on postwar domesticity that it seemed, even among many feminists, hard to believe.[14]

Despite a flood of scholarly, popular, and self-help books on father-daughter incest published in response to the recent accusations, few historians have investigated the subject. Historians Elizabeth Pleck and Linda Gordon published groundbreaking books on domestic violence, including incest, in 1987 and 1988, respectively. In 1989 Ann Taves edited the diary of Abigail Bailey, the wife of a Revolutionary War officer, which describes Bailey's horror as she realizes that her husband is sexually assaulting her daughter.[15] However, with the exception of some histories in which incest was an important but ancillary issue, fourteen years passed before historians Irene Quenzler Brown and Richard D. Brown published a book-length account of Ephraim Wheeler's trial for father-daughter incest in early America.[16] Two years later Rachel Devlin published her nuanced work on father-daughter relationships in postwar America, which included an exploration of how psychoanalysts and social scientists reconceptualized the occurrence of incest and influenced postwar Americans' views about it.[17] Much of the other scholarly, professional, and popular literature published in response to the accusations seemed broad ranging but in fact focused primarily on debates over psychoanalytic theories or feminist politics, not the empirical reality of incest.[18] Without any social scientific context in which to make sense of such startling claims, interest in the topic diminished as the debate reached a stalemate.

Unspeakable provides this context. It stretches back in time to the late seventeenth century and moves forward to 1940, concluding, in the Epilogue, with a discussion of the recent debates. It focuses on the period between 1890 and 1940, when doctors and reformers grappled most intensely with the new data about gonorrhea infection and struggled over its interpretation. It was during this period, I argue, that public knowledge of the occurrence of father-daughter incest, and Americans' views about the types of men who seemed capable of committing it, changed. Doctors' involvement in this process temporarily abated when penicillin, which provided doctors with the first effective treatment for gonorrhea, became available at the end of World War II. With the availability of "rapid treatment," doctors no longer considered infected girls a public health problem, and the need to prevent infection by identifying the source of the disease lost its urgency. The number of articles on the etiology of the disease published in medical journals shrank until the 1960s and 1970s, when the medicalization of child abuse, skyrocketing rates of venereal disease, and second-wave feminism all combined to renew interest in girls' infections.

Chapter 1 uses nineteenth-century newspaper reports and court cases to demonstrate both the widespread occurrence of father-daughter incest, especially in respectable white families, and how knowledge of its occurrence was disseminated and understood in America. It also shows how, in the last quarter of the nineteenth century, race, class, and ethnicity began to shape ideas about what kind of man Americans thought capable of engaging in such behavior. At the end of the nineteenth century, medical and scientific breakthroughs provided doctors with the tools to more reliably diagnose gonorrhea in girls. But they did not use their improved diagnostic authority to assist victimized girls by confirming that they had been assaulted and bolstering criminal prosecutions against their assailants. Subsequent chapters explain why that did not happen. Chapter 2 traces the history of gonorrhea vulvovaginitis in Anglo-American medicine and law. Chapter 3 recounts American doctors' growing realization in the late nineteenth century of the extent to which gonorrhea had infected girls and their initial attempts to define infection as a by-product of immigration and ignorance. Chapter 4 focuses on the Progressive Era and interwar responses to the epidemic, which were marked by recasting the source of infection from sexual contacts with men to everyday contacts with mothers and household objects. Chapter 5 examines how the focus on poor sanitation eventually moved from a child's home to public spaces, especially school lavatories. This view designated infected girls as infectious agents and resulted in their increased surveillance and even incarceration. Chapter 6 turns to newspaper and courtroom accounts of father-daughter incest in the first half of the twentieth century to measure the effect of the medical profession's intervention into and authority on the issue. The Epilogue summarizes postwar views about both incest and gonorrhea and then discuss the reemergence of father-daughter incest into public consciousness in the late 1980s and 1990s.

The medical history of gonorrhea vulvovaginitis is a window through which we can see evidence of both the occurrence and the denial of father-daughter incest in twentieth-century America. However, I am not arguing that every diagnosis of gonorrhea vulvovaginitis was correct or that every infection was proof of father-daughter incest. Rather, I examine the history of medical views about gonorrhea vulvovaginitis to ask why late nineteenth- and early twentieth-century Americans found the possibility of widespread father-daughter incest among the white middle and upper classes too disturbing even to contemplate. White males cemented their social superiority in the early twentieth century in many ways. One was how white male professionals responded to the possibility of incest within their own socioeconomic class.

Early twentieth-century doctors did not intend to deny that father-daughter incest ever occurred. They freely acknowledged that it occurred in those homes where they expected to find it—among the poor, people of color, and immigrants. But an epidemic of sexually transmitted disease among girls from every type of family challenged their worldview, and they resolved this challenge by suspending their scientific beliefs in order to maintain ideological integrity. How Americans have understood, discussed, and denied father-daughter incest sheds light on the role that sexual behaviors, and assumptions about them, play in shaping Americans' beliefs about one other. Those views have provided justification for the way the social hierarchy ranks various groups, including women and girls, poor people, and immigrants and men of color.

This book also investigates how knowledge—whether professional or popular—is constructed and how gender affects which types of knowledge achieve social authority. Because authoritative knowledge, such as medicine and law in the twentieth century, can affect Americans' perceptions of what seems "true," "natural," or plausible, struggles over knowledge can determine whose reality society affirms and whose it marginalizes or even denies. The book demonstrates the power that "science" and "reason" can exert in controlling—whether by acknowledging or denying—a woman's or girl's ability to find social affirmation of her lived experience, particularly when that experience flies in the face of deeply ingrained national ideologies. Accusations of father-daughter incest among the white middle and upper classes threaten the status quo. They contradict ideologies about race, class, and gender that support and rationalize the male-headed family structure and the power and privileges that flow from it. And in twentieth-century America, they led well-meaning and earnest professionals to deny the obvious evidence before them.

Adult-child sexual contacts most often include oral contact, fondling, and rubbing, acts that do not leave lacerations, bruises, or scars.[19] Contemporary doctors therefore most often diagnose child sexual abuse based on the patient's history, not on any physical evidence.[20] Doctors have identified a number of physical indicators that raise the suspicion that a child has been sexually abused. These include infection with a sexually transmitted disease, or STD; genital injuries, bruising, scratching, and bites inconsistent with history; bloodstains on underwear; anal, genital, gastrointestinal, or urinary area pain; bedwetting and bowel issues; vaginal discharge; and pregnancy.[21] However, the American Academy of Pediatrics advises that few girls display any physical evidence of sexual assault.[22] Even in those cases in which a man penetrates (whether with his penis, his finger, or an instrument) and injures a girl's genitals, such wounds usually heal within hours or days, long before a physician may examine her, and leave no lasting marks.[23]

An untreated sexually transmitted disease can infect a child for months and even years, and the academy instructed physicians in 1998 that an STD may be the only physical evidence of sexual assault.[24] Children are not commonly infected with gonorrhea, and doctors do not routinely test children for STDs. When they do, usually to investigate an allegation of sexual assault, about 5 percent of the children test positive. Gonorrhea is the most common diagnosis.[25] This diagnosis is exceptionally useful forensic evidence because, unlike bacteria that result in systemic infection, *Neisseria gonorrhoeae* cause a local infection only at the point where they enter the body.[26] Gonorrhea infections in boys and girls appear in the rectum and throat, but even today, gonococcal vaginitis, a vaginal infection in prepubescent girls, "is the most common form of gonorrhea in childhood."[27]

The primary symptoms of gonococcal vulvovaginitis in prepubertal girls include genital itching, painful urination, and a yellow or green bloody or crusty vaginal discharge of purulent (puslike) consistency that may stain underwear and bedding in the first stages of infection.[28] Not all infected girls become symptomatic, and prepubertal girls may not test positive for gonorrhea until two weeks after contact, older girls in three to seven days.[29] Girls infected with nongonococcal vaginitis, caused by poor hygiene, local irritation, or streptococcus, may exhibit similar symptoms. But girls whose symptoms include a green or purulent vaginal discharge most often test positive for gonorrhea.[30]

In 1998, both the Centers for Disease Control and Prevention (CDC) and the American Academy of Pediatrics advised doctors to assume that "gonorrhea in children [is] sexually transmitted."[31] Doctors must rule out sexual assault before attributing the source of infection to some other type of contact. Mothers do not pass genital gonorrhea to their babies, and the bacteria do not spread by nonsexual or casual contact.[32] Because the epithelial lining of girls' genitals remains thin until puberty, doctors have speculated that prepubertal girls might be susceptible to *N. gonorrhoeae* from casual contacts with fomites, objects soiled with bacteria.[33] Some girls may have been infected by contacts with fomites when, for instance, nurses in the nineteenth and early twentieth centuries used one unsanitized rectal thermometer to take the temperature of every girl on a hospital ward. But this type of contact was neither casual nor indirect, and doctors remedied these types of practices early on.[34]

More important, *N. gonorrhoeae* survive so briefly outside a warm, moist area that although fomite transmission is possible, it is highly improbable. For the bacteria to remain virulent enough to transmit the disease, an object freshly covered with discharge would have to come into intimate contact with a girl's vagina within minutes. And although doctors cannot rule out the possibility, in the American medical literature, not one girl's infection has been traced conclusively to an object, whether a toilet seat, towel, or bed linen.[35]

Infection *has,* however, been linked to incest. In its most recent treatment guidelines, the CDC identified circumstances that constitute a "strong indication" for STD testing. These include vaginal discharge or pain, genital itching or odor, urinary symptoms, and genital ulcers or lesions; a suspected assailant who is infected or at high risk of STD infection; a sibling or another child or adult in the household or immediate environment who has an STD; the patient or a parent requests testing; and evidence of genital, oral, or anal penetration or ejaculation.[36] When family members agree to be tested (to trace the source of infection or to identify other persons in need of treatment), in nearly half of the cases an infected girl's parents or sisters also test positive.[37] Physicians who investigate the source of a girl's infection often discover that a family member or close family friend has had sexual contact with the child.[38] Because every state requires doctors to report cases of suspected child abuse to the police, a diagnosis of gonorrhea may have "serious social and medicolegal consequences."[39]

Yet few data exist with which to measure the prevalence of child sexual abuse, particularly father-daughter incest. As the U.S. Supreme Court noted in 1987, "Child abuse is one of the most difficult crimes to detect and prosecute . . . A child's feelings of vulnerability and guilt and his or her unwillingness to come forward are particularly acute when the abuser is a parent."[40] Reporting agencies estimate that at least two-thirds of the victims of sexual assault never contact the police, and many agencies that collect such data do not delineate father-daughter incest from other types of sexual abuse.[41] Even if the incest is reported, there is no uniform interstate system that tracks reports made to welfare or law enforcement agencies, and in some states the police do not routinely investigate reports filed with welfare agencies.[42] Still, authorities were able to substantiate nearly 150,000 cases of child sexual abuse in 1992, a number equal to 15 percent of all reported cases of child mistreatment (including all types of abuse and neglect). In 1998 the number of substantiated cases of sexual abuse dropped to almost 104,000 and has continued to decline, although researchers cannot yet explain why.[43]

Most children who are sexually abused are girls who were assaulted by their father or other male head of household. When researchers at the Crimes Against Children Research Center at the University of New Hampshire studied the aggregate data from twelve states collected for 1997 by the National Child Abuse and Neglect Data System, which measures and tracks all types of child maltreatment, they concluded that male caretakers—fathers, stepfathers, and a mother's unmarried partner—were responsible for 92 percent of child sexual assaults reported to the police. Boys and girls suffer nonsexual assaults and kidnapping by caretakers at almost equal rates, but 80 percent of sexually victimized children are girls.[44] Findings in a study published by the U.S. Department of Justice in 2000 showed that the largest category of victims reporting child sexual assault comprises girls under the age of 12. Most were assaulted in their own homes by an adult male who was either a family member or close family acquaintance. Yet the Justice Department concluded that crimes against children had the lowest arrest rates for all categories of sexual assaults.[45]

A startlingly high number of women report having been sexually assaulted in childhood, including by their fathers.[46] When psychologist David Finkelhor, a professor of sociology and director of the Crimes Against Children Research Center at the University of New Hampshire, reviewed retrospective studies of child sexual assault (surveys of adults asking about their childhood experiences), he concluded that rates of child sexual assault may be at least three times higher than reports suggest. In his review of nineteen studies published in the United States and Canada from 1980 to 1991, Finkelhor found that as many as 62 percent of women surveyed claimed to have been sexually assaulted as girls. He concluded that the most reliable estimate is that at least one in five American women has a history of childhood sexual abuse. As many as half of the women respondents identify their abuser as a family member, most often their father or male head of household. The risk of someone outside the family assaulting a girl sexually before adolescence is slight but rises through her teen years.[47]

The percentage of women reporting a history of father-daughter incest is even higher among women who are incarcerated, seek psychiatric treatment, engage in prostitution or severe drug use, report a history of rape, or are runaways.[48] Women outside nonclinical settings may underreport because they are unwilling or unable to acknowledge that it happened.[49] Still, as many as 25 to 30 percent of college women surveyed report histories of incest and child sexual abuse.[50] There is little research suggesting that race, class, or ethnicity affects a girl's risk of sexual assault or incest, though such factors may affect her willingness to disclose the abuse or of governmental authorities to investigate it. Retrospective surveys confirm that rates of child sexual assault are relatively equal when compared across race, class, and ethnicity.[51] Race and ethnicity may affect the types of physical and emotional symptoms of incest a woman suffers.[52]

The most tragic consequences of father-daughter incest are neither ideological challenges nor gonorrhea infection, however widespread. In contrast to uncertainties about the prevalence and incidence of child sexual assault, researchers are in "relative unanimity" about its psychological consequences.[53] Child sexual assault so profoundly disturbs a girl's sense of bodily integrity and her developmental process that she may suffer from both immediate and long-term physical, psychological, and emotional problems.[54] The age at which the assault occurred or began can make a difference, as can the identity of her assailant. When a girl's own father is her assailant, she may experience more severe psychopathologies than when a stranger is the assailant. Fathers have not only a unique relationship to girls who are their daughters but also unique access, particularly when the girls are young. Fathers generally victimize their daughters more than once, often repeatedly over a period of weeks, months, and even years. Such long-term abuse compounds the psychological suffering a girl experiences from the initial breach of the father-daughter relationship.[55]

Psychological symptoms of sexually abused girls include "sexualized behaviors" not seen in girls without a history of sexual assault. These include sexualized play with dolls, putting objects into their anus and vagina, public masturbation, seductive behavior including requests for sexual stimulation from adults, and age-inappropriate sexual knowledge. Their behavior often includes self-injury (cutting), cruelty, delinquency, running away, aggression, and regressive behavior such as bedwetting and tantrums. Health care professionals often diagnose victimized girls with post-traumatic stress disorder (PTSD), low self-esteem, high levels of anxiety, fear, depression, withdrawal, sleep issues and nightmares, school and learning problems, and neurotic mental illness.[56] Girls receiving treatment in clinical settings for other reasons may display the same symptoms at similar rates, except for sexualized behaviors and PTSD.[57]

Left untreated, many of these symptoms continue throughout life, and at each stage of development new symptoms may emerge. In puberty and adolescence, a girl who has been sexually assaulted is at increased risk for early and unsafe sexual activity, including less birth control efficacy, and a younger age of menarche, voluntary intercourse, and birth of first child. The effects on their sexual behavior range from aversion to preoccupation.[58] And teens may shift from cutting to suicidal behaviors and substance abuse.[59]

Women who were victimized as girls suffer from many psychiatric disorders, higher rates of medical problems, lower self-esteem and sense of self-worth, and altered body image. The abuse may even affect a woman's desire to have children and how she perceives her value as a long-term mate.[60] Victimized girls often develop eating disorders, including anorexia and obesity, as they grow up.[61] And the risky sexual behaviors in which they often engage, including unprotected sex, multiple partners, and exchanging sex for drugs or money, put women sexually assaulted in childhood at significantly higher risk for STD infection, including human immunodeficiency virus.[62]

CHAPTER ONE

Incest in the Nineteenth Century

NINETEENTH-CENTURY AMERICANS had a surprising familiarity with accusations of father-daughter incest. But over the course of the century, ideas about the type of man white genteel Americans thought capable of committing incest changed, as did views about the need to protect and nurture children and the consequences of male intemperance. This chapter begins to examine father-daughter incest by looking at how newspapers and court records identified and discussed its occurrence in all types of homes. These documents often reflected Americans' differing views about social respectability and class status, particularly before the Civil War. After the war, race and ethnicity came into sharper focus as white males struggled to protect their social power and white women developed a stinging critique of men's ability to protect women, including their own daughters.

Beginning in the late eighteenth century, the discovery of father-daughter incest in towns across America provoked indignation and a swift response. Word spread so quickly that local men, who apparently did not pause to consider the truth of the allegation, raced to the perpetrator's home, eager to tar and feather the accused before casting him out of town.[1] Accused men who reached police custody, including white men of "good circumstance," found themselves under heavy guard against lynch mobs—and long odds.[2] The mob that grabbed A. A. Stegall, a white man accused of incest, out of a Texas jail "hooted and shouted" and "laughed and leaped with joy" as they hung him.[3]

Men fortunate enough to see their day in court sat in rooms "crowded to suffocation," with 200, 500, or even 1,000 people jostling for seats, transforming a fact-finding hearing about a family nightmare into a public holiday that usurped business as usual.[4] If a nervous defendant chose at the last minute to avoid a trial by pleading guilty and throwing himself on the mercy of the court, throngs of "eager spectators" reacted as though a "wet blanket" had dropped over them.[5] Even after a jury acquitted

Albert H. Essex, a respectable, well-known New England merchant, the mob that swarmed the courthouse—so thick that it took Essex an hour to leave the building—showered him with stones and chased him to the train station. Some of the crowd followed Essex onto the train and hounded him until, they claimed, he jumped off.[6]

Newspaper editors had few qualms about reporting father-daughter incest, sometimes in terms so routine as to suggest that editors considered its occurrence neither extraordinary nor surprising enough to be newsworthy. In the late eighteenth century, American newspapers began to augment their standard coverage of incest in European society and literature with reports of father-daughter incest closer to home.[7] On July 28, 1794, the *Hartford Gazette and the Universal Advertiser* matter-of-factly reported that local officials in Danbury, Connecticut, had arrested Moses Johnson, "a noted thief," for having raped his two eldest daughters. The one-sentence story ran on page 3.[8] In the three weeks before Johnson was tried, at least seven newspapers republished the story among columns of sundry other events.[9] A reader of the *Norwich Packet* might have missed the brief notice, which was printed above a more detailed account of the damage wrought by a summer thunderstorm.[10] Only the editors of the *Columbian Centinel* and the *Oracle of the Day* thought Johnson's crime worthy of elaboration. They placed an exclamation point after the first sentence and lamented, "O Nature! How art thou at times degraded."[11] When Johnson was convicted of attempted rape, the court's sentence was harsh: life at hard labor—an opportunity, the six papers that carried the story drolly noted, for Johnson to perfect the art of nail making.[12] But public interest in Johnson did not end with the conclusion of the case. When, on his way to prison, he wrote a letter to his wife that proclaimed both his love for her and his innocence, the *Columbian Centinel* published it for all of Boston to read, on page 1.[13]

Many newspapers both reflected and encouraged the "profound sensation" that rippled through communities that discovered father-daughter incest occurring in their midst.[14] When the *Portland Gazette and Maine Advertiser* reported in 1805 that a jury had convicted Ephraim Wheeler of the rape of his 13-year-old daughter, it declared the "particulars [of the crime] too horrid for newspaper publication."[15] Perhaps, but even if they failed to deliver the salacious details themselves, newspapermen told their readers where to find them. Publishers, booksellers, and businessmen who astutely measured the public's appetite for sordid detail heavily promoted their ability to deliver. The editor of the *Northern Post* printed a letter by one enterprising citizen with the foresight to have hired a man to take notes at Wheeler's trial. He promised that the product of their labors, a "Report of the Trial" consisting of "100 octavo pages, *more or less,*" would soon be available.[16] Sixteen different boxed advertisements for trial reports, on sale by subscription before the trial and in book shops after, appeared in at least eleven newspapers published from New York City to Maine. Some ads ran for as many as nine weeks.[17]

Advertisements for trial reports left no doubt as to the case's appeal. As one emphasized, "The particulars of this unnatural transaction are so shocking in their nature that public utility not only warrants but demands a publicity of facts." To entice anyone still uncertain whether the eighty pages promised would be worth the twenty-five-cent subscription fee, the notice teased that "the Trial occupied 13 hours," an extraordinarily long duration in that era, which subscribers might reasonably have imagined to have been filled with electrifying testimony.[18] After Wheeler's execution in the winter of 1806, newspapers affirmed both the significance of the case and readers' interest in it by reporting that "not less than 5000" people turned out to witness the hanging.[19] Yet even Wheeler's death failed to sate the public's appetite for lurid detail. That spring, the Springfield Bookstore was one of many businesses that capitalized on the public's long-standing interest in criminal executions. It advertised for sale a pamphlet entitled *Narrative of the Life,* which also included a report of Wheeler's trial and the sermon preached at his execution.[20]

Even if we assume that newspapers inflated the number of people absorbed with the trials of Moses Johnson and Ephraim Wheeler (whose execution occurred during an intense political debate about capital punishment), how can we account for the sustained interest their crimes provoked? It might appear that the discovery of incest captured public consciousness because the crime so rarely occurred, and someone who searched the historical record looking only for arrests and prosecutions might conclude this was the case. In her meticulous search of Connecticut court records for the 150 years before 1789, historian Cornelia Hughes Dayton found only three criminal prosecutions for father-daughter incest.[21] In her study of rape in the nineteenth-century South, Diane Miller Sommerville also found only three cases of father-daughter incest, two of which involved stepdaughters, all occurring before the Civil War.[22] In 1852, New York City reported only one prosecution and conviction for incest, the same rate of prosecution for the crime of voting twice in the same election.[23] Three years later, the only person charged with having committed incest in Brooklyn was Ellen Vaughn, who was sentenced to sixty days in the penitentiary for "feloniously intermarrying with James Thompson," presumably a relative by blood or marriage.[24] In 1886 the rate of prosecution for incest in Brooklyn was unchanged—one.[25] In Los Angeles the following year, the sheriff arrested two people for rape, seven for attempted rape, two for incest, and twenty-three for indecent exposure.[26] In 1888 the *Los Angeles Times* complimented the police for their "creditable showing" of five hundred arrests in May, which netted mostly drunks and gamblers but also one arrest for incest, none for rape, and two for insanity.[27]

There are few criminal records from which to determine the incidence of father-daughter incest and no indexes from which to calculate the number of men who were arrested and prosecuted nationally. But even if such indexes existed, they would contain only an incomplete accounting. Some accused men fled before the sheriff could arrest them.[28] When the sheriff caught up with Lewis Gorman, he shot himself to avoid capture.[29] Other men committed suicide in prison while awaiting trial or sentencing.[30] Richard Goodwin, an "old negro," waited until his trial began and then "drew his 'Barlow' knife" in the South Carolina courtroom and "began to hack away at his own throat" as the spectators sat in stony silence and watched him die.[31]

Although there are some records of cases that went to trial and resulted in a conviction, there is no way to know how many other cases ended in a verdict of not guilty, a plea agreement, or a conviction on another or lesser charge. No one officially recorded testimony in many courts, and there is no system for locating the transcripts that random counties preserved.[32] Appellate court opinions are more readily available but cannot be used to estimate the number of men prosecuted. Only defendants found guilty filed appeals, and not all of them would have had the legal right or the personal will or means to do so.

Evidence from other sources, however—newspapers in this chapter and medical discourses in the next—shows that Americans regularly detected and publicly acknowledged father-daughter incest. In 1802, Dietrich von Bulow, a German national who published an account of his tour of the early Republic, reported that venereal disease was common among Americans and that adultery, incest, and murder were "rife in every part of the union."[33] Prominent Americans endorsed von Bulow's observations about Americans' disgraceful habits, which provided fuel for their own political ends, and saw that his work found a wide audience. John Quincy Adams translated von Bulow's book *Travels in America*, which newspapers across the eastern seaboard, including the *Port Folio*, the premier Federalist literary and political journal, published serially.[34] Yet not every American relished von Bulow's account. The enraged editor of the Philadelphia newspaper the *True American* complained that the "churlish and misanthropic" foreigner had caricatured Americans' behavior and character for the sake of scandal, charging that, "without the title page, [his travels] would appear a description of an uncivilized horde of Hottentots, or Cannibals."[35] But though other papers repeated the *True American*'s attacks on the "viscous Prussian," von Bulow had backed up his charges with a list of specifics, such as the countryman near Lancaster, Pennsylvania, who "had children by his daughter," incidents von Bulow claimed were "generally known, detailed in all the newspapers, and never contradicted by any one."[36]

Searching microfilm, electronic databases, and archival collections nationally, I was able to confirm von Bulow's impressions. I found a small number of trial transcripts, a larger number of appellate court cases, and a vast number of newspaper articles reporting father-daughter incest, many of which also describe a seemingly endless pool of contemporaries drawn into public discussion of the crime: family members, neighbors, doctors, lawyers, judges, jurors, policemen, newspapermen, reformers, clergymen, and churchgoers from Bangor to San Diego.

Taken together, these sources show that incest was neither rare nor the behavior of a few or easily identifiable miscreants. As noted in the Introduction, I found more than five hundred different newspaper reports of father-daughter incest occurring in the United States published between 1769 and 1899, most between 1817 and 1899. Because newspapers sometimes followed a case over a period of days and more than one newspaper might report or reprint the same story, the total number of stories published about these incidents nearly doubled, to more than nine hundred. A few involved stepfathers and daughters, or adopted daughters, but I excluded countless additional cases that reported incest but did not specify the relationship of the parties or those that involved grandfathers and granddaughters or men who served as putative father to sisters-in-law, nieces, or other girls who lived under their roof.

Some stories contained only a brief notice and no details. But others tracked the course of the investigation and reported extensively on the trial, and I located trial transcripts and appellate court records for some of these claims. I am not suggesting that every accusation was true. For most reports there is no way to determine the veracity of either the accusation or a newspaper's account, whether the reports reflect the frequency with which incest occurred, or how fully or fairly they convey the details of the charge, defense, and public response. And juries found some men not guilty, appellate courts overturned many convictions, and newspapers occasionally reported that a daughter's claim proved to be false or that she had retracted her accusation, sometimes after her father had already been convicted or even committed suicide.[37]

What is clear from the newspaper and court reports is that incest was neither so unusual as to seem outlandish on its face nor too shameful for public consumption. Until the end of Reconstruction, nearly all reports of father-daughter incest involved native-born white Americans, and accused men came from every walk of life. In the century before psychoanalytic theory defined incest as a psychological or pathological event, Americans were upset to learn of its occurrence. But they did not necessarily recoil from the news with shock or disbelief. The pages of nineteenth-century newspapers were teeming with stories of murder, deceit, betrayal, and mayhem in all types of American families, and stories of men sexually assaulting children, mostly girls but also boys, abounded.[38] In this context, news of father-daughter incest was neither a stretch of the imagination nor any more implausible than the regular accounts of violence and lust with which they shared column space on the pages of both local and national newspapers.

Newspapers sometimes mentioned incest only because it was a factor in another legal proceeding. Stories reporting that a wife had filed for a divorce sometimes identified her discovery of father-daughter incest as the grounds.[39] Investigations into murder sometimes also led to the revelation of incest. Trying to determine who had killed Chauncey Knapp, a Minnesota farmer, the police learned that he had fled from home four months earlier after being caught trying to rape his stepdaughter. When Knapp returned, his brothers-in-law and nine other men from the community took him out to the barn and offered him five hundred dollars to leave permanently. When he refused, the men dragged Knapp into a pond and, after repeatedly asking whether he would leave and ducking him when he refused, drowned him.[40]

Other stories reported that men had murdered wives and daughters who threatened to reveal the incest or daughters who refused to submit to their assaults.[41] Some men were charged with having procured an abortion or committing infanticide in a desperate attempt to hide babies they had fathered with their daughters.[42] In 1871 at least eleven newspapers reported on the trial of Joel W. Perkins, a Connecticut man who confessed that for ten years he had repeatedly raped every one of his six daughters. A jury convicted him of murdering three of the six babies that resulted.[43] In the last quarter of the century, a few accounts portrayed the daughter as the protagonist: young women who killed fathers they claimed had assaulted them; daughters who married their fathers or stepfathers, sometimes bigamously; and assertive young women who mocked the police or reformers who tried to save them.[44] The police charged one seemingly remorseless young woman, whose mother's "horribly mutilated" body was found in a river in Leavenworth, Kansas, with having hired the man who killed her. The daughter had apparently wanted her father all to herself.[45]

Why did newspapers so frequently publish such stories? Genteel Americans expected "heinous" and "beastly" behavior from the lesser sorts: dirt-poor farmers, men of color, rough-mannered immigrants, and dissipated drunks and scoundrels. Reports of incest in socially marginalized families provided genteel Americans with proof that their social biases were justified, and they were useful to the moral, temperance, suffrage, and civic reformers who used these tales of degradation to promote their own causes.[46] Newspapermen and judges rarely exhibited any sympathy for men accused of incest, referring to them as "unnatural fathers" and "inhuman brutes," names they readily applied to assorted villains, fiends, and ne'er-do-wells.[47] In 1883 the *National Police Gazette* suggested that "monsters" who commit incest should receive "some nice, neat Chinese punishment," like being chopped into pieces or flayed alive.[48] Reporting on the suicide of a wealthy and influential farmer charged with incest, the headline in the *Gazette* approvingly proclaimed, "He Did Right."[49] And if a young woman tried to bring her father to justice by killing him herself—well, all the better.[50]

After Reconstruction, newspapers on both sides of the Mason-Dixon reflected the struggle over a rapidly shifting social landscape by trumpeting allegations of incest as proof of the hypocrisy and moral degeneracy of the other side.[51] Southern newspaper editors often placed stories about African American men accused of raping girls, whether their own daughters or white girls, on the front page, and they applauded the lynch mobs that went after them.[52] In 1886, Charles Hudson, an African American man who had languished three months in a Missouri jail, appeared in court and pleaded guilty to incest with his 18-year-old daughter, after which he returned to his cell to await sentencing. The sheriff stated that Hudson then saturated his clothing and bedding with coal oil and set himself on fire. Although the sheriff's detailed account seemed to contradict his claim that he wasn't aware of Hudson's actions until it was too late to save him, newspapers carried his description of the "Negro Pyre" on page 1.[53]

News of father-daughter incest was also good business. In 1887, when the New York Court of Appeals affirmed a father's conviction for incest, it omitted the customary recital of facts, fearing that to repeat the "disgraceful history . . . would imperil the calmness and cleanness which belong to a judicial record."[54] But the publishers of the *National Police Gazette*, a scandal sheet, found a market for salacious family intimacies among the more than five hundred thousand men across the country who passed the paper to one another in taverns, boardinghouses, and barber shops.[55] Historian Howard Chudacoff describes the content of the *Gazette*, which was printed in pink paper after 1878, as a view of a "reality that existed below—but not too far below—prevailing standards of respectability."[56] News accounts of incest provided a superficially respectable way for publishers to disseminate—and for readers to consume—titillating content under the pretext of denouncing it.

In 1867, the *Gazette* reported on the "considerable excitement" among "the few who are in on the secret," an incest accusation against a "wealthy merchant, one of the first business men in this city, a member of the Board of Trade, and . . . one of our first churches, who is the head of a highly respectable family." The *Gazette* described the victim and her circumstances in idealized terms, as "an interesting young girl, about 14 years of age, who was adopted into the family when quite young, and has received the same care and attention . . . as though she were their own daughter." Nodding first to conventional morality and romantic family sentiments by reprimanding the "inhuman wretch, who should have been the protector and guardian of his ward's virtue," the reporter then delivered a primer on seduction: "[He] commenced using those arts and persuasions which only such human fiends know . . . to induce her to gratify his unholy lusts. At first she was startled and grieved, but gradually and almost imperceptibly he succeeded in arousing her slumbering passions and removing her conscientious scruples, until . . . she fell a victim to the base passions of her seducer."[57] The purple prose sandwiched between exclamations of disgust betrayed the purported news value of the gossipy story.

When they crowded together in excited courtrooms or enjoyed the *Gazette* in homosocial venues where they drank beer or got a shave, American men shared the thrill of "unprintable" details with one another. Men who read about incest and elbowed their way into courtrooms to watch lawyers tease out explicit details from reluctant and shamefaced girls and young women may have felt that their position in the spectators' box demonstrated their superiority to the defendant and his depraved sexual desires. When men condemned incestuous behavior, they also asserted their own virtue as manly citizens. But newspaper accounts that describe the pleasure of male spectators absorbed in the pornographic details of incest suggest otherwise. As the *Tacoma Daily News* noted repeatedly during its coverage of the trial of Thomas Bower, who was caught in a hotel bed with his unrepentant teenage daughter, the judge had to lock the courtroom doors to stop the crush of men pushing to join the lucky few inside who "eagerly listened" to "unpublishable testimony."[58] The *Chicago Tribune* more pointedly described the spectators at one hearing as "an audience eager greedily to seize upon every disgusting detail . . . and manifesting all that love for the scurrilous and depraved which such an exhibition is likely to produce."[59]

The sensational manner in which newspapers sometimes reported accusations among genteel families reinforced the notion that incest was not a behavior expected of men whose reputations and claims to social privilege rested on their apparent mastery of self-restraint. The *National Police Gazette* took special delight in reporting on such cases, which tied neatly into both its editorial posture of exposing the hypocrisy of respectability and its business model of stoking the class resentments of its readers with a naughty story.[60] As the *Gazette* intoned when it reported a bastardy charge filed by the stepdaughter of an "excellent citizen" and leading member of a prominent church, the "greatest rascals assume the heaviest coat of pretended righteousness."[61]

In April 1846, when the editor of the *Gazette* learned that Daniel Burnett, a wealthy 52-year-old New York City butcher, was arrested for the "unnatural and revolting *crime of incest with his own daughters!*" he made no attempt to conceal his glee.[62] Burnett's son, a Wall Street broker, filed the charges on behalf of his adult sisters, which the *Gazette* pronounced the talk of the town, an assessment the trial judge confirmed when he noted that a charge of incest against a man who "has been long and most favourably known" had intensified public interest in the case.[63] The *Gazette* seized the opportunity to profit from the scandal by hiring someone to provide its readers with an "exclusive" transcript of the first hearing in the case, a probable cause hearing. Most probable cause hearings are routine matters during which the prosecutor demonstrates that he or she has evidence sufficient to at least minimally prove the prosecution's case. Burnett's probable cause hearing lasted nearly a month, weeks longer than most murder trials at the time, and ended with the judge dismissing the case for lack of evidence.[64] But because the *Gazette* knew that its readers found much pleasure in the "wonderful and surprising details" of the scandal, it immediately compiled its coverage of the case into a twenty-four-page booklet that was "neatly bound in ornamental covers, and accompanied by a spirited engraving." The booklet went into three printings, and the *Gazette* dutifully reminded readers of the case whenever other respectable men tried to deflect an accusation of incest with evidence of their good character.[65]

Most newspapers, however, matter-of-factly reported accusations against genteel white Americans, who constituted the single largest group of defendants before the 1890s. For every "hardened wretch" or impoverished cooper charged with incest, two wealthy doctors, merchants, farmers, or other churchgoing men were also charged with having raped their daughters.[66] Whether the accused was the lowest or loftiest man in town seems to have had little effect on whether his neighbors deemed the story credible. The social prominence of the unnamed man described in the *Gazette*'s "insider's" story had not led the paper to dismiss the allegation as improbable. Rather, the reporter had reminded readers that underneath the father's social trappings he might be "guilty of an outrage which stamps him as a villain of the deepest dye."[67]

Similarly, when the Reverend William S. Douglass, a Maine preacher, was arraigned in 1846 on charges that he had raped his two daughters, the court set his bail at seven thousand dollars, a sum that ensured he would remain in prison.[68] At trial, Douglass defended himself first by besmirching the reputations of his daughters and then by offering the testimony of "several brother clergymen" who "swore to his unimpeachable character." Still, the jury needed only a "short absence" to find him guilty.[69] In another case, the court accepted the guilty plea of the Reverend F. A. Strale, a 36-year-old Presbyterian minister who taught at the Female Seminary in Great Bend, Pennsylvania, for having attempted to rape his stepdaughter. But the judge scoffed at Strale's request for mercy—based on his "literary attainments"—and sentenced him to five years.[70] The neighbors of W. H. Hurd, a prominent Sioux City politician and businessman, did not waste time considering his guilt or innocence. As soon as they heard that Hurd's daughter had accused him of incest, they set out to lynch him.[71] And when an Ohio newspaper reported in 1834 that a woman from a "wealthy and respectable family" had discovered her husband's "diabolical course of incest" with her three daughters, it encouraged townspeople to shun and detest him "as one who entails upon society a blighted and blasting influence . . . more destructive, than the raging epidemic or depopulating plague."[72]

All incestuous fathers broke the social contract, for which they received a damning public rebuke. But race and class shaded the consequences. Genteel white men who raped their daughters displayed only their own individual weakness. When a genteel man's neighbors discovered the truth, their anger reflected the insult of having been deceived by a "brute in human garb" whose manhood was only superficial and therefore unworthy of respect.[73] Men who initially stood by an accused friend withdrew once they realized the enormity of the charge, and they stood silent when asked to pay his bail.[74] Having been unmasked as a fraud, incestuous fathers were expelled from their social class, a move that both affirmed the values of genteel society and restored it to wholeness. But whereas a genteel white man suffered mostly from a loss of social status, poor men and men of color, whom genteel white Americans believed defined the lower sorts by their lack of manly ability, represented their class. This uneven response to what everyone agreed was inexcusable behavior cost some men their freedom and others their lives but left the social hierarchy intact.

Unlike prostitution, incest was not an activity in which an otherwise respectable man might discreetly indulge. Genteel Americans valued respectability because it was the result of considerable effort. Leading a life of self-restraint was a never-ending process, and any number of missteps could ruin a man's reputation as both a father and a man. A man caught sexually assaulting his daughter had not merely stumbled and committed an embarrassing personal gaffe. Whereas a man could swear off drink and redeem himself from intemperance, the disclosure of incest tore away completely the curtain of respectability to reveal a man whose behavior had sunk so far below acceptable conduct that he forfeited the respect and company of genteel society and the privileges that followed.

Since the colonial era, law and custom had granted fathers extraordinary social privileges over those whom the state placed under their governance and control: wives, children, slaves, servants, and apprentices. In return, fathers were charged with protecting and nurturing their dependants.[75] The rape of any woman or child reflected poorly on the manliness of the perpetrator. He had abandoned his duty to protect women from harm, displayed a lack of sexual restraint that resulted in a woman's ruin, and insulted his victim's father, husband, or male protector, who had a right to sue for monetary damages.[76] When newspapers reported that a man had committed incest "*upon his own daughter*," they underscored the line he had crossed that magnified the venality of his conduct, destroyed the family's privacy, and required public intervention.[77]

Before sentencing Thomas Johnson, a widower convicted of having repeatedly raped his 16-year-old daughter Lavinia, the judge listed Johnson's failings as a man. At least four papers carried the story, including the *Farmers' Repository*, which printed the judge's entire opinion. The court first reminded Johnson that women have a right at all times to be protected from improper assaults. A man who could force himself on a crying woman without taking pity on her weakness and innocence was, the judge told him, a "monster in human shape, possessing the form and figure, without the heart and feelings common to man." The court continued to rebuke Johnson's manliness by pointing out that by committing incest Johnson had failed in "the most sacred and solemn obligation . . . to bring up this child in the paths of innocence . . . and to be her faithful adviser, guardian, and protector." Berating Johnson for having become instead his daughter's "worst enemy," the court finally sentenced him to fourteen years at hard labor.[78]

As the number of cases of father-daughter incest implies, ideologies about manliness and restraint were not enough to protect girls from predatory fathers. Idealized notions of manhood and respectability did work, however, to delay the exposure of incest to public, and even household, view. Ephraim Wheeler was a poor, uneducated man who had a history of not supporting his family and beating his wife. Yet even a man who had uncontrovertibly failed to earn the respect of his neighbors still had paternal authority at his disposal. Wheeler used his to rape his daughter a few feet into the woods on an empty stretch of road.[79]

Like Ephraim Wheeler's daughter, victimized girls frequently confessed that they had submitted to the incest and kept it secret because their fathers had threatened to beat or kill them and other family members—with guns, knives, whips, or clubs—if they did not.[80] George Skyles told his 14-year-old daughter that he would cut her heart out if she told.[81] Doke Tribble, who lived outside Nashville with his three children, repeatedly raped his older daughter, who had scars from the knife wounds he inflicted on her when she tried to resist. When his younger daughter also resisted him, Tribble first threatened to kill her with a large knife and then raped her in front of her sister and brother, telling them that if they exposed him he would kill them, too.[82] Some men made good on their threats.[83] John Jones, a widowed Chicago meat packer who had placed three of his seven children in an orphanage because he could not support them, poisoned his eldest daughter when she threatened to disclose the incest.[84] When 19-year-old Ollie Kitzelman refused her father's demand to "cohabit" with him, he shot her.[85]

Other men tried to avoid the public scrutiny that an unmarried pregnant daughter or bastard child invited by murdering their daughters or newborn babies.[86] Some young women tried to escape their fathers' assaults by marrying a suitable partner, a move that sometimes only deepened their fathers' rage.[87] When Edward Parr discovered that his daughter Maggie had married "Ready" Gamble, Parr hunted her down. He tried twice to kill her, succeeding the second time by plunging a shoe knife into his daughter's body at least twenty times.[88]

Men who had achieved higher social status than that of Ephraim Wheeler also exploited their paternal authority both to abuse their daughters and to escape the consequences. Some men used their money and influence to purchase their freedom through financial settlements, to flee, or to jump bail, leaving their families to bear both the scandal and the debt.[89] Attorney Hiram A. Potter tried to avoid suspicion by relocating with his daughter from Kansas to Arizona, where he opened a new practice.[90] J. T. E. Johns tried to avoid prosecution by paying a young man to marry and move away with his daughter before investigators could interview her.[91] Other men, who feared the loss of social status more than death, committed suicide to "avoid the disgrace" of a daughter's imminent public disclosure.[92]

Newspaper accounts also repeatedly discussed the situations and the responses of mothers who discovered that their husbands were sexually assaulting their daughters. Such knowledge could be costly; some husbands threatened, beat, or killed wives who stood in their way.[93] And some genteel wives and daughters resisted or delayed publicly disclosing the incest because maintaining the appearance of respectability, and the privileges that accompanied it, was important to them as well as to the husbands. Yet many mothers took effective action to protect daughters. They contacted their extended families for help, filed divorce or criminal complaints, and corroborated their daughters' testimony in court.[94] Mothers and daughters often also found strong support not only from other women but from other men who were members of the family or household, including uncles, grandfathers, sons, brothers, sons-in-law, and boarders.[95]

Still, a genteel mother who publicly disclosed the incest in order to protect her daughter might receive a withering rebuke, rather than support, from her community. When Mrs. Thomas Jeffries discovered that her husband had raped their 13-year-old daughter Leora, who became pregnant, Mrs. Jeffries immediately told her relatives, who reported the crime to the district attorney. Thomas admitted what he had done, but his confession did not end the suffering he inflicted on his family; Leora died from complications during childbirth. However, Mrs. Jeffries's neighbors viewed her decision to make the scandal public worse than the immensity of her loss. The only leniency the Willstown Baptist Church showed her was the opportunity to resign her membership before being read out of the church.[96] Fearing a fate like that of Mrs. Jeffries, mothers with the means to do so tried to both protect their daughters and avoid scandal by sending their daughters to live out of town.[97]

Respectability and the appearance of ordinariness could deflect any suspicion of incest, sometimes for years. The cases that newspapermen considered the most vexing emphasize this point. When a "well-to-do" Indiana farmer was indicted for incest in 1869, the accusation seemed so remarkable that newspapers from Kentucky to South Dakota reported it.[98] The *Union Dakotaian* recounted the grand jury testimony that led to an indictment against a 65-year-old farmer for the rape of all six of his daughters and at least one granddaughter.[99] The incest had continued for so long that the eldest daughter testified that she had no idea "whether her children were the offspring of her husband or her father."[100] The prosecutor was so flabbergasted by the accusations that he considered the possibility that the eldest daughter, his own witness, was deranged, a suggestion her neighbors firmly put to rest. Yet what the newspaperman considered most surprising was "that the old man . . . has for years been a member of the church, and has always been upright and honorable."[101] Had the farmer been a known criminal or drunkard, the prosecutor could have easily woven the daughters' accusations into existing rationales about the type of man unable to meet local standards of manly respectability. A community's distress arose not so much out of an unwillingness to believe that a genteel man was capable of such a thing as out of its own inability to discern his perfidy, and the progeny of it, which had been circulating in public view.

Incest, Law, and Criminal Prosecution

Daughters, wives, families, and communities responded in a variety of ways to the discovery of father-daughter incest, sometimes looking to the law for protection and redress. But whether a community responded by vigilantism or an orderly criminal trial, legalistic responses failed to address the gap between domestic ideals, which protected and promoted the existing social hierarchy, and the reality of the lived experiences of victimized girls, whose narratives challenged the integrity of paternal authority and the privileges that flowed from it.[102] White Americans from the colonial era through the nineteenth century generally viewed the family as having one unitary interest, which a presumably capable male head of the household both defined and represented.[103] Absent a monarchy and a system of inherited status, the family was a central organizational element of the United States, and fathers were responsible for maintaining order and discipline within their "little commonwealth."[104] In this system, the "state did its utmost to support [the families'] rulers in the proper exercise of their authority," though the definition of what contemporaries considered "proper" often changed.[105]

The distribution of power within American families has also been more dynamic than is implied by the consistency of its dominant pattern: a male commanding a household of dependents.[106] The control the male head of family exercised over various members of the household (including slaves, children, and wives) steadily, if unevenly, decreased over time. But an allegation of father-daughter incest instantly ruptured the community's presumption that a father had capably exercised his paternal authority. It signaled a breakdown of paternal responsibility and order, and because these formed the foundation of the social system, this breakdown demonstrated an urgent need for community intervention.[107]

Recognizing that some men would fail in their responsibilities, the British colonists enacted laws prohibiting incest. Most colonies adopted the language of the Old Testament, either directly or as interpreted by English ecclesiastical law, which declared sexual intercourse between certain family members a crime.[108] The first such law, in the 1639 Articles of Confederation between the Massachusetts, Plymouth, Connecticut, and New Haven plantations, made incest a crime punishable by death.[109] But the goal of most colonial legislatures was not to punish fathers who had coercive sexual relationships with their minor daughters. Rather, they hoped to avoid complicated inheritance issues that arose when kin related by affinity married, such as a widower marrying his wife's sister.[110]

Still, by 1900 every state had approved laws prohibiting various forms of incest, including that between father and daughter.[111] Yet because the language and intent of the laws in most states were vague, courts wrangled over whether they applied only to consensual marriages between adults and not to cases of child sexual assault.[112] As late as 1878 the Supreme Court of Louisiana asked in frustration, "But what is incest? It has not, like murder, a fixed and definite meaning everywhere."[113]

As a consequence, few men who sexually assaulted their daughters were punished before the 1890s under a criminal law prohibiting incest. The U.S. Census for 1880 counted only 121 inmates convicted of incest among the national prison population of 58,000, and some of these may have been guilty of marrying an adult relative rather than assaulting a child. By 1890 the prison population had increased to more than 82,300, and the number of inmates serving time for incest had nearly doubled. Yet the 222 inmates serving time for incest, which included 8 women (most likely marital partners), accounted for less than 1 percent of convicted felons.[114] Nearly twice as many (390) prisoners were incarcerated for adultery, 1,300 for rape, and more than 7,300 for murder.[115] Other data confirm the census numbers. One Michigan county, for example, did not prosecute a single case of father-daughter incest over the entire nineteenth century.[116]

Appellate court records, which are more accessible and better preserved than those from lower courts, confirm that many men who were prosecuted and convicted of incest were not always punished. My search of electronic legal databases found 150 appeals of criminal incest cases in the nineteenth century.[117] Of these appeals, 104 involved fathers and daughters: 82 biological fathers and 22 stepfathers. Appellate courts reversed the father's conviction in almost half (46) of the cases.[118] In California, the supreme court heard only 8 cases involving father or stepfather and daughter incest and reversed convictions in half.[119] Texas reported the greatest number, 34 cases, but most states reported only 1 or 2, and 9 states did not record a single criminal appeal for incest for the entire century.[120] I did not include additional criminal cases involving child sexual assault that lacked the formal relationship required to prosecute a man for incest, such as men who assaulted minor sisters-in-law living under their roof.[121] And I omitted court cases involving murder, slander, and divorce that arose out of the revelation of father-daughter incest.[122]

The dearth of prosecutions and convictions and the overturning of many convictions on appeal seem to suggest that men enjoyed virtual immunity from punishment for committing incest. But a man who sexually assaults his daughter may break any number of laws, and prosecutors also charged fathers under laws that classified rape (defined as forcible sexual intercourse), statutory rape (sexual intercourse with a girl under the age of consent), attempted rape (no penetration), and sodomy as criminal behavior.[123] Some of these crimes also carried a higher penalty than did most incest statutes, including capital punishment. But the harsher penalties and higher burden of proof required for a conviction under a rape statute may have decreased the likelihood of conviction and so discouraged prosecutors from pursuing some cases.

Prosecuting incest generally required the state to prove only that there had been sexual contact between father and daughter, and the defense had few options other than to challenge the daughter's credibility or, at a time when public records were sparse, her paternity. A prosecution for forcible rape, however, also required that the victim prove that she had not consented to the contact. Some states considered incest the proper charge if the female appeared to consent to the sexual contact, and rape the appropriate charge if the contact occurred without the female's consent or expressly against her wishes.[124] Those states that considered a willing female an accomplice to incest also required additional witnesses to corroborate the girl's account of what happened because they viewed the testimony of one accomplice against another as insufficient to prove the crime.[125]

Legal consent is especially problematic in cases of adult-child sexual contact because it assumes some parity of strength and reason between victim and assailant. In cases of alleged child sexual assault, the prosecution must overcome substantial evidentiary hurdles, such as whether a child is legally competent to swear an oath to tell the truth or can explain in sufficient detail what transpired.[126] A child may not have understood the nature of a man's advances; may have appeared to consent only because she was afraid of, or taught to be obedient or to defer to, adult males; or may have been too physically immature or weak to meaningfully resist. Such evidentiary rules were designed to protect the defendant's right to a fair trial, and prosecutors may have declined cases that involved very young girls or those without independent witnesses. Most newspaper articles and court records involve teenage girls and young women, victims old enough to have had the wherewithal both to seek help and to satisfy a judge that they were competent to testify.

Nineteenth-century doctors were pivotal witnesses in rape cases, and they set a high standard of proof on the issue of consent, insisting that it was physically impossible for a man to rape a woman who "resisted to the utmost." A defendant would go free unless the victim could show the type of physical injuries that demonstrated she had fought so hard that her assailant could accomplish the rape only after resorting to severe violence.[127] Some men exploited this loophole. Mid-nineteenth-century New Yorkers sometimes defended themselves against a charge that they had sexually assaulted a girl by claiming that, in the absence of physical injuries, they had not "hurt" the girl.[128] A girl's father might have been even keener to avoid visible injuries that could lead to questions or raise the suspicion of incest.

But any man, especially a girl's father, might have found physical coercion unnecessary if a girl lacked the physical strength or ability to rebuff him or the emotional maturity and courage to defy him. In an era when many girls lived in virtual isolation and did not regularly attend school, some were too ignorant or mentally impaired to protect themselves.[129] In a front-page story in the *Santa Fe New Mexican* in 1899, a bewildered pregnant teenager told police she had submitted to her father's sexual demands because she thought it was legal for her to fill her deceased mother's place in his affections.[130]

Even though men who assaulted their daughters exploited their dominance over their families, most courts refused to consider the impact of paternal authority on a daughter's ability to resist. Whether the family was rich or poor, urban or isolated, the circumstances in which men chose to sexually assault their daughters followed repeated patterns.[131] In most cases, the incest was not an isolated event but repeated over days, weeks, months, and years.[132] Some men attacked more than one daughter.[133] A few assaulted their daughters in the presence of the girls' mothers, but most waited until their wives had died, were absent from the home, or were too ill to effectively protect their daughters.[134] C. D. Sharpe raped his daughter the day after her mother died, and James Watson took advantage of his wife's trip to the 1893 Columbian Exposition to assault his stepdaughter.[135] That some men waited until their wives' temporary or permanent absence from the home suggests not only that the presence of the girl's mother could deter a man from sexually assaulting his daughter but also that fathers had the ability to restrain themselves from committing incest, at least until they judged the circumstances most favorable for them to succeed without getting caught.

Even in the absence of a maternal protector, many girls resisted their fathers as best they could. Some men had to use their physical dominance or weapons to force a daughter into submission. Courts recognized such coercion, both in accomplishing the assault and in keeping it secret.[136] In 1886, the Virginia Supreme Court recognized that a girl was trained to view her father as "her only protector and guardian [and] to yield obedience . . . to [him]."[137] But other jurists did not consider whether a girl's ability to repel her assailant might be diminished if she was attacked in her own home by her father. Most courts followed the logic of the Arizona Supreme Court, which in 1895 went so far as to declare the fact of a father-daughter relationship irrelevant to the issue of consent in a prosecution for rape.[138]

To avoid the pitfalls of the issue of consent, a prosecutor could charge the father under a "statutory rape" law, removing the issue of consent entirely. However, before the 1880s and 1890s, only a few states recognized an "age of consent," an age below which the law presumes that a girl is developmentally incapable of consenting to sexual contact, regardless of her apparent willingness.[139] In 1787, for instance, the New York legislature declared "carnal knowledge" of a female under 10 years old to be a crime punishable by death, a penalty it later reduced to life in prison. In the early nineteenth century, Illinois and Massachusetts also specified child rape as a capital crime, and other states mandated prison terms from ten years to life.[140] North Carolina hanged Franklin Smith in 1867 for having raped the daughter of his common-law wife, a girl under the age of 10, whom he also infected with a venereal disease.[141]

Infecting his daughter with venereal disease may have been a factor in the severity of Smith's sentence. In most states the punishment for incest or statutory rape did not include capital punishment. Men convicted of sexually assaulting their daughters were seldom executed unless they were also guilty of having murdered their daughter or her baby.[142] But they still paid a steep price. Trial courts, filled with members of the defendant's community, readily convicted and imposed harsh sentences, often life in prison.[143] The court sentenced George Godfrey to life in prison for raping his 8-year-old daughter.[144] When Charles Thompson sought the court's mercy by pleading guilty to having raped his daughter, he still received a life sentence.[145] New Mexico sentenced Francisco Vallegas to prison for sixty years.[146] James Neal Winthrow was nearly 60 years old when he was convicted, but the court still sentenced him to thirty years.[147] And though Henry Thompson was sentenced to only ten years, he served it in solitary confinement.[148]

Appellate courts, however, overturned many convictions. Appellate judges, who try to reconcile law with ideology rather than facts, could not, or would not, believe that American men like themselves were capable of committing incest. Some convictions surely rested on flimsy evidence or flaunted due process. Juries were often so eager to convict that they rendered their verdict at the first opportunity, sometimes not even bothering to leave their seats to deliberate.[149] In 1820 a jury took only twenty minutes to find Thomas Dickerson guilty of raping his stepdaughter and to sentence him to death.[150]

However, prominent jurists were often more concerned with upholding assumptions about social class than with the integrity of the legal process. In 1849, for instance, a Texas jury convicted Mr. Tuberville of incest with his 14-year-old daughter. Tuberville was a respected member of the community, but his marriage and social status collapsed after his wife learned of, and reported, the incest. But the justices of the Texas Supreme Court could not reconcile the accusation of incest with Tuberville's reputation as an honorable man. It was not that the court thought incest unimportant. Rather, in language similar to that used by the *National Police Gazette*, the justices condemned incest as "shocking to the moral sense of every civilized being, so degrading . . . to human nature, reducing man from his boastful superiority of a moral, rational being to a level with the brutal creation."[151] Yet precisely because it viewed incest as the behavior of people "less than human," the court refused to "believe it possible to have been committed in this age and country" by a civilized white American.[152]

To acknowledge that a man of Tuberville's standing had committed incest, not only would the court have had to accept Mrs. Tuberville's testimony over that of her husband's, but it also would have had to admit that the differences between white gentlemen and other males might not be as great as it liked to imagine. In the antebellum South, where the supposedly natural superiority of genteel white males over all others was a key justification not only for slavery but for the intricacies of the social hierarchy, this was impossible.[153] To avoid an outcome so dissonant from their beliefs, the Texas justices chose instead to conflate white male character with national virtue and strove to protect both by overturning the verdict.

They ruled that there was no evidence "that our country has been degraded by the commission of so loathsome, so heartsickening an offense in our midst."[154]

In appellate court opinions, the *idea* of incest affirmed the inherent superiority of genteel white males by representing a fictional border they did not cross. At a time when evidence of a defendant's character and reputation was an important element of a criminal defense, the rationale of the Texas court insulated an entire class of white males from suspicion. In return for the privileges of his gender, race, and class, a genteel man was supposed to respect his marriage vows, protect his family—especially the innocence of his daughters—and perfect his self-restraint.[155] Incest violated all these standards. By setting Tuberville free, the Texas justices sought not only to restore the reputation of one southern gentleman but also to mend the rift between ideology and reality that the verdict against him had exposed.

Appellate court opinions emphasized the squalor of the lives of socially marginalized defendants, pointing to the men's failure to provide for their families as evidence of their unfitness as men, a justification for the state to intrude into and remove the father from his home to a prison. Appellate justices noted the primitive locations in which the assaults occurred: cotton fields, hog pens, and barns, locations that also refuted any suggestion of consent. Emphasizing the distance between gentility and the families in which incest occurred, justices ominously recited the grim physical conditions in which defendants lived, including the number of people sleeping together in a one-room house or "cabin" where family members and strangers might share only one or two beds.[156] In an 1891 appeal filed by a widower who lived with his four children in a one-room log hut that was 13.5 square feet, the Nebraska Supreme Court remarked that incest was not surprising in a "family steeped in wretchedness."[157]

By the end of the century most men imprisoned for incest were from socially marginalized backgrounds, which confirmed both their exclusion from manly respectability and the proclivity of the "lower classes" to engage in such heinous behavior. Before the end of Reconstruction in 1877, the vast majority of newspaper reports of father-daughter incest involved respectable, financially comfortable, and even well-to-do families. In the last quarter of the century, however, three factors reshaped incest as a marker in the broader cultural conversation among genteel white Americans about race, ethnicity, and manliness. The first was the struggle over how to remake southern culture and the role of racial difference in American society; the second was the response of native-born white Americans, particularly in the North, to the influx of immigrants; and the last was a shift from the nineteenth-century idea of manliness to twentieth-century masculinity. As part of the process in which manliness was parsed into masculinity, genteel white Americans began to believe that male privilege was attached less to a man's qualities, such as self-restraint and discipline, and more to a man's social status, particularly his race or ethnicity and class.

As they reconceptualized gender as a natural function of race and class, white Americans elevated ideology over lived experience. They promoted the idea that respectable white men were intrinsically unable to engage in behavior like incest, unlike other, less civilized, "primitive" males, who they imagined retained a natural proclivity for "beastly" conduct. As this process took hold, the credibility of allegations against respectable white males appeared increasingly improbable, while allegations against socially marginalized men seemed not only credible but expected. And genteel Americans could nod to assumptions about the type of men who committed incest as proof that the social position of both the privileged and marginalized was deserved.

Of the 214 men serving prison sentences for incest in 1890, only 3 claimed to have had professional occupations. Most reported having worked in agriculture (82), unskilled labor (55), or manufacturing (48).[158] Kathleen Parker characterized most of the defendants in her sample from Michigan as having such poor social skills and bad judgment that they seemed self-destructive.[159] Many cases came to public attention when a girl became pregnant or gave birth and the local prosecutor inquired as to the baby's paternity, a public intervention that a well-off family may have had the means to avoid.[160] As Nell Irvin Painter observed in her essay on the sexual abuse of slaves, "Then and now, family violence and child sexual abuse are usually concealed, and the people with the most privacy, the wealthy, are better at preserving their secrets than poor people, who live their lives in full view of the rest of the world."[161]

In search of protection against an abusive father, mothers and daughters from poor and immigrant families may have had little choice but to report his behavior to the authorities. But genteel girls and women with more extensive financial, familial, and community resources did not necessarily have to publicly disclose the abuse in order to stop it, and they may have weighed the consequences of public disclosure differently than did less fortunate women. A genteel wife may have been unwilling to complain about a type of domestic misconduct that would almost certainly result in scandal, publicize the fact that her daughter had been ruined, lead to her husband's arrest and possible incarceration and the resulting loss of earnings and property, and destroy the family's social cachet. And in an era when women were painfully aware that they had few legal rights, genteel wives may have felt pessimistic about the outcome of any court proceedings they initiated against their husband.

The late eighteenth-century journal entries of Abigail Bailey demonstrate these concerns. Abigail lived on a New Hampshire farm with her husband, Asa Bailey, a well-regarded man of property and knowledge who had served as an officer in the Revolutionary War.[162] In 1788, Abigail was 44 years old, had been married to Asa for twenty-one years, and had given birth to fourteen children.[163] That year, she began writing about her growing awareness that Asa was sexually assaulting their 16-year-old daughter Phebe. Even though Phebe was too ashamed to discuss the matter, Abigail was "fully convinced of the wickedness" and was distraught over how to protect her.[164] When Abigail mustered the courage to confront her husband, he taunted her, reminding her that she "was under his legal control."[165] Phebe resolved the immediate problem by moving away, but Abigail, who considered Asa's breach of his marital vows irreparable, had to enlist her brothers, her neighbors, and members of her congregation to help her to protect her other children and to obtain a separation from Asa and a fair property settlement. Asa felt so confident of his ability to determine his family's future that he ignored them all and continued to mock Abigail, even after she finally had him imprisoned. But Abigail recognized when she finally had the upper hand. Visiting Asa in prison, she reminded him that he would shortly face criminal prosecution for incest, then a capital crime in New Hampshire, and, after years of deception, he quickly agreed to her terms and vanished from her life.[166]

In contrast, Hannah Wheeler, a mixed-race woman living in Massachusetts in 1805, lacked any financial comfort, and her family's social status left her with few options with which to respond to her daughter Betsey's statement that her father had raped her. Ephraim Wheeler was a

43-year-old farm laborer who had been orphaned as a child, could not sign his name, and earned a living too meager to support his family. Hannah had separated repeatedly from Ephraim because he beat her, and when she learned of the incest, Hannah sought community intervention. She filed charges on behalf of her 13-year-old daughter and testified in support of Betsey at Ephraim's trial. Not a single witness came forward, as was common at the time, to testify to Ephraim's good character. The jury quickly convicted him, and the judge sentenced Ephraim to death.[167]

Historians Irene Quenzler Brown and Richard D. Brown, who uncovered the case, were amazed that the "community took a daughter's accusations against her father so seriously that it sent a man to the gallows."[168] The Browns found only two other incest cases in the Berkshire County sessions of the Supreme Judicial Court of Massachusetts between 1790 and 1810, and none in which a man was executed for incest. Public sentiment against the death penalty was high, but the governor refused to stop Ephraim's execution, even though Hannah and Betsey had added their signatures to a petition signed by one hundred men asking that he commute Ephraim's sentence to life in prison.[169] The Browns attributed Ephraim's fate not only to his marginal social standing but also to Betsey's persistence in maintaining her accusation against him.[170] Yet Betsey's testimony would have carried little evidentiary weight absent Hannah's unwavering public support for her daughter. Mothers like Hannah frequently instigated criminal proceedings, but a conviction often rested on their willingness to also provide corroborating testimony.[171] The Wheelers lacked the social power that Abigail Bailey had exploited to manipulate her husband, and Hannah shared none of Abigail's concerns about a fair division of their property. Hannah had nothing to lose by prosecuting her husband immediately and no other apparent means of protecting her daughter.

Genteel women may have had more choices for protecting their daughters, but they were constrained by the punishing social consequences that might follow. When they chose to remain silent about the incest, genteel mothers became complicit, however inadvertently, in maintaining social constructions about gender and respectability that they knew to be unfounded. But even if she had wanted to, how could a genteel woman have acknowledged a reality so anomalous to the ideals of manliness, womanhood, and domesticity? Even today, psychiatrist Judith Lewis Herman argues that incest accusations present such a sharp challenge to conventional beliefs that they are taken seriously only when a political movement like feminism exists to support a radically different view of social relations.[172] Until the late nineteenth century there was no social movement that could support, affirm, or explain a counternarrative as provocative as an accusation of incest against a respectable man.

The Child Protection Movement

In the last quarter of the nineteenth century, the child protection and temperance movements recognized the gap between ideology and lived experience and worked ambitiously, if not always effectively, to narrow it. Both movements promoted the notion that children had a right to physical safety from abusive parents, and they shaped new remedies that improved the ability of girls to receive legal protection and redress from incestuous fathers. Charitable interest in child neglect and abuse developed into an organized child protection movement after 1868, when reformers created organizations to investigate, catalog, and publish information about crimes, including sexual assaults, against children. The intense urban poverty that accompanied immigration, urbanization, and industrialization frightened white native-born Americans. Self-styled "child savers," who viewed all foreign-born or poor parents with suspicion and disdain, believed that regulating the behavior of the underclasses and Americanizing immigrant and poor children were key to reducing urban ills and reestablishing order.[173]

Historians who have documented the zeal and selectivity with which child protection agencies identified child abuse and its remedies have raised troubling questions about reformers' methods and motivations.[174] Yet despite its many shortcomings, the child protection movement succeeded both in justifying the idea of state intervention into domestic life and in changing the ways that Americans, including victimized girls, conceptualized the meaning of childhood. In Boston, poor mothers and daughters sought help from the Massachusetts Society for the Prevention of Cruelty to Children (MSPCC), where they found advocates with both the clout and the determination to intervene effectively to end the incest.[175] In Manhattan the New York Society for the Prevention of Cruelty to Children (NYSPCC), organized in 1875, was instrumental in identifying and prosecuting cases of child sexual assault. By the 1890s the district attorney ceded all authority over allegations of child sexual abuse to the society, which used its resources, including doctors who quickly developed an expertise in examining girls who claimed to have been sexually assaulted, to investigate accusations and decide whether to prosecute.[176]

But the activities of the child savers worked inadvertently first to hide and then to erase father-daughter incest in homes outside their purview. Child protection advocates assumed that genteel parents naturally met the ideal standards of nurturing and protecting their children and so did not pry into their domestic lives.[177] For their part, genteel families responded to the ideas about childhood that protective workers espoused and became increasingly "child-centered." As the notion of children's inherent innocence and the need for a protective, nurturing environment took hold, genteel parents sought to protect their own children from corruption.[178] No one questioned their ability to do so, particularly in an era in which respectable white women espoused their intrinsic high morality as central to successful childrearing and domestic stability.[179]

Child protective agencies focused instead on the problems confronting children from socially marginalized families, whose parents they viewed as intrinsically immoral or too ignorant or "foreign" to know and do better. When Elbridge T. Gerry, the founder of the NYSPCC, argued in 1895 that incest was on the rise, he linked its occurrence to the influx of immigrants, whom he called the "offscourings of the criminal classes of Europe." Gerry claimed that immigrants sought asylum in the United States both to avoid prosecution at home and to "enact here the crimes with which they are familiar in their own land."[180] Because child savers such as Gerry believed that crimes like incest were "not so much as to be named among Christians," they readily acknowledged incest only where they expected to see it, among the socially marginalized families over whom they wielded virtually unchecked social control. In turn, this one-sided process of data collection provided them with ample evidence that seemed to affirm their belief about a natural association between social class and respectable behavior.[181] As a result, even though the numbers of men who were prosecuted for child sexual assault and incest rose at the end of the century, the vast majority of them fit an increasingly constricted profile defined by class, race, or ethnicity.

Female Moral Reform Movements

In the early nineteenth century, female moral reform societies initiated a critique of male sexuality and its relation to social power that activists later incorporated into the women's rights, suffrage, temperance, and purity movements. Female reformers were deeply offended by male sexual license and understood male privilege as so pervasive that it subordinated not only women's sexuality but also women's social and economic power. Women who joined moral reform or anti-vice societies in the 1830s often focused on prostitution, which they viewed as both the literal manifestation of male sexual power and the perfect symbol of unequal gender relations. But some women reformers who were less interested in individual moral uplift or religious virtue than in social and legal reforms broke with their peers over strategies and goals. Rather than target fallen women for moral reform, they identified men as the aggressors in prostitution, and, hoping both to shame individual men into changing their behavior and to rouse public antipathy against them, they published the names of men who visited brothels.[182]

Antebellum female temperance reformers took a similar approach. They discussed intemperance as a problem primarily of the poor and working classes, and they condemned the "brutish" men who had not mastered self-restraint. Antebellum temperance reformers avoided any open discussion of topics as unseemly as father-daughter incest, but it was implicit in their references to domestic violence.[183] By the 1860s, women in various reform movements had become less reticent about incest. Officials at the Lancaster State Industrial School for Girls and the New York House of Refuge, for instance, noted incest as a matter of course when they inquired into the life histories of new inmates.[184] But reformers began to publicize domestic violence and incest only after the Civil War, when Elizabeth Cady Stanton and Susan B. Anthony included such incidents as examples of "outrages" against women.[185] In the 1870s, activists Lucy Stone and her husband, Henry Blackwell, the editors of the *Woman's Journal*, published reports of "crimes against women," including child sexual abuse, to support their demands for expanded women's rights and suffrage.[186] But these were still only isolated voices, and no one before the 1880s developed a critique of male sexual power as it pertained to the sexual abuse of girls within their own homes.

The first reformers to do so were the members of the Woman's Christian Temperance Union (WCTU). At the height of its power in the 1880s and 1890s, the WCTU organized a successful national campaign for age-of-consent legislation. When WCTU president Frances Willard learned that girls constituted a high percentage of the victims of sex crimes, she concluded that the individual men whom society had entrusted with the task of protecting girls—their fathers or male guardians—were inadequate to the task.[187] And at a time when economic opportunities drew vulnerable adolescent girls and young women away from the apparent protection of their homes and into anonymous urban centers, for which they were unprepared, female reformers feared for the moral and physical safety of young women outside their homes. Historian Sarah Deutsch argues that, to shore up their own moral authority, which rested on the value of the domestic "shelters" they created, white middle-class women in the late nineteenth century emphasized the vulnerability of women and girls to the lecherous, unscrupulous, and nameless men they might encounter on city streets.[188] As respectable women across the country became convinced that both the law and the social structure left girls exposed to the unchecked rapacity of male avarice and deceit, the WCTU pulled together disparate critiques of male power and wove them into a national and hugely successful campaign for age-of-consent legislation.[189]

Historians have confirmed Willard's assessment of the sexual dangers confronting nineteenth-century girls. Rape, prostitution, and reformatory records show not only that men from every class sought out sexual contacts with children but that girls "were preferred sexual objects of some men in nineteenth-century New York."[190] Men could find girls in brothels across the city, and they were willing to pay fifty dollars, a sum equal to what an adolescent girl might otherwise earn in an entire year, for a single sexual contact.[191] Some men simply took what they wanted. Between 1810 and 1876, one-third of all rape and attempted rape cases prosecuted by the New York District Attorney's Office involved girls aged 12 and under. The girls were not precocious adolescents; most were between the ages of 1 and 9.[192] According to historian Stephen Robertson, after the 1880s public exclamations of disgust in New York City against men who raped children reflected a growing intolerance for abuses against children.[193] But in the Jim Crow South, where whites frantically sought to retain white male supremacy by spreading the lie that African American men were incapable of restraining their sexuality, white male legislators tellingly decried age-of-consent legislation as impinging on white males' "human liberty."[194]

Trying to understand how so many men could behave so badly, late nineteenth-century reformers pointed to alcohol as a "scapegoat to explain and excuse disreputable male behavior."[195] Postbellum temperance writers increasingly found a cause-and-effect relationship between intoxication and a man's loss of will power and moral responsibility for his actions. Because it interfered with the exercise of male volition, a crucial element in demonstrating one's ability to control and protect his dependents, inebriation connoted "a failure of manhood."[196] The belief that alcohol impaired manly volition was so pervasive that men, too, accepted it. When they got caught molesting children, including their own, some blamed their misconduct on drunkenness, a defense that no one seemed to question.[197] Child protection workers readily accepted this explanation from a class of men about whom they already had a low opinion, and the MSPCC estimated that two-thirds to three-quarters of its cases, including incest, involved alcohol.[198]

Yet judging from the newspaper articles and court cases I found, incest was almost never a one-time occurrence. And newspaper articles and court testimony describe a degree of planning and a strategic use of threats and violence that reflect a more sober and calculated intent than might be displayed by a sloppy, drunken accident. Few newspaper articles mentioned alcohol, and then usually only in the context of a father's reputation as a drinking man, not as the proximate cause of the incestuous attack.[199]

Even if alcohol played only a minor role in father-daughter incest, however, alcohol as a metaphor for male corruption resonated with late nineteenth-century women. Temperance women—even those married to an abstinent man—used temperance to air "their concerns about social dislocation, their frustrations with an irresponsible male culture, and their own vulnerability."[200] WCTU reformers criticized a social structure that, in their view, facilitated the circumstances under which child sexual abuse and incest could occur, an insight that, as I argue in later chapters, Progressive Era reformers would so thoroughly obscure that it remained dormant until rediscovered by second-wave feminists in the 1980s. The WTCU aimed broadly—they wanted to change the social context in which men exercised authority, and they looked to the law for an authority effective enough to make this change.[201] By adding the sexual abuse of children to its agenda, the WCTU refashioned incest from a personal failure of manly responsibility to a social and political issue that signaled the need for a dramatic and national change of course.[202]

Women responded enthusiastically to Frances Willard's critique of male domestic abuses and sexual excess. The WCTU organized an influential and determined social movement of 150,000 women members—ten times the membership of the major suffrage organization—that identified gender relations, not individual men, as in need of reform.[203] Emboldened by Willard's leadership and their collective authority, female reformers stormed not only into saloons but into courtrooms, where they both demonstrated their resolve and disrupted business as usual. When a judge rebuffed the efforts of Sarah Sanford, an agent for the Humane Society in Oakland, California, and the wife of a politically active Oakland businessman, to exclude men from the courtroom during an incest trial in 1895, Sanford shrewdly "packed the room with rows of grey haired women, effectually crowding out the horde of curious men who flocked to hear the details of the case."[204] In 1899, a reporter in Vermillion, South Dakota, was impressed when, on the first day of the trial of a teamster charged with incest, more than two hundred WCTU women showed up to observe the proceedings. He claimed it was the largest number of women ever gathered in a Vermillion courtroom, a fact that affected his view of the case. When the reporter wrote that the testimony had begun to convince the male spectators that the charge was false, he was careful to add that the WCTU women seemed to have reached the same conclusion.[205]

The determined presence of female reformers no doubt confounded the expectations of those men who looked forward to enjoying the erotic spectacle of an incest trial. But the presence of so many reform-minded women also sent a larger political message. The WCTU was using its growing popularity to its political advantage in a national campaign to either raise or establish an age of consent in every state. Unlike the ambiguous incest statutes, age-of-consent or statutory rape laws unequivocally defined sexual contacts with girls below a certain age, including one's daughter, as a crime.[206] With the removal of consent as a defense, the only factual issues left for trial would be the girl's age and whether there had been sexual contact, a boon to criminal prosecutions on behalf of victimized daughters, who could now circumvent the uncertainties of both the incest laws and the issue of consent.[207]

The WCTU's aggressive campaign to enact new age-of-consent laws heightened public awareness and political intolerance of adult-child sexual contacts. By 1900, thirty-two of the forty-five states had enacted legislation that either established or raised the age of consent (which in some states had been as low as 10) for girls to at least 16.[208] And the numbers of prosecutions for child sexual assault in general, and father-daughter incest in particular, rose substantially. In its newspaper, the *Union Signal*, the WCTU claimed that the first defendant to be convicted under Michigan's new statutory rape law was the stepfather of a victim.[209] Appellate courts in many states reported their first incest case in the 1880s. Of the 150 appellate court opinions on incest that I identified, nearly all (120) occurred in the late 1880s and 1890s. After Michigan raised the age of consent to 16 in 1887, convictions for incestuous assaults doubled.[210]

Yet even though women from the genteel and emerging middle classes had campaigned vigorously for age-of-consent legislation, they did not use the new laws when incest occurred within their own homes. Most defendants were still drawn from the poor and working classes, perhaps because age-of-consent reformers, like the child savers, continued to identify only the poor and working classes as exposed to corruption and so in need of protection.[211] But the unwillingness of genteel women to identify their own daughters as in need of protection is at odds with the passion with which they endorsed the WCTU's rallying cry for home protection. Temperance literature, whether fictional morality tales or moral suasion articulated as polemic, may have provided an acceptable forum for talking about an unpleasant reality. The extensive use of the incest plot in WCTU literature, underscoring the organization's focus on safeguarding girls from domestic, rather than public, dangers suggests that genteel women considered incest, if not an everyday reality, at least a real concern.[212]

But the rhetoric of home protection was double-edged. Genteel women who participated in reform activities assumed that, as fortunate members of society, they had a responsibility to uplift the women and families they viewed as beneath them. Women who viewed gender relations in terms of female moral superiority against a dangerous male public may have been unable to recognize, let alone act on, threats to their daughters' purity within the home. If a woman's husband could so blithely trample the respectable and moral domestic environment into which a mother had presumably put her heart and soul, how could she admit such a failure to herself, let alone to the world? As Sarah Deutsch pointed out in her discussion of men's abuse of domestic servants in this period, middle-class and elite women were not going to criticize the morals of the men of their own class—their husbands, sons, and brothers—because doing so would "bring into question the efficacy of their own, female, moral authority in the home."[213] In an era when genteel women understood their social power as rising out of their natural ability to guard the values and choices of their husbands and children, a daughter's unequivocal ruin at the hands of her father demonstrated that her mother had failed utterly.[214] What kind of woman could fail to discern that her husband was a heinous beast, that the most revolting sort of sexual immorality had taken root, in many cases for years, in her own home, or that one or more of her daughters had been ruined—all within the purview of her watchful maternal eye and moral influence?

The fact that genteel women did not turn to the statutory rape laws for protection may indicate that they perceived the WCTU's critique of male authority and political activism, coupled with the new level of state intervention permitted into the family, as a sufficient rebuke to abusive husbands and fathers. In this context, some women may have felt emboldened to disclose the abuse to family and friends who could help them to achieve an extrajudicial resolution. Others turned to the divorce rather than the criminal courts. Perhaps some accepted the argument that suffrage was the best solution to their domestic problems. Still, it was only after women insisted on talking about incest and organized to protect their own and other women's daughters that legislators felt significant political pressure to vote for laws that removed legal barriers that had confounded prosecutors seeking legal redress for victimized girls.

Just as a politicized analysis and public discussion of the topic began to circulate widely, however, doctors discovered that gonorrhea was widespread among American girls, including white girls from respectable families. Reformers could have used this new scientific data, which they realized was evidence of child sexual assault at best and of father-daughter incest at worst, to strike a fatal blow against male privilege. But this did not happen. The idea of home protection took on new meaning at the end of the century instead as native-born white Americans became anxious about shifting race relations, the uncontrolled growth of large urban areas and the decline of small communities, and the changing complexion of a country filling rapidly with dark-skinned immigrants and their babies. Viewing themselves as under siege and in danger of losing their social power, Progressive Era white, genteel Americans closed ranks around race and class. Looking backward, they clung to the ideal of the male-headed nuclear family as vital to ensuring domestic safety and American civilization. A social movement of sufficient strength and sophistication that could support the notion that father-daughter incest was widespread among the white middle and upper classes would not emerge until almost a century later. And even then it was daughters, not mothers, who revealed the domestic trauma.

CHAPTER TWO

Medicine and the Law Weigh In

When Americans sought evidence that might corroborate a girl's allegation that she had been sexually assaulted, they often looked to doctors for an opinion as to whether any physical evidence of assault existed. In the nineteenth century, doctors considered genital injuries, marks of violence (bruises, lacerations), semen stains, and venereal disease—particularly vaginal gonorrhea—as evidence of sexual contact in both girls and women. However, unless a girl's injuries were so severe that she required immediate medical care, days, weeks, and even months might pass before a physician examined her for evidence of sexual assault. The symptoms of vaginal gonorrhea, particularly a foul-smelling, yellowish green purulent discharge, were often the only physical evidence visible to physicians. But in the era before doctors understood that bacteria are the agents that lead to infections such as gonorrhea, they had no tools to help them make the diagnosis. And doctors were not eager to give a medico-legal opinion that might lead to the criminal prosecution and even execution of a girl's alleged assailant. This chapter explores medical views about the etiology of girls' infections in Anglo-American medicine and law, beginning with a late eighteenth-century case in Manchester, England. The case seems innocuous on its face, but the most eminent physicians writing on the topic in Britain and America repeated it and struggled over its meaning—within the profession, in the courts, and with mothers who suspected their daughters had been assaulted and infected.

The Jane Hampson Case: A Cautionary Tale about Doctors' Duty to Protect Men

In Manchester, England, on February 10, 1791, 4-year-old Jane Hampson complained to her mother of painful urination. When Mrs. Hampson examined Jane, she was surprised to find her daughter's genital area "highly inflamed, sore, and painful." The next day Mrs. Hampson took the girl to the Manchester Infirmary, which admitted her "on account of a mortification of the female organs" and treated her with leeches, "other external applications," and "internal remedies."[1] Jane remained in the hospital for nine days, during which time her condition worsened until she died. At that point, the hospital began to consider the possibility that Jane had been murdered.

Mrs. Hampson told the doctors that just before Jane became ill, she had "complained of being very much hurt . . . during the night" by a 14-year-old boy with whom she had shared a bed. Suspicious that the boy had taken "criminal liberties," the hospital asked Dr. Michael Ward, one of its surgeons, to investigate the cause of death. Ward concluded that Jane had died as a result of "external violence," meaning that the boy had sexually assaulted Jane and infected her with gonorrhea, the complications of which had led to her death. The boy was charged with murder and fled to avoid arrest (and also, though the account does not say so, perhaps to avoid a medical examination).[2]

Before the boy was found, however, Dr. Ward encountered several other girls suffering from vaginal symptoms like Jane's. But, as Ward felt he had "no reason to suspect" that these girls had also been the victim of "injury or guilt," he concluded that they were ill with typhus, then a common disease.[3] We now know that lice and fleas are the primary carriers of typhus, the symptoms of which include fever and chills, body rash, abdominal pain, and delirium.[4] But in 1792, British doctors believed that typhus was a disease of poverty that struck people whose constitution was already weakened by unhealthy and unsanitary living conditions that left them vulnerable to infection from factors that included the "warm moist state of the atmosphere" and "animal and vegetable effluvia . . . emitted from flagrant putrid water."[5] Doctors also identified a more severe type of typhus, "putrid fever," which heaped on the unfortunate sufferer the additional symptoms of diarrhea, constipation, black bile, and hemorrhaging.[6] Horrific as these descriptions are, patients who were treated promptly with a combination of purges and emetics enjoyed a good prognosis.[7]

Twenty years earlier, Dr. Thomas Percival, one of the hospital's socially progressive surgeons, had boasted that, even with a 50 percent death rate among children under age 5, the residents of Manchester, a factory town already afflicted with coal soot, poor sanitation, and "appalling conditions of child labor," still enjoyed better health than the people of London.[8] But the population of Manchester had nearly tripled in the forty years since the hospital had been founded, and the demand for services far exceeded the infirmary's nearly three-hundred-bed capacity. Worse, a series of "epidemic fevers," particularly through the winter of 1789 and spring of 1790, had drained both the public health and the hospital's resources.[9]

Dr. Ward reasonably attributed the spate of new infections in girls to typhus, one of the "epidemic fevers" Manchester had just endured, a view his colleagues shared. Dr. John Ferriar, who would become the city's most highly regarded physician, was a colleague of Ward's and a close friend of Thomas Percival's, with whom he collaborated both in modernizing the infirmary and in initiating major public health projects they hoped would improve the living conditions of the poor.[10] In 1792 Ferriar published *Medical Histories and Reflections*, which included Percival's detailed accounts of both the epidemic fevers of 1789 and 1790 and the typhus that lingered through the summer of 1791. In a chapter on the fevers, Ferriar described a particularly deadly variant: "In the course of the last twelve months, I have met with several instances of putrid fever, in young girls, accompanied with . . . a gangrenous state of the labia pudenda. The parts were greatly tumefied [swollen], and extremely painful. It was a very fatal complaint."[11] Ferriar did not mention Dr. Ward by name, but he clearly endorsed Ward's diagnosis.

In light of the number of girls he treated after Jane's death, and with the weight of his more eminent colleagues behind him, Ward reconsidered his diagnosis of Jane Hampson. His reasoning was similar to that of a French physician who had recently been faced with similar circumstances and who warned, "As the great number of such cases attests the existence of some general cause, operating upon many individuals at the same time[,] . . . the production of the disease cannot be attributed to one so special and extraordinary [rape], since if violence were inflicted upon so many children at the same time, the circumstances could not escape general observation."[12] Ward became convinced that it was more probable that Jane suffered from typhus, like the other sick girls, than that they had all been sexually assaulted. He informed the coroner that his initial diagnosis had been wrong and, in view of the capital charge pending, "designedly" made his new view of her case public. The accused boy soon surrendered and appeared for trial. If he felt any anxiety, the judge allayed the boy's fears as soon as he opened the case. Turning to the jury, the judge told them that there was insufficient evidence to convict, that a trial would entail "much indelicate evidence," and that he hoped "they would acquit . . . without calling any witnesses." The jury "immediately complied," and the defendant was dismissed.[13]

Jane Hampson's story seems too sketchy and unremarkable to have amounted to anything more than an obscure medical anecdote. But Dr. Percival viewed the chain of events—from a mistaken diagnosis to the dismissal of a murder charge—as a cautionary tale about medical practice. From the 1790s until 1890, doctors in Europe and America worked to increase their professional authority and prestige by arguing that science should replace superstitious or folk beliefs in a rational society. This was a difficult task in the prebacterial era, when physicians knew that their ability to diagnose many diseases, including gonorrhea, often relied on whatever history a patient gave them. And patients complicated the task; doctors knew that people often lied instead of admitting to "immoral" behavior. But in making a decision as to whether a girl's symptoms were proof that she had been sexually assaulted, any history given by the patient was also the event to be proved. Doctors and a new category of professionals called medical jurists, who might be trained in science, medicine, or law, sought to broaden their social influence by interjecting their expert views into criminal proceedings. But medical jurists and doctors found themselves in an unexpectedly precarious position when it came to the issue of gonorrhea infections in girls and child rape, which they understood in the context of their social biases and which threatened to undermine their claims to professional expertise.

Three years after Jane Hampson's death, Dr. Thomas Percival circulated copies of *Medical Jurisprudence; or, A Code of Ethics and Institutes Adapted to the Professions of Physic and Surgery* (1794), his attempt to resolve internal disputes at the infirmary by codifying a standard of professional conduct.[14] The British medical community already held Percival in high regard. He was a prolific writer, a fellow in the learned Royal Society (whose past presidents include Samuel Pepys and Isaac Newton), and in 1781 he co-founded the influential Literary and Philosophical Society of Manchester and served as its president for many years.[15] Doctors praised Percival's attempt to formulate a set of medical ethics, and, having found a receptive audience for *Medical Jurisprudence* beyond the walls of his own institution, Percival published an expanded and renamed version in 1803, *Medical Ethics: Or, A Code of Institutes and Precepts Adapted to the Professional Conduct of Physicians and Surgeons*. It became the first standard British textbook on the topic.[16] An American edition followed in 1821, and twenty-five years later the American Medical Association (AMA) used it as the basis for its first Code of Ethics. The AMA still credits Percival with having made the "most significant contribution to Western medical ethical history subsequent to Hippocrates."[17]

But why did Percival include two accounts of Jane Hampson's story, his own and Ward's narrative of the case, entitled "Uncertainty in the External Signs of Rape," in a book about medical ethics?[18] Percival set his discussion of Jane's case in a chapter about the role of the physician in a case of alleged rape. Percival admired Sir Matthew Hale, the seventeenth-century jurist who wrote the major English text on criminal law, and he recognized that emotions in cases of child rape could easily prejudice a man's right to a fair investigation and trial. If a woman or girl claimed to have been raped, doctors checked her for physical evidence, including "appearances of violence on examination &c." and any genital bruises, lacerations, or disease that might corroborate her claim.[19]

Though he abhorred the crime of rape, Percival did not view the examining doctor's role as assisting the putative victim in the service of justice. Rather, he emphasized the doctor's part in helping the defense detect "malicious prosecutions" or false accusations. Percival believed that protecting the accused was a singularly heavy responsibility, and he endorsed Hale's axiom, that an accusation of rape is "easy to be made, and harder to be proved; but harder to be defended by the party accused, though innocent."[20] Percival warned physicians to give their opinion about any corroborating physical evidence "with the utmost caution" because "even external signs of injury may originate from disease, of which the following examples . . . in Manchester, are adduced on very respectable authorities."[21]

In other words, before diagnosing gonorrhea, a doctor should be certain that the girl's physical ailments are not symptoms of a disease like typhus. But Percival's only example was the Jane Hampson case, and he altered Ward's version of events by claiming that the boy was freed after a full trial at the Lancaster assizes, implying that his reprieve from the gallows was more harrowing than the quick relief he actually obtained.[22] Yet Ward's account echoes Percival's dire view, warning that a doctor should "consider well the important duty he has to discharge, both to an individual, and to the community: And that he makes himself responsible for the consequences which may result from the influence of his judgment in the minds of the jury."[23]

Neither Percival nor Ward identified gonorrhea by name when discussing Jane's illness or diagnosis. Yet throughout the nineteenth century, doctors and medical jurists in Britain and America treated the Hampson case as a notorious incident proving the ease with which a doctor could misdiagnose gonorrhea infection in girls and send an innocent man to the gallows—even though Jane's assailant had neither been tried nor executed. To medical jurists, the willingness of practitioners to diagnose gonorrhea in girls proved that they failed to appreciate the gravity of the diagnosis and justified their charge that unskilled doctors were contributing to a miscarriage of justice. Medical jurists' complaints had merit, but their response to the problem failed to offer practitioners any information with which they might improve the accuracy of the diagnosis. Rather, their advice about making the diagnosis in a case of suspected child rape became increasingly convoluted and contradictory.

The Jane Hampson Case in Britain and America

George Edward Male, a physician at Birmingham's General Hospital, was the first to take up the case. In 1816 he published *An Epitome of Juridical or Forensic Medicine*, the "first, major original" English book on medical jurisprudence, which earned him the title "father of English medical jurisprudence."[24] Male viewed the murder charge filed against Jane Hampson's bedmate as an example of the ruinous consequences of a misdiagnosis. Male joined Percival and Ward in urging doctors to be extremely guarded when diagnosing a girl with gonorrhea, even if the infection seemed to be accompanied by other well-accepted physical signs of rape, including "symptoms of defloration."[25]

Caution was appropriate, but Male suggested a standard of diagnostic rigor that early nineteenth-century physicians could not meet. In the prebacterial era, doctors were keenly aware that they lacked the tools to diagnose gonorrhea—in a man, woman, or child—with certainty and that their ability to make a differential diagnosis was often tied to the patient's willingness to provide an accurate history, particularly about his or her sexual activities. As the London surgeon, naturalist, and lecturer Dr. John Hunter wrote in 1786 about gonorrhea in women, "The appearances of the [genital] parts often give us but little information . . . I know of no other way of judging . . . but from the circumstances preceding the discharge."[26] In legal cases in which sexual contact was the issue to be proved, the single most important piece of diagnostic information—a history of sexual contact—was also the fact to be legally proved. It is easy to imagine a doctor influenced by circumstances strongly suggesting rape and mistakenly diagnosing a common vaginal infection as gonorrhea.

However, the link between vaginal inflammation in girls and gonorrhea was so widely known and accepted that mothers like Mrs. Hampson, who saw their daughters' genitals unusually inflamed, reached the same alarming conclusion. But their concern only stiffened doctors' determination to protect the accused. In 1817 an American medical journal reprinted an article written by Dr. Kinder Wood, a member of the Royal College of Surgeons with a practice in London. Wood railed against the "disgusting" frequency with which parents of girls suffering from an infection of the genitals claimed their daughters had been raped.[27] He claimed to have treated twelve girls, all between the ages of 1 and 6, who had suffered from "affection of the pudendum," which he considered a symptom of typhus.[28] But Wood began his lengthy discussion of individual cases not with an account of one if his own patients but with Jane Hampson, stating that "this disease has been frequently considered in court as evidence of violence and venereal infection; inflammation, ulceration and discharge having always had particular attention in a consideration of the evidence." Wood criticized such conclusions as wrong, and saying that Ward's account of Jane Hampson's case had "not hitherto attracted sufficient attention," he reprinted it in its entirety.[29]

Yet if doctors and medical jurists believed that such diagnostic errors were as common as they feared, they never actually argued that a diagnosis of gonorrhea was so unreliable that it should be categorically excluded as medico-legal proof of child rape. Instead, they acknowledged the utility of the diagnosis while also impugning the skills and ethics of those doctors willing to make it. For example, a year after Kinder's paper was published in America, Thomas Cooper published *Tracts on Medical Jurisprudence*, reprints of four European medico-legal essays, including George Edward Male's *An Epitome*, intended to aid doctors in making accurate diagnoses.[30] Cooper, who had studied law, medicine, and science, had been a contemporary of Percival and Ferriar in Manchester but left England in 1795 for America, where he became a citizen and a member of the Pennsylvania bar and later a judge, university professor, and college president.[31] In his history of nineteenth-century American medical jurisprudence, James C. Mohr argues that Cooper believed the goal of medical jurisprudence was to protect "defendants against wrongful conviction" and that he published *Tracts* so that Americans would have access to the most useful works on the topic.[32]

The inclusion of Jane Hampson's story in the seminal works of medical ethics and jurisprudence published in Britain and America, and its retelling and sanctioning by influential and learned scientists steeped in Enlightenment thought, must have sent a powerful message to practitioners in the early Republic. A century before American medical schools would begin to design a comprehensive program of study, a misdiagnosis that ended in the execution of an innocent man might have not only ended a physician's practice but also damaged the still thin reputation of the profession itself. But why did doctors choose the Hampson story to illustrate their point rather than a case in which a supposedly innocent man had actually been executed? Percival may have chosen the Hampson case because it was the first such case he had either encountered or could document. But subsequent authors, many of whom claimed to know of "many" such cases, usually only cited Hampson.

Their choice is curious. Nothing in Percival's or Ward's accounts suggests that Ward had acted recklessly or lacked competence; the fact that the hospital selected him to investigate Jane's death implies just the opposite. When he had reason to believe that his initial diagnosis was mistaken, Ward acted quickly and publicly to obtain the defendant's release, a course of action that an official history of the Manchester Infirmary, published in 1904, still considered worthy of praise.[33] More important, in their retelling of the case, medical jurists omitted the fact that Jane had complained that the boy had hurt her, the central allegation that had raised the suspicion that she had been assaulted. Nineteenth-century medical jurists in Britain and America who discussed physical evidence of child rape instead took a bewildering tack: they promoted the notion that gonorrhea was key evidence for a successful prosecution while also insisting that such evidence was unreliable, sonorously invoking the Hampson case as Exhibit A.

Diagnosing a Loathsome Disease

Doctors did not doubt that men raped children, including their own daughters. At a time when doctors expected virtuous women to resist a sexual assault "to the utmost," they recognized that girls, who lacked both the physical strength to repel an assailant and the emotional ability to understand and effectively respond, were easy prey.[34] But doctors became uncomfortable when asked to give a medical opinion about whether a girl had been raped, particularly when physical symptoms suggested gonorrhea infection. If a physician found that the girl exhibited a purulent vaginal discharge, he then examined the accused. If the defendant also showed signs of infection, the doctor then tried to discern when the girl's discharge first began. If it began before the alleged assault, or more than five days after, she may have acquired the infection elsewhere. In that case, her symptoms had no further bearing on the issue of the defendant's guilt but reflected instead on the girl's character, suggesting that she was sexually promiscuous and not credible. The defendant might then be released, without any determination about his guilt or innocence. In such cases the physician, whose opinion might ruin the girl's reputation for chastity, was relieved of the greater burden of having to give an opinion on which a man's life might be at stake.

In the prebacterial era, doctors had few tools with which to diagnose or treat many illnesses, but they expressed a marked reluctance about diagnosing gonorrhea in girls. A number of obstacles stood in their way. For one, medical knowledge about gonorrhea in women was so slight that doctors did not even agree on whether women were susceptible to the disease.[35] Yet even when physicians agreed that gonorrhea infected both females and males, they could not agree on its symptoms, particularly in girls. Doctors generally agreed that the symptoms of gonorrhea vulvovaginitis, infections in prepubescent girls, included vaginal discharge; genital irritation, redness, and sores; and a painful, burning sensation when urinating. But doctors also knew that a number of other gynecological and childhood illnesses, including pinworms, pelvic inflammatory disease, and leucorrhoea (nonsexually transmitted vaginitis or "the whites"), produced similar symptoms.

Physicians who turned to medical texts for help in making a differential diagnosis found contradictory advice. In 1848, Dr. Charles D. Meigs, professor of midwifery and diseases of women and children at Philadelphia's Jefferson Medical College, founded only twenty-four years earlier, published his lectures, *Females and Their Diseases*. Meigs was also a director of the American Philosophical Society and had achieved some notoriety for his debate with Dr. Oliver Wendell Holmes Sr. over the cause and treatment of childbed fever.[36] The best advice Meigs could offer his students was to distinguish gonorrhea from vaginitis by the degree of pain a woman suffered. Extreme pain (probably due to complications such as pelvic inflammatory disease) signified gonorrhea.[37] Similarly, Dr. Robert Liston, a well-known London surgeon and lecturer, thought that pain while urinating distinguished gonorrhea from leucorrhoea.[38] Others considered the color, odor, or texture (ropy versus mucous) of the discharge key, but they disagreed about whether gonorrhea discharge was green, white, or red (mixed with blood). Liston, for instance, claimed that gonorrhea discharges were yellow with a black border whereas leucorrhoea discharges were white or yellowish, with no border.

Doctors were stymied, in part, because they lacked any sort of technology that might have assisted them in making the diagnosis. Before 1879, when Dr. Albert Neisser identified *Neisseria gonorrhoeae* as the bacterium that produces gonorrhea infection, few doctors had access to a microscope, still a new device that was out of the financial reach of most practitioners and hospitals. New York City's Mount Sinai Hospital, for instance, which opened as the Jews' Hospital in 1855, purchased its first microscope in 1867 for general staff use. Even so, the hospital did not make space for a lab in which to use the microscope until 1903, when it converted a coat closet that held space for two people.[39] Before Neisser's discovery, however, access to a microscope would have been of little help. As Baltimore physician John Morris complained in 1878, even with a microscope, a doctor trying to diagnose gonorrhea had no idea what he was looking for.[40]

Five years after Neisser's discovery, Danish bacteriologist Hans Christian Gram developed his innovation, the Gram stain, the first microbiological test to detect *N. gonorrhoeae*. The Gram stain, which is still the most commonly used diagnostic technique in microbiology, involves smearing a slide with biological material, staining it various colors, and then examining it under a microscope.[41] Medical journals quickly publicized Gram's technique, and by 1886 Penn Medical University professor Dr. Seth Pancoast had incorporated it into the revised and enlarged edition of his *Ladies' Medical Guide*, in which he advised women that "the discharge of gonorrhoeae can only be detected from that of vaginetis [*sic*] by the aid of the microscope. No physician should dare pronounce the discharge gonorrhoeal without such microscopic examination."[42]

Although Pancoast's appreciation of the new diagnostic technique proved prescient, he overstated physicians' confidence in it. Doctors raised concerns about the reliability of the Gram stain through the early 1900s, some of which were warranted.[43] Doctors' inexperience with both the microscope and the staining method, coupled with the likelihood of error and contamination in conditions such as those existing at Mount Sinai, no doubt resulted in countless misdiagnoses. But even aside from the uncertainties inherent in the use of new technologies, the lack of medical knowledge about gonorrhea is striking. Why was medical expertise about a disease that Hippocrates had identified in ancient Greece still so scarce? The paucity of information about gonorrhea infection, particularly in women and girls, reflects, in part, the long-standing belief shared by many nineteenth-century British and American doctors that people with venereal disease were not suffering from physical disease but from immorality.[44] Until the early twentieth century, many American doctors and hospitals refused to treat venereal disease patients, who they believed were paying the wages of sin and were thus unworthy of medical attention. At best, venereal disease patients were placed in segregated units, and in many cities the only medical care available was in a degrading public charity of last resort, an almshouse or penal institution.[45]

As a result, few doctors had experience in diagnosing and treating patients with venereal disease. In antebellum New York City, for instance, the New York Hospital did not permit a person to be admitted more than once for treatment (which, for men, included applying leeches and cold water to the penis) for venereal disease.[46] Desperate New Yorkers sometimes feigned the crime of vagrancy so that they could be committed to the Penitentiary Hospital on Blackwell's Island, the city's frightening poorhouse, for treatment.[47] In the 1830s or 1840s, the Boston Dispensary became that city's first institution to provide health care for venereal disease patients, even though doctors knew that, if untreated, the complications of venereal disease could lead to prolonged illness, physical and mental disability, and death.[48] Treating soldiers during the Civil War brought venereal diseases into sharper focus among Boston's physicians, but as late as 1894 the trustees of the Massachusetts General Hospital debated whether it was appropriate for the hospital to accept a bequest that was to be used to underwrite treatment for venereal disease patients.[49]

Moral judgment was most acute when the patient was a woman. Doctors often assumed that venereal disease infection among women was limited to prostitutes, whom doctors blamed for spreading the infection to men and for whom they expressed little sympathy.[50] For example, in 1824 the Pennsylvania Hospital refused to treat prostitutes, whom it attempted to exclude by requiring women seeking admission to the venereal disease ward to show a certificate of marriage.[51] Such policies contributed to physicians' ignorance of women's gynecological conditions, an ignorance that some genteel women applauded. They viewed a gynecological examination by a male physician as improper under any circumstances and especially when the patient was a girl, whom the exam would, by necessity, ruin.[52]

Yet however much physicians wished to believe that gonorrhea infected only sinful rakes or the "lower classes," enough doctors treated infected patients with such frequency that when nascent medical societies set fees for practitioners, they included treatment of venereal disease among the list of ordinary services.[53] And doctors who wrote medical textbooks revealed the regularity with which they encountered and treated venereal disease among their genteel and presumably worthy peers. In *A Treatise on Gonorrhoeae and Syphilis*, published in 1859, Boston's Dr. Silas Durkee, who treated venereal disease patients for thirty years, remarked that physicians disliked the task of having to inform a respectable patient that he or she was infected with a venereal disease.[54] Durkee described the panic and despair with which genteel men and women reacted to the diagnosis, particularly unfaithful marital partners who, having been caught, expected the physician to help them to solve the "difficulties which spring up within the domestic circle in connection with this embarrassing subject."[55] Venereal infection was grounds for divorce, and though he would be well paid for a medical opinion related to a court proceeding and his testimony in court, no reputable physician wanted his name associated with a marital scandal certain to provoke the opprobrium of respectable society.[56] Among genteel society, in which sexual restraint was essential for continued admission, an infected person faced an uncertain future. If the patient's spouse learned of the diagnosis or the condition became grist for the rumor mill, the patient risked losing not only his or her marriage but also his or her social and financial standing.

With so much at stake, doctors knew that their respectable patients would rather fabricate a medical history than admit to immoral behavior. Dr. John Hunter wrote in 1786 that a patient's medical history in cases of gonorrhea "is not always to be trusted to, for very obvious reasons."[57] Dr. Liston marveled at the propensity of patients to "give very ridiculous accounts of the way in which their clap was contracted . . . They will exert their ingenuity to the utmost, in order to deceive their surgeon."[58] And Durkee, who complained that patients seemed not to tire of trying to convince doctors they had become infected after eating spicy foods or riding horseback without a saddle, complained that "many persons, who are perfectly reliable on all other occasions, will not hesitate to deceive their medical advisors concerning impure connections, even when they know that deception . . . will operate against their interests, perhaps for life."[59]

Imagine, then, how fraught the conversation must have been when the patient was the daughter of a respectable or even prominent family. Doctors worried that parents might feel insulted by the news that their daughter, whose purity, after all, it was their duty to guard, was infected with a venereal disease. Dr. John Morris ruefully told his colleagues at the Baltimore Academy of Medicine that one genteel family terminated his services after he diagnosed their daughter with gonorrhea. They found a doctor who diagnosed the child with vaginitis instead, and they retained him to treat the girl.[60] Similarly, in 1887, the editor of the *Medical and Surgical Reporter* commented on a letter he had received from Dr. J. S. Prettyman Jr. of Delaware about a case of gonorrhea he had diagnosed in a 9-year-old girl, "the youngest member of a family of excellent social position." The family had called Prettyman after another doctor diagnosed the girl with "cutaneous disease" and unsuccessfully treated her with vaginal creams. Prettyman bravely diagnosed the girl's trouble as gonorrhea and excused his predecessor's diagnosis by observing that doctors "did not expect to see gonorrhoea in children so young." Yet Prettyman made his diagnosis without "interrogating [the girl] in regard to the source of infection" because he believed that doing so would humiliate a girl of her social position.[61]

Only the slow but steady implementation of microbiologic testing after the 1890s reduced a physician's crippling reliance on the history and character of his patient or her family to reach a diagnosis.[62] Still, microscopes remained an extravagance in an era when hospitals relied on the generosity of charitable benefactors in order to provide even basic medical services, and many physicians had neither access to one nor the training to use it.[63] Not until 1912, after the state legislature passed a law requiring physicians to report venereal infections, did the New York City Health Department address the issue by organizing a central pathology laboratory where doctors could send cultures to be examined microscopically at no cost.[64] Until then, the identification of gonorrhea remained entirely within the purview of the physician's diagnostic knowledge and skills, his interpretation of the patient's history and reputation, and his willingness to make the diagnosis.

Diagnosing Gonorrhea Vaginitis

In nineteenth-century America, a "medical examiner" might be an expert in forensic medicine or simply any available physician close to the scene of a crime. In a case of suspected child rape, the decision to prosecute often depended on the medical examiner's opinion as to whether the victim was infected with gonorrhea and, if so, whether she acquired it from her alleged assailant.[65] In the prebacterial era the crucial diagnostic question was whether a vaginal discharge was a symptom of gonorrhea or another disease that had not been sexually transmitted. But a differential diagnosis was little more than guesswork, a process that offered physicians wide latitude with which to contextualize their findings within their own biases or self-interest. For guidance, doctors relied on textbooks written by medical jurists, a field in which few doctors or lawyers had been trained and in which almost none had expertise.

In 1831 the prestigious Guy's Hospital in London named medical scientist Alfred Swaine Taylor its first lecturer in medical jurisprudence, a position he held for forty-six years. In 1836 Taylor began writing *Elements of Medical Jurisprudence*, which became the standard nineteenth-century British work on the topic. Over the next twenty years, Taylor claimed to have consulted on five hundred medico-legal cases, and by the mid-1850s he had become the country's leading medical jurist.[66] In the preface to the first American edition of *Medical Jurisprudence*, published in 1845, Taylor defined medical jurisprudence as the science that applies "every branch of medical knowledge to the purposes of the law."[67] In Taylor's view, medical jurists placed science in the service of lofty goals, representing neither the prosecution nor the defense; medical jurists worked "for the country."[68]

Nineteenth-century American medical jurists expressed similar aspirations. James Mohr contends that "the champions of medical jurisprudence crusaded self-consciously as the agents of applied science."[69] Medical jurists may have been trying to improve their professional stature by using scientific authority to sway public opinion, but Mohr views their contribution in broader terms. Because they advocated logic and science in place of "what they considered to be outdated attitudes, medieval precedents, folk perceptions, and irrational policies," Mohr argues that antebellum medical jurists played an important role in the formation of a modern American worldview that "confirmed an American version of rationalism, rooted in logic and science."[70] As American doctors wrestled with the complexities of diagnosing gonorrhea in girls, medical jurisprudence did prove influential, but its power to persuade owed more to social bias than to logic, science, or enlightenment.

The first standard American authority on medical jurisprudence, published in 1832, illustrates the conflict. New York medical professor Theodoric Romeyn Beck wrote the *Elements of Medical Jurisprudence* after concluding that the "duodecimo volume" of English-language material on medical jurisprudence "did not do this important subject justice."[71] The response to *Elements* seems to have proved him right. Shortly after the book's publication the Medical Society of the State of New York elected Beck its president, and by midcentury he had become "the best-known expert in medical jurisprudence in the world."[72] Alfred Taylor, whose own book would not displace Beck's as required reading in the Yale Medical School curriculum until 1869, claimed that *Elements* had inspired him to enter the field.[73] George Edward Male praised *Elements* in the influential *Edinburgh Medical and Surgical Journal*, calling it "one of the best works ever to appear in this or any other country."[74] The nine-hundred-page book, which grew to more than thirteen hundred pages by the fifth edition, dominated medical jurisprudence, in both medical schools and courtrooms, for the next thirty years.[75]

Like his British colleagues, Beck emphasized the pitfalls facing doctors who investigated allegations of child sexual assault. And he discussed Jane Hampson's case to prove his point. In the third edition, published in 1829, Beck began his discussion of gonorrhea vulvovaginitis by reiterating that infection was important medico-legal evidence of child sexual assault. He stressed that it was more reasonable to assume that a child who claimed to have been assaulted had contracted gonorrhea from her assailant than, coincidentally, from some nonsexual source: "*It is extremely improbable that diseases which occur so rarely should happen to appear in a child to whom violence was offered, unless that violence had some effect in producing it.*"[76]

Still, Beck warned that gonorrhea was not absolute proof of child rape, and he diminished the importance of genital injuries as corroborating evidence by arguing that they could be a symptom of disease rather than assault. Reiterating Percival's account of the Hampson case, Beck warned "that disease has produced the appearance of external injury, and led to suspicions against innocent persons."[77] Like George Edward Male, Beck argued that doctors could consider gonorrhea infection to be corroborating evidence only in those cases in which the girl also suffered genital injuries consistent with the use of force so severe that it could rupture the perineum (the area between the vulva and rectum).[78]

But doctors knew that men who sexually assaulted girls rarely engaged in attacks brutal enough to cause such injuries. In 1859 Dr. William R. Wilde asked twelve of the most esteemed medical jurists and surgeons in Britain whether they had ever seen a case in which a girl's perineum had been ruptured. Only Alfred Taylor responded affirmatively, and he could cite only two cases, one of which occurred in India.[79] A girl might not exhibit any physical injuries if a man only rubbed his penis against her genitals or penetrated her with a finger or object. Yet, as often happens in medical discussions of gonorrhea in girls, Beck omitted any discussion of assault by these means, even though medical jurists considered such possibilities in cases of the rape of women.

For instance, when Michael Ryan, a British physician, lecturer, medical jurist, and member of the Royal College of Physicians, published his *Manual of Medical Jurisprudence*, in London in 1831 and in Philadelphia the following year, he noted that a woman who is raped may not suffer genital injuries because "the [male] organ may be so small."[80] But as to girls, Ryan said only that "when defloration of any young female has recently taken place, the signs are very evident."[81] Ryan accepted genital lacerations and contusions as convincing evidence of child rape, but doctors knew that these injuries healed quickly in young girls, often before the symptoms of gonorrhea appear.[82] Unless a doctor examined the child immediately after an assault, and again in three or four days, he might not detect both genital trauma and vaginal discharge. But after reading Ryan or Beck, even a practitioner confident in his skills might hesitate to diagnose gonorrhea.

Ryan's discussion illustrates the contradictions that marked medical jurists' discomfort with the subject. He was the earliest British medical jurist to state directly that "venereal infection is a proof of violation when it coincides with the time at which the crime is alledged [*sic*] to have been perpetrated . . . and, above all, if the accused is affected with the disease."[83] But he clouded the issue with caveats that substantially narrowed the number of cases that could be attributed to sexual assault. First, Ryan claimed that only "venereal" gonorrhea was acquired by sexual contact, whereas "simple" gonorrhea could be acquired by eating foods such as asparagus. He attributed infections in patients who had eaten any one of a list of thirty foods, used spices, or had a history of constipation or bladder infections to "simple" gonorrhea.[84] Next, Ryan warned that a "simple mucous discharge" could be mistaken for venereal gonorrhea, and he repeated verbatim Percival's account of Jane Hampson's death. Referring to one case in which he believed that a man had been wrongly convicted, Ryan claimed that "cases like the present are unfortunately of too frequent occurrence," and he quoted Sir Astley Cooper's claim that he knew of thirty cases in which an innocent man had been convicted because simple gonorrhea had been misdiagnosed as venereal gonorrhea.[85]

In this manner Ryan offered practitioners a way in which to diagnose gonorrhea without having to also consider the possibility of sexual assault, thereby avoiding the troubling legal implications of the diagnosis. Ryan emphasized the physician's responsibility to the defendant, and he turned, as did many medical jurists, to the "impressive language" of Sir Astley Cooper's medical lectures: "Children . . . are frequently the subjects of a purulent discharge . . . [The mother] goes to a [poorly trained] medical man . . . and he says, 'Good God! Your child has got a clap.' (A laugh). A mistake of this kind, gentlemen, is no laughing matter . . . I can assure you a multitude of persons have been hanged by such a mistake."[86] Astley Cooper was by all accounts a popular and highly regarded member of London society. He served as vice president of the Royal Society and as an officer of the Royal College of Surgeons, performed surgery at Guy's Hospital, and lectured in medicine at Guy's, St. Thomas's Hospital, and the Company of Surgeons (later the Royal College of Surgeons), where he was also on the board of examiners.[87] When Cooper died in 1841, the *London Times* reported not only on his funeral but on the memorials that followed for years after his death.[88] The lengthy eulogy in the *Times* celebrated a long list of Cooper's accomplishments, including having been King George IV's choice when he needed surgery to remove a head tumor, having received the "largest fee ever known to have been paid for an operation," and, at a time when medical students paid for their education by paying the instructor directly for each lecture, having increased attendance at his lectures from fifty to four hundred, the "largest class of medical students ever known in London."[89]

Cooper worked right up until his death, and his influence on British and American medicine is both wide and deep.[90] But however much physicians and medical jurists repeated Cooper's claim about the multitude of innocent men hanged, none, except for Dr. William Wilde, attempted to document his claim, and Wilde knew the truth to be exactly the opposite. In 1859, when Wilde sought to overturn the rape conviction and death sentence of Amos Greenwood, he acknowledged that hanging a man for child rape would have been an extraordinary occurrence in Britain.[91] Men who raped girls and infected them with gonorrhea were rarely sentenced to death. As early as 1694, gonorrhea infection was important evidence in trials for child rape heard at the Old Bailey, and not all of those cases ended in guilty verdicts. Because so few men attempted penetration, courts dismissed many of the rape charges (which required proof of penetration) but, cognizant of the evidence of gonorrhea infection, permitted the prosecutor to file charges of attempted assault instead, a noncapital offense.[92]

Wilde believed the circumstances in Greenwood's case were similar to Jane Hampson's because the girl who claimed that Greenwood had raped her had also suffered from mortification of the pudenda and died. In a letter he wrote to the judge in which he reported Percival's conclusions about the Hampson case, Wilde claimed that since then only one man in Ireland and none in England had been executed for "the murder of a child by rape, or rape and disease conjointly."[93] When Wilde solicited comments about the case from prominent doctors and jurists, Dr. William Acton replied, "I think we should all aid in solving these odd cases in young girls, and rescue the man, if we conscientiously can; for I believe *many have been transported or hanged on imperfect medical evidence.* The case you propose for my opinion is similar to a few I have seen."[94] Yet Wilde's own research had uncovered only one case, and Acton had cited none.

Still, medical jurists intent, perhaps, on bolstering their claims to authority and expertise without alienating their peers or revising their social biases blamed ordinary practitioners for the supposed miscarriages of justice they insisted regularly occurred. Dr. Ryan's complaint is typical: "Every well-informed physician and surgeon is conversant with the purulent discharge of female children of scrofulous and delicate habits . . . such discharge is seen almost every day in dispensary and hospital practice among the poor . . . and it is often mistaken by ignorant practitioners for gonorrhoeae."[95] "Scrofulous" referred to a constitutional weakness that characterized people suffering from scrofula, a form of tuberculosis affecting the lymph glands.[96] And like venereal disease, scrofula carried a stigma that by the mid-nineteenth century implied moral corruption.[97] According to Dr. William Nisbet, a fellow of the Royal College of Surgeons of Edinburgh, in 1795 scrofula was "more frequent in Britain than any other disease," so much so that there were "few families in which it does not make its appearance in one form or another." But Nisbet explained that because practitioners knew that the diagnosis was "apt to give offence," they usually broke the news to patients by using one of its "less alarming denominations," meaning that many people were unaware they were afflicted.[98]

Physicians viewed scrofula and gonorrhea as simply different sides of the same coin, the detritus of poverty. But the medico-legal implications of the diagnoses differed significantly. Medical jurists often repeated Dr.

Wilde's discussion about two "filthy" girls he had examined for evidence of sexual assault. Even though Wilde had not examined the girls immediately after the alleged assault, he argued that because they displayed no marks of violence "it was simply a case of a disease . . . which is very common among children who are strumous [scrofulous], or badly cared for . . . It is usually found in low life."[99] Sir William Robert Wilde (husband of the popular Irish poet, nationalist, and salon hostess Sperenza and father of Oscar) was a Dublin ear specialist and a fellow of the Royal College of Surgeons of Ireland who made a name for himself by, among other things, disputing the medico-legal implications of vaginal discharges in cases of child sexual assault.[100] Wilde devoted a large part of his medical practice to the poor, and his diagnostic skills reflected the socially progressive view that the conditions of poverty caused the poor to be more susceptible than the wealthy to a host of illnesses.

But he was disinclined to admit the validity of any woman's allegation of sexual impropriety against a respectable man. Wilde had notoriously fathered three children before his marriage, and when a former patient and jilted mistress charged that he had raped her while she was anesthetized, the ensuing scandal and lawsuit ruined both his reputation and his finances.[101] Perhaps unsurprisingly, he viewed allegations of child rape with a high degree of skepticism, insisting that entire classes of men were defenseless against women who wished to blackmail them: wealthy men, men to whom a woman was financially indebted, or men against whom a woman held a grudge.[102] Wilde analyzed the case of the scrofulous girls in this context, stating that he had been "horror-stricken" to learn that this particular defendant had been "accused of such a crime."[103] The social standing of both the victims and the assailant influenced Wilde's opinion as to whether the girls were infected with gonorrhea. And even though his personal experience may have understandably affected his judgment in such matters, Wilde's logic would prove enduring among physicians in Britain and America, as class-based views about gonorrhea infection as evidence of child sexual assault intensified in the medical literature over the remainder of the century.

Still, when an American edition of Alfred Taylor's *Medical Jurisprudence* was published in 1845, it included gonorrhea infection among the four types of forensic evidence that supported the conclusion that a girl or woman had been the victim of "impure intercourse" or rape: marks of violence in the genital area; marks of violence on the bodies of the victim or defendant (or both); semen stains; and gonorrhea.[104] Taylor differed from his predecessors, however, in the value he placed on genital marks as a sign of child rape. Whereas earlier medical jurists required genital marks to substantiate an accusation based on gonorrhea infection, Taylor recognized both that any number of nonsexual or accidental activities can bruise or injure children's genitals and that victimized children do not always suffer obvious genital trauma.[105]

However, as with Wilde, Taylor's suspicions about the usefulness of genital injuries as a diagnostic tool arose not from his familiarity with the ease with which active children suffer injuries but from an alarming gender bias. Taylor had studied with Sir Astley Cooper and claimed that women were so eager to exploit men and financially profit from false claims that many would "purposely produce . . . genital injuries . . . on young children," an opinion his colleagues shared.[106] Capital cases require vigilant skepticism, and Taylor's emphasis on the duty of a physician to investigate the evidence carefully was appropriate. But he did not contextualize his warnings with additional information that might have been useful to practitioners, such as an estimate of the frequency with which women made malicious accusations or how to discern intentionally inflicted genital injuries from those a girl might be expected to suffer from a sexual assault. Rather, Taylor doggedly warned physicians, at every step of the investigation, about concluding that a girl had been raped.

Thus, even as he advocated the importance of gonorrhea infection as more reliable medico-legal evidence than genital injuries of dubious origin, Taylor discouraged physicians from making the diagnosis, stating, "We should be well assured, before giving an opinion, that the discharge is of a gonorrhoeal, and not simply of a common inflammatory character."[107] Like other medical jurists, Taylor reminded practitioners that "the existence of a purulent discharge from the vagina has been erroneously adduced as a sign of rape . . . The parents or other ignorant persons . . . often look upon this as a positive proof of impure intercourse." And if that warning was not sufficient, he added, without citing Jane Hampson by name, that "some . . . individuals have thus narrowly escaped conviction for a crime which had really not been perpetrated."[108] Yet just as Taylor persistently discouraged doctors from diagnosing gonorrhea in a girl who claimed to have been victimized, he also overlooked the element of Jane Hampson's case that had raised suspicions that her bedmate had sexually assaulted her in the first place—her complaint that the boy had hurt her.

Taylor had no advice for practitioners on how to avoid making a similar mistake, other than to refrain entirely from making the diagnosis. But his rule about the importance of gonorrhea infection as an indicator of rape became the standard among medical jurists for the remainder of the nineteenth century. Still, his equivocation about the sexual origins of vaginal discharges invited speculation, which London practitioner Dr. Burke Ryan soon provided. In a talk before the Medical Society of London in 1851, "On the Communicability of Gonorrhoeae, in Reference to Medical Jurisprudence," Ryan argued that it was possible for girls to acquire gonorrhea from nonsexual contacts.[109] Ryan, who had never published a treatise or undertaken any research on the topic, had no proof: he merely repeated speculation he had heard—probably from a servant—about the source of infection of two girls. Yet a two-column abstract of his talk in the *London Medical Gazette* had an immediate and durable impact. Doctors and medical jurists embraced and repeated his speculation so often that by 1940 medical textbooks in England and America treated it as fact.

Burke Ryan had examined two sisters, ages 1 and 4, both of whom he diagnosed with gonorrhea. The girls' symptoms included a profuse pus-like discharge, high fever, swollen genitals, and severe pain when urinating. Ryan expressed no doubt about his diagnosis, even though both girls were apparently free of any genital injuries, no one claimed they had been raped, and their mother denied being infected. The girls' mother claimed she had no inkling that her daughters might be infected until a neighbor commented on the girls' discharges. Ryan's account omits any mention of the girls' father or whether he was also infected. But Ryan sought out another woman in the house, whom someone claimed to have "observed washing herself in the same vessel used for washing the children, and using the same sponge to her private parts as was used for them."[110] The woman, whom Ryan did not identify, admitted that she was infected with gonorrhea and that she and the girls had used the same water closet and sponge.

Ryan got the idea about the nonsexual sources of the girls' infections not from the textbooks of medical jurists like Alfred Taylor or Michael Ryan but from this unnamed woman, who "thought it more probable that the eldest child having sat upon the same vessel as herself, to pass water, was thus infected, and that the second took it from the sponge used by both."[111] Ryan thought this the most probable explanation, too, and when he repeated it to the girls' mother, she responded with relief. Had she also used the same sponge, she explained, her husband would never have believed that she had become infected in such a manner, in other words, without sexual contact. But why did they believe that the girls had become infected in such an otherwise incredible manner? And why did Ryan and the girls' mother so readily accept the infected woman's account?

Ryan did not explain, but he acknowledged that no precedent existed to support his suggestion and that none of the medico-legal authorities considered it possible for a child to acquire gonorrhea without "impure connection." Still, he argued that the unique role of the physician in the medico-legal process made it all the more urgent for doctors to act "in the cause of humanity." What was the cause? Ryan turned to Jane Hampson, reminding his colleagues of their responsibility "to take care that no innocent person suffered, for had this disease not been so easily traced to its source . . . then the same fate might perchance befall the boy whom Beck mentions as having been condemned to die on account of the death of Jane Hampden [*sic*]."[112]

Ryan may have been right to suspect that a warm, wet sponge and toilet seat soiled very recently with gonorrhea discharge might, under perfect circumstances, have fostered the bacterium and spread infection. However, Ryan had not investigated the possibility of sexual contact; nor had he gathered any evidence to support his speculation about the vessel and sponge. Perhaps because he wished to avoid offending or insulting parents who might have been socially important or at least respectable, he had neither examined them to learn whether one or both might also be infected nor wondered whether any of the parties, including the woman who reported the story, had a motive to lie. Still, he concluded that "there is no reason why people should not contract disease from the seats of water closets."[113]

Burke Ryan had strayed so far from even the few authorities that existed on the subject matter that one might have expected his speculation to be soundly rejected and forgotten by the medical community. But his speculation had a profound and long-standing effect on how doctors viewed the etiology of girls' infections. Rather than criticize Ryan, doctors in Britain and America quickly embraced his "proof," which they repeated as fact over the next hundred years. In the ninth edition of *Taylor's Principles and Practice of Medical Jurisprudence*, for instance, published in 1934, editor Sidney Smith declared that "Ryan long ago traced the origin of a discharge in two children to . . . a sponge."[114] How had Ryan's speculation become so influential?

Only four years after his talk at the London Medical Society, the next standard work on American medical jurisprudence, Wharton and Stillé's *Treatise on Medical Jurisprudence*, endorsed Burke Ryan's speculation. In 1855, attorney Francis Wharton, a well-known legal writer, and Dr. Moreton Stillé, a lecturer for the Philadelphia Association for Medical Instruction, published a synthesis of the most recent European (British, French, and German) and American literature on medical jurisprudence.[115] Wharton and Stillé began a lengthy discussion of medical evidence in cases of child rape by explaining that they would rely on the work of "celebrated" Prussian medical jurist Johann Ludwig Casper, whom they praised for grounding his research in "authentic cases" instead of "theoretical discussions."[116] Casper was a public health officer and lecturer in Berlin who revolutionized German medical jurisprudence in 1857 when he published *Practical Textbook of Legal Medicine: Based upon Personal Experience*, which emphasized clinical data.[117] Wharton and Stillé believed that Casper's clinical experience was particularly useful in cases of alleged child rape because "there is no subject upon which it is more necessary for the physician to be guarded in his opinion than this."[118] They discussed numerous cases of false claims but also acknowledged that girls were often victimized because they made easier targets than did women.[119] Wharton and Stillé pointed to the number of cases that Casper had investigated personally and argued that similar numbers of child rape cases could be found in any country.[120]

Like Taylor, Wharton and Stillé argued that gonorrhea infection was convincing evidence of child sexual assault, but they also severely limited the types of cases in which a physician might comfortably make the diagnosis. However, unlike Taylor, Wharton and Stillé argued that gonorrhea could be diagnosed only in a child who also showed "marks of violence," physical indicators of rape or attempted rape that included a swollen labia, an inflamed and painful vaginal entrance, a purulent discharge, and pain when urinating and defecating.[121] As straightforward as their rule appears, Wharton and Stillé admitted that lasting or severe injuries were rare. Even more confusing, the "marks of violence" that Wharton and Stillé describe as competent evidence of assault are also the symptoms of gonorrhea vulvovaginitis. How could a practitioner know whether genital irritation, for instance, was due to sexual contact or to the chafing caused by the purulent discharge itself?

The crucial issue was whether a girl could acquire gonorrhea in the absence of sexual contact, and Wharton and Stillé approached it by discussing one of Casper's cases. In a trial for child rape, Casper had testified that the victim, an 8-year-old girl, was infected with gonorrhea and had been raped. However, because the defendant showed no signs of infection, the judge asked Casper whether "the common use of an *unclean chamber vessel* could possibly be the means of conveying the gonorrhoeal disease." Casper answered that he thought it possible, but not in a case in which the child had obvious genital injuries.[122] From this statement, Wharton and Stillé concluded that "there can be no doubt of the occasional transmission of gonorrhoeae by other means than sexual intercourse; but it is important for the physician to keep in mind the fact, that *in the case of children* at least, the presumption is entirely in favor of the ordinary mode of infection [sexual contact], unless the signs of violence . . . do not exist."[123]

In other words, a physician should presume, as Beck had advised, that a purulent discharge from a girl who showed other signs of rape was gonorrhea, but in the absence of physical marks he should neither diagnose gonorrhea nor conclude that the girl had been raped. Yet these symptoms rarely overlapped, as Wharton and Stillé knew; they had also reported Casper's statement that injuries to girls' genitals usually heal within forty-eight hours. Even Casper admitted that he had never examined a victimized girl sooner than three weeks and in some cases up to a year after the alleged assault, meaning that he would never have examined a girl who was suffering from both gonorrhea infection and genital injury unless the injuries were so severe that they were permanent.[124]

As much as they favored clinical evidence over theoretical speculation, the only other support that Wharton and Stillé offered for the possibility of nonsexual transmission was Burke Ryan's speculation, which they repeated as fact: "Dr. Ryan, nevertheless, examined two children who were infected with gonorrhoeae by using a sponge belonging to a servant girl who had the disease."[125] And Wharton and Stillé seemed to have had additional information or made an assumption about an important factor that the *London Medical Gazette* had not reported: the woman who told Ryan she was the source of infection was the family's live-in servant. Even if Wharton and Stillé's characterization of the infected woman is incorrect, they may have assumed that it was more likely that, if anyone in the household was infected, it was a servant and not her respectable employers. Perhaps the presence of a servant in the household also explains Ryan's lack of curiosity about whether his patients' father was infected. A servant, whom respectable people in Britain and America considered a source of both moral and physical contagion, may have seemed so obvious the source of infection that Ryan did not see the need to consider any others.[126]

American Practitioners Encounter Gonorrhea Vulvovaginitis

In the "Queries and Replies" section of the April 30, 1870, issue of the American *Medical and Surgical Reporter*, Dr. Jasper I. Hale asked the editors whether they had "ever seen a case of a specific gonorrhoea occurring in a female child as young as 5 years old." Hale had a patient who had been "purposely inoculated," but he had no idea how to treat her because he had never, either in his practice or in his "limited course of reading . . . met with such a case." Neither had the editors, who could recall only one "somewhat similar case."[127] Beginning in the 1870s, the American medical literature began to document the surprise of doctors who encountered gonorrhea among girls from genteel families, many of whom reported no history of sexual contacts. Doubtless in part to avoid the embarrassment that the diagnosis presented to both physician and parent, as well as to explain the infection's appearance in homes where they did not expect to encounter it, doctors became increasingly interested in the possibility of infection from nonsexual or fomite contacts.

Although doctors expressed their puzzlement over how the girls had become infected, they rarely wrote about father-daughter incest as a serious possibility. It was not the case that doctors had no reason to consider whether the girls had been sexually assaulted. American doctors often complained about mothers who, after discovering stains on their daughters' underpants or bed linens or a purulent discharge and redness around their genitals, became suspicious that someone had sexually assaulted their daughters.[128] But, like Sir Astley Cooper, they trivialized women, this time anxious mothers. Rather than acknowledge the important role women played in detecting child sexual assault, physicians dismissed all mothers as either ignorant or greedy, insinuating that their suspicions were always and obviously unfounded. Doctors' attitudes may have been a part of a more general development in nineteenth-century medical practice. Historian Jonathan Gillis argues that, to elevate their status in the early nineteenth century, doctors insisted that their diagnostic powers arose from their professional skill in discernment; everything they needed to know about a patient they could learn from their own superior powers of observation. One result was that after 1850 doctors were hostile to any observations or ideas that a mother, who had watched her daughter and the progression of her illness closely, might think useful to share.[129]

But why did mothers react with such distress to symptoms that doctors so easily shrugged away? If a girl complained of painful urination and chapped genitals, why would her mother worry that she had been sexually assaulted instead of assuming that she had simply acquired a common childhood disease? "Ignorant" mothers seemed to have known enough about sexually transmitted disease, which was, after all, rampant and unsusceptible to treatment in the nineteenth century, to realize that their daughters might have been raped. In New York City in 1822, for example, when Mrs. Jameison saw that her daughter's bed linens were spotted and discolored, she immediately suspected that the 10-year-old had been assaulted. Her daughter, Mary Ann, had not made any such claim, but when Mrs. Jameison questioned her, Mary Ann said that, when her father had sent her out to buy beer, Hugh Flinn, the grocer, had raped her. Mrs. Jameison confronted Flinn, who denied any wrongdoing, but she did not drop the matter. She took her daughter to a doctor, who told her that Mary Ann was infected with gonorrhea. Mrs. Jameison's next stop was the police station, where she filed charges. Flinn was arrested, and the doctor who examined him diagnosed Flinn with gonorrhea, too. At trial, three doctors gave contradictory testimony about whether the coincidence of infection was sufficient to corroborate the charge of rape, but the jury quickly found Flinn guilty.[130]

Like countless other women in the nineteenth century, Mrs. Jameison did not file charges until a doctor had confirmed her suspicions. Yet in the sixth edition of his *Treatise on Gonorrhoeae and Syphilis* (1877), Dr. Silas Durkee warned doctors about mothers who noticed a purulent discharge, which "may awaken alarm in the mind of the mother, and sometimes suspicion may be excited that foul play has been practiced upon the child . . . when such is not the fact."[131] But the problem for both victims and defendants was not "ignorant" mothers running amok or practitioners who should have given less weight to mothers' narratives than to their own diagnostic powers. Medical jurists and doctors who cast aspersions at infected girls' mothers and readily made confusing and unsupported pronouncements about the etiology of girls' infections unnecessarily complicated a diagnostic problem that was already genuinely troubling. Doctors' ability to diagnose gonorrhea was tenuous in the prebacterial era. But had they not been so eager to act on their gender biases, they may not have rallied so willingly around untested speculation. Having imagined the possibility of nonsexual transmission to girls, doctors had only themselves to blame when they found their authority questioned in courtroom debates over their skills, a process that made them look inept, unscientific, and lacking in the authority they strove to demonstrate.

Still, in the absence of any clear physical evidence of sexual assault or in those cases in which a girl neither complained that she was assaulted nor seemed to have suffered a traumatic event, there would have been nothing in the patient history suggesting that a vaginal discharge might be gonorrhea. Wharton and Stillé had argued that in the absence of physical marks a doctor should not diagnose gonorrhea, but late nineteenth-century American doctors increasingly accepted Dr. Burke Ryan's "proof" that girls were susceptible to gonorrhea infection from nonsexual or casual contacts. Detached from the implication of incest or sexual assault, doctors could diagnose the infection without having to parse the motives of anxious mothers or to address the issue of sexual assault, which, one doctor claimed, caused "visions of tedious law suits . . . [to] flit before him."[132] But just because they wished it so did not make it true. Doctors remained unable to explain satisfactorily how, besides sexual contact, girls became infected, a process about which they confessed to be perplexed long after the discoveries about bacteria revolutionized medicine in the last quarter of the century.

In 1878, Dr. Isaac E. Atkinson, who, the following year, would begin a career as a professor and then dean of the medical school at the University of Maryland, published the first original article on gonorrhea vulvovaginitis in an American medical journal.[133] Two years earlier, Atkinson had been one of six physicians to organize the American Association of Dermatology, whose members often specialized in the treatment of venereal disease, and he later served as both vice president and president of the Maryland Medical and Chirurgical [Surgical] Society, the state's professional medical association.[134] In contrast to Wharton and Stillé's statement that nongonorrheal discharges in girls were uncommon (part of their rationale for thinking it more reasonable than not to conclude that a vaginal discharge accompanied by genital marks was a symptom of gonorrhea), Atkinson stated that physicians routinely encountered girls with purulent discharges.

Citing both the "popular disposition to attribute such maladies in young children to criminal causes" and the importance of the medical opinion in a case of suspected child sexual assault, Atkinson would be the first of many to urge his colleagues to reach "a correct etiological understanding."[135] Atkinson discussed six cases of vulvovaginitis he had treated in the Baltimore charitable institution where he was attending physician. After questioning the girls, he concluded they had become infected from sexual contacts with one another. Atkinson was particularly swayed by one girl's confession that "some of the larger girls were in the habit of titillating the genitals of the smaller ones with their fingers, buttons, sticks, etc."[136] Atkinson did not suppose that the girls had become infected through means other than sexual contact, and he recognized that fingers and other instruments could spread infection without causing injuries.

But doctors who treated children living in respectable homes may have been less inclined to identify sexual contact as the source of infection. In the next American medical journal article on gonorrhea vulvovaginitis, Dr. John Morris's 1878 talk before the Baltimore Academy of Medicine, Morris advocated a more nuanced approach than Atkinson's.[137] Unlike most of his colleagues, Morris seemed unfazed by the task of making a differential diagnosis, and he expressed no hesitation in diagnosing gonorrhea in girls he believed had acquired the disease from nonsexual contacts. Morris could view the matter so calmly because he had seamlessly married his social biases with his medical views, leaving no room for any awkward conflicts to arise that might have required him to choose between them.

Of the cases he discussed, Morris conceded that sexual contact caused infection in only one—that of a 4-year-old girl molested by a "tramp" in the privy of "a tavern kept by a German."[138] In the case of a girl from a well-to-do family, Morris considered a different set of facts: that a "colored" servant in the house had gonorrhea, that he had also treated another girl who was the companion of the infected girl, and that all three used the same water closet.[139] In a third case Morris claimed that two girls became infected after using the same water closet as an infected "apprentice boy." In each instance, Morris placed the source of the disease on a person outside the nuclear family and of lower social position. Identifying the water closet as a shared object avoided the necessity of considering sexual contacts entirely.

In Morris's scheme, socially marginal "tramps," apprentices, and people of color were responsible for spreading the disease to genteel white girls. But Morris had also treated girls whose entire families were infected, and in those cases he named the girls' fathers as the person responsible for bringing the infection into the household. Implying that the father had been guilty of philandering, however, was as far as Morris was willing to go; he drew the line at suggesting incest. He imagined instead that shared bed linens and towels spread the disease from father to daughter, and he blamed the frequency with which girls became infected on the "prominence of the external sexual organs" in prepubertal girls. As Morris saw it, "The father of the family contracts the disease; he infects his wife; the little girls sleep with their parents on the bed linen, or towels become tainted, and, as a consequence, the female children are infected. I have treated a great many cases of gonorrhoea contracted in this manner."[140]

Yet Morris offered no reprimand to those family men who had, presumably, visited prostitutes and brought gonorrhea into their homes. He reserved his condemnation instead for people with "beastly habits"—all of whom lived not only outside the household but outside the United States. Morris claimed, for instance, that women in "Continental and Oriental countries" acquired rectal gonorrhea infections from "unnatural practices." But he believed that American women acquired rectal gonorrhea only when their own vaginal infection spread because of poor hygiene. Americans, he declared, did not engage in "unnatural practices."[141]

Assumptions about sexual behaviors as a function of a person's race, class, and nationality informed Morris's diagnostic skills: respectable Americans did not engage in behaviors like anal intercourse, incest, or sodomy. When the evidence suggested otherwise, Morris's discomfort is apparent. In a story about two male college friends infected with rectal gonorrhea, Morris added a footnote in which he attempted to explain away the unseemly implication about their character. He recounted how two of his friends at Princeton "slept, as was the custom at that time, in the same bed" and that, "as the weather was cold, they slept closely together, 'spoon fashion.'" However, "Mr. E." was infected with gonorrhea, and, as it turned out, shortly after spooning, so was Mr. E's companion. How could this have occurred? Doctors knew that men had sexual contacts with other men and that they could transmit venereal disease during these encounters.[142] But Morris blithely stated that "the cloth which Mr. E. used to swathe his penis became detached in the night, and greatly to his horror, he discovered in a few days that his friend had contracted gonorrhoea in the rectum."[143]

As easily as Morris conjured the dropped cloth, American doctors would soon join some of their European counterparts in identifying other common household objects—the chamber pot, bedding, washcloth, or linens—that might also be doing double-duty.[144] Against a growing tide, one experienced medical examiner remained unmoved. Dr. Jerome Walker insisted that gonorrhea in girls was a sexually transmitted disease and that contemporaneous gonorrhea infection in both the victim and alleged perpetrator was compelling evidence of assault.[145] Walker, who worked as a school physician and was active in Brooklyn community affairs, examined children for the borough's Society for the Prevention of Cruelty for Children (SPCC).[146] In 1886 he published an article in the *Archives of Pediatrics*, which had been founded only two years earlier as the first American journal in the new field of pediatrics.[147] In his article, Walker discussed his investigations between 1882 and 1885 into twenty-one cases in which a girl alleged she had been sexually assaulted.

In his work for the SPCC, Walker had discovered that "apparently respectable men, as well as ordinary disreputable characters[,] outrage children; that even fathers, step-fathers, and brothers will do it."[148] Unlike some European doctors, who seemed unable to comprehend why a man would sexually assault a child, Walker argued that men did so simply because they had the power to do so and to get away with it. Like earlier medical jurists, Walker explained that girls were easier prey than women because they are "more readily obtained and influenced, and because the risk of conception is not run, and because . . . children will be so influenced by threats and fear of exposure that they will not tell they were outraged."[149] All the girls Walker discussed were between the ages of 2 and 14, and except for one man charged with incest, all the defendants were charged with rape or indecent assault. Eleven of the girls, more than half, accused a family member: six natural fathers, three stepfathers, one foster father, and one uncle. Two of the girls who claimed to have been assaulted by their fathers were also infected with gonorrhea.

In the nineteenth-century medical literature on gonorrhea and child sexual assault, Walker's openness to the evidence is remarkable. He accepted gonorrhea as proof of assault even without signs of genital injury. He criticized his colleagues for believing that child sexual assault always resulted in substantial physical injuries. He argued that few men attempted penetration, that most attempted only to rub their genitals against a girl, and that medical jurisprudence textbooks overstated the typical severity of girls' genital injuries.[150] Walker noted that even when the assailant did attempt penetration visible physical injuries were rare, either because penetration was not completed or because the man lacked the physical capability to injure the girl: "We are to bear in mind, as physicians, that the size of a man does not necessarily determine the size of his instrument."[151]

Except for one girl he described as extremely dirty, Walker diagnosed gonorrhea in every girl whose alleged assailant was also infected.[152] But he was not yet using bacteriologic testing, and he complained that his inability to make a conclusive differential diagnosis hampered his ability to recommend prosecution. In one such case, Walker examined a 2-year-old girl with a vaginal discharge but no genital injury. The girl's parents were separated, and she shared a bed with her father, who showed no symptoms of infection. Walker took into account a claim by the girl's mother that two years earlier an older daughter who had also shared the father's bed had contracted gonorrhea. However, because the alleged assailant was not also infected and the victim had no apparent genital injuries, Walker felt unable to state with certainty that she was infected.

Walker never suggested that girls could have become infected from nonsexual contacts; nor did he consider dirt or poor hygiene a cause of gonorrhea.[153] But he complained that a "shrewd" lawyer could exploit the doubt inherent in the diagnosis.[154] Such was the case for William Moore, an Ohio man convicted in 1867 of incest with his 5-year-old daughter, Fanny. The trial court had been persuaded by medical testimony that father and daughter were both infected with gonorrhea. However, the examining doctor's admission that he could not be absolutely certain that Fanny's infection was gonorrhea rather than leucorrhea was enough for the Ohio Supreme Court to reverse Moore's conviction.[155] Walker acknowledged that such experiences had made him reluctant to make the diagnosis in some cases. And he lamented the result: a more lenient sentence or a favorable plea bargain for the defendant.[156]

Yet even as intent as Walker was about obtaining justice for victimized girls, the issue of whether the discharge was leucorrhea or gonorrhea was his last, not his first, concern. He may have been more willing than most of his colleagues to acknowledge the occurrence of incest, but Walker also shared the medical jurists' sense of responsibility for protecting men from false accusations. "Owing to the natural sympathy of both judge and jury for ill-treated children, a prevalent respect for innocence and purity, as well as a current dislike for unnatural crime," he reasoned, "a man charged with indecent assault, rape, or an attempt at rape upon a child, though he has good legal talent to defend him, stands a poor chance of acquittal if a reputable doctor swears that the child has been tampered with."[157] Walker's concerns were justified in the era before bacteriologic testing became standard. The way in which he dealt with his concern, however, exhibited the same misogynist bias that had long influenced the analyses of medical jurists.

Before he examined the physical evidence, Walker first investigated whether the girl or her family had a motive to lie. Only after he had conclusively ruled out any reason for the child to lie, considered the probability of the victim's and assailant's respective stories, and confirmed that the defendant had a motive for engaging in the assault did Walker consider the physical evidence.[158] In each case, he scrutinized the motives of every woman involved: the victim, her mother or guardian, and any witnesses. Because there was no physical evidence in most cases, Walker relied heavily on his impression of whether the people he interviewed were telling him the truth, taking into account the "feeble-mindedness" of the victim and the respectability of each person involved.[159]

But across the country on the Pacific coast, mothers, doctors, and the lay public showed none of Walker's hesitation. On May 2, 1888, just over a century since Jane Hampson's death in Manchester, a doctor in Los Angeles confirmed Mrs. Peter Butti's suspicion that her husband had sexually assaulted their 8-year-old daughter and "inoculated her with a loathsome disease." Mrs. Butti had not wanted to believe that her husband could have committed incest, but, as the *Los Angeles Times* reported, "it was only when the child's condition became so bad that a physician was called that she learned the truth" and called the police. The *Times*, noting that the "Butti Brute" was a "disreputable-looking Italian," called the matter "one of the worst cases of total depravity in the criminal history of the county."[160]

Yet even though the *Times* did not identify the disease by name, it did not shy away from publishing a second story on the Butti arrest the next day. It reported that Dr. Kannon, who was treating the "little victim of her unnatural parent's lust," had confirmed that both father and daughter were suffering from "the disease."[161] The *Times* again did not identify "the disease"; nor did it offer any explanation about why the diagnosis carried a strong insinuation of guilt. It did not need to. Just as the phrase "9/11" needs no explanation for contemporary Americans, nineteenth-century Americans understood gonorrhea infection as proof of sexual contact, including among children.

Little had changed over the century: mothers still recognized and became alarmed when their daughters displayed symptoms of gonorrhea, and they turned to medical and criminal authorities to help them protect their daughters. Doctors' misogyny accentuated their misgivings about the crudeness of their diagnostic abilities and supported their reluctance to make the diagnosis or to identify cases of child sexual assault. But within a decade after Dr. Jerome Walker's article, medical examiners would reject his emphasis on discerning the motives of all interested parties as irrelevant to making the diagnosis. Advances in bacteriology and technology would enable doctors to test a girl's vaginal discharge to learn whether she was infected with gonorrhea.

The objective clarity of bacteriologic analysis should have soothed doctors' anxieties about rendering an opinion in a case of alleged child sexual assault. But the result was not, as Walker might have expected, a more aggressive role for medical examiners in detecting and prosecuting incest or an increased appreciation for mothers sufficiently knowledgeable to recognize the signs of disease and assault. It was exactly the opposite.

CHAPTER THREE

Gonorrhea and Incest Break Out

TURN-OF-THE-CENTURY technological breakthroughs provided doctors with the means to diagnose gonorrhea vulvovaginitis to a reliable standard of certainty. These advances should have resolved their anxiety over diagnosing gonorrhea and child sexual assault and signaled the triumph of medical progress in the service of society. When physicians began to use the new technology, however, they were shocked to discover that gonorrhea vulvovaginitis was widespread among American girls. How could such a thing have happened? The possibilities were startling: either men from every class were regularly sexually assaulting girls, particularly their own daughters, or girls became infected so easily that they could acquire the infection from bed linens and bathwater. If doctors attempted any research to discover the answer, none of the major medical journals published it, and none of the country's eminent pediatricians mentioned it. Doctors who wrote about the etiology of girls' infections reflected the social rather than scientific concerns of the period. In a decade marked by lynchings in the South, unprecedented rates of European immigration in the North, the end of the Indian wars in the West, and a new focus on incursions against brown-skinned people overseas, doctors spoke of sexual behaviors—and assumptions about them—as the new borders between native-born, genteel, white Americans on the one side and an ignorant horde of savages whose presence threatened to destroy the nation on the other.

If sexual behaviors had, in fact, resonated only within private life, doctors may not have begun speaking publicly about topics that the Comstock laws and respectable mores deemed inappropriate. But though condemning child prostitution among the working classes was one thing, explaining how girls from good families had become infected with a sexually transmitted disease was another. On the cusp of moving from the margins to the center of American health care, doctors used their professional authority, if not their medical skills, to twist the etiology of girls' infections into existing narratives that fit more seamlessly with what doctors believed than with what they had discovered. In the process, doctors attributed fewer infections in girls to sexual assault. And father-daughter incest, which Americans had discussed so openly for most of the century, began to recede from public view.

Uncovering an Epidemic

When the American Medical Association (AMA) held its forty-fourth annual meeting in Milwaukee in 1893, gynecologist Charles P. Noble of Philadelphia read a paper naming gonorrhea one of the top five causes of diseases of women. The *Medical News* did not publish the text of Noble's paper, but in the discussion that followed Dr. Hoff of Ohio responded indignantly to Noble's statement "that many young girls have had gonorrhea." Hoff protested that, "while this might be true in Philadelphia, it will not hold good in Ohio."[1] In the 1880s, American medical journals began to publish translations of articles by German physicians about so-called epidemics of gonorrhea vulvovaginitis. German doctors attributed the upsurge of infections, in part, to the popularity of a superstition "in many countries . . . that a man infected with clap can be cured [by] intercourse with a virgin."[2]

American doctors did not react to the articles from Germany with alarm, even though Dr. Charles O'Donovan, professor of diseases of children at the Woman's Medical College of Baltimore and attending physician at its dispensary, claimed to have diagnosed so many cases of gonorrhea among African American girls that he considered the disease endemic in the city's black community.[3] Doctors like Hoff would never have imagined that white American men had any more in common with African Americans than with superstitious European peasants. Rather, medical journals affirmed doctors' comfortable assumptions about racial difference and disease as they smugly reported that gonorrhea infections in girls were "fortunately . . . much less frequent in this country than they appear to be on the continent of Europe."[4]

But new evidence soon proved at least some of their assumptions wrong. In 1894 Boston pediatrician and Harvard Medical School professor Dr. John Lovett Morse, who would write one of the first American textbooks on pediatrics, began to bacteriologically test purulent vaginal discharges from the girls he treated. When he found that five girls (presumably white, as he did not state their race) within a five-month period tested positive for gonorrhea vulvovaginitis, he was so surprised that he reported the "outbreak" in the *Archives of Pediatrics*, the major journal in the field.[5] Three years later, when Dr. L. Emmett Holt published *Diseases of Infancy and Childhood*, which would become the authoritative twentieth-century textbook on childhood diseases, he also noted that bacteriologic testing had demonstrated that infected girls were "very much more numerous" than doctors had assumed to be the case.[6]

One reason doctors considered the numbers so high was that they were diagnosing girls outside the circle where they expected to find them, among the relatively small numbers of victimized girls championed by child protection societies. Over the next ten years, doctors diagnosed gonorrhea vulvovaginitis so often that by 1904 an epidemiologist writing in the *Journal of Infectious Diseases* declared it "epidemic" among girls.[7] That same year the Babies' Hospital in New York City, which had implemented preadmission testing for girls, discovered that at least 5 or 6 of the 125 girls seeking admission each month were infected, including 5 in just one day.[8] These figures led Dr. Holt, the hospital's leading pediatrician and one of the country's most respected physicians, to conclude that the actual incidence of infection, not merely doctors' ability to detect it, was "steadily rising."[9]

Other physicians agreed, and they citied even more alarming figures. In 1909 Dr. Flora Pollack, who treated girls at the Johns Hopkins Hospital Dispensary, estimated that at least a thousand girls in Baltimore became newly infected each year, and she visited police stations and met with community groups to press for more aggressive prosecution of their assailants.[10] During the winter of 1911–12, the fifty-bed Children's Venereal Disease Ward at Chicago's Cook County Hospital, which Jane Addams called the "most piteous . . . of all children's wards," placed girls three to a bed and turned away many more.[11] And in just one month, September 1926, the Vanderbilt Clinic on Manhattan's West Side, the city's major provider of outpatient treatment for girls infected with gonorrhea, saw 218 new cases.[12]

But knowing how to diagnose the disease did not enable doctors to stop its spread. By 1927 the *American Journal of Diseases of Children* ranked gonorrhea vulvovaginitis the second most common contagious disease, after measles, among children.[13] Yet two years later the New York City Department of Health called it "the most neglected and poorly managed condition seen in medical practice."[14] Part of the neglect may have arisen from a combination of factors, including the historically low professional status of doctors who specialized in venereal disease and the lack of data collected to track its incidence. Before 1911, no state laws required doctors to report gonorrhea infections, and no agency collected statistics on infection.[15] The New York City Health Department began to collect data on venereal infection after a 1913 law required it to do so, but even then the only doctors who made reports were those employed by the department.[16] By the 1920s, only Washington, D.C., and Massachusetts had developed useful systems for tracking infection—and they consistently determined that girls accounted for approximately 10 percent of reported infections among females.[17]

Yet even if the number of infected girls approached 100 percent, doctors still had sound reasons to neglect it. The complications of gonorrhea vulvovaginitis were mild compared with the many other diseases that contributed to a frighteningly high rate of mortality among urban children.[18] Doctors routinely saw half of their hospitalized patients under the age of 5 die, and rates of death among infants and children in New York City at the turn of the century were comparable to those that Thomas Percival had boasted about in Manchester, England, a century earlier.[19] Diphtheria, for instance, which was on the rise and endemic in New York City in the 1890s, struck nearly 5,000 people in 1893, killing almost 1,400 children.[20] In contrast, many girls infected with gonorrhea vulvovaginitis were asymptomatic, and with its low incidence of death and debility, and without any reliable treatment at their disposal, doctors and public health officials may have considered it more of a nuisance than an emergency.

The larger problem was the fact of widespread infection itself. When they realized that infection had spread to girls outside the small group who either lived in tenements or claimed to have been sexually assaulted, doctors wondered whether they had been mistaken not only about the frequency with which girls became infected but also about how they became infected. Did the prevalence of infection suggest that girls were susceptible to infection by some means other than sexual contact? If so, doctors needed to quickly explore fomite or other types of nonsexual contacts to which girls might commonly be exposed. If not—if gonorrhea vulvovaginitis was, as medical jurists had taught, sexually transmitted to girls—then doctors had uncovered a repellent social problem of staggering proportions. Doctors agreed they needed to study the issue, but the magnitude of its implications for men of their race and class paralyzed them rather than provoked them into action.

In 1896 Dr. Herman B. Sheffield, an instructor at the New York School of Clinical Medicine and attending physician at the Hebrew Sheltering Guardian Society Orphan Asylum, reported an epidemic of sixty-five cases of gonorrhea vulvovaginitis at the orphanage.[21] Ten years earlier Sheffield's predecessor, Dr. William M. Leszynsky, an authority on public health, had reported an epidemic of vaginal discharges among the girls at the orphanage. He assumed the girls were infected with leucorrhea.[22] But bacteriologic testing provided Sheffield with the type of diagnostic certainty that Leszynsky had not been able to claim, and he was confident of his diagnosis. Yet even as Sheffield admitted that "there is still a great deal of dissension" as to the etiology of girls' infections, he did not believe that any of his patients had been sexually assaulted. He thought it improbable that so many girls could have simultaneously acquired a sexually transmitted disease. After reviewing the European medical literature on epidemics, he argued that the disease spread through bathwater, noting that the orphanage commonly bathed twenty to thirty girls together in one large tub.[23] One American medical journal, however, had mentioned such a possibility only to mock it as a "racket," like the "water-closet explanation," and predicted that doctors would soon relegate the notion "to the realms of fiction."[24] Sheffield shared these concerns and called on gynecologists, pediatricians, general practitioners, and medical jurists to "dispel any and all doubt as to the real nature of infectious vulvo-vaginitis" by making "a more careful study" of the etiology of infections in girls.[25]

But no such study was forthcoming. It is easy to understand why Sheffield considered it improbable that a sexually transmitted infection would break out simultaneously among so many institutionalized girls (he raised no suspicions about the staff). Yet doctors also ignored the likelihood of sexual contact, particularly incest, even when fathers and daughters who were both infected also shared the same bed. By 1900 Herman Sheffield had moved to the Metropolitan Hospital and Dispensary for Women and Children, New York City's public hospital. When he had written about the epidemic at the Hebrew Orphanage, Sheffield had thought it possible for girls to pass the infection with their hands. But now that his patients were living with their parents and not in an institution with other girls, Sheffield did not discuss the possibility that his patients' fathers might have transmitted the infection in the same manner. He refused to consider sexual assault unless a girl also had obvious genital injuries, which Sheffield knew rarely occurred.[26]

How, then, did Sheffield explain the means by which a girl living at home, under her father's care and protection, became infected? Looking again at reports by European doctors that American medical journals published in the 1880s and 1890s, he argued that "little girls sleeping with [infected] parents or elder brothers" became infected from "soiled bedclothes, cotton-pads, or rags."[27] Sheffield's commitment to the notion that objects "mediated" the "accidental" transmission of the bacteria from parent to child was so strong that he seemed to overlook the possibility of incest in even the most provocative patient histories. In one case, a man asked Sheffield to examine his 2-year-old daughter, who had a fever of 101.5 degrees and was constantly crying, especially when urinating, and whose genitals appeared "frightfully inflamed and swollen and bathed in greenish-yellow pus."[28] The father told Sheffield that two weeks earlier he had returned home from a four-month business trip and that, because he had "strained" himself (a euphemism for acquiring gonorrhea), he had told his wife he would abstain from sex. Sheffield diagnosed both father and daughter with gonorrhea, and although he did not note whether he had also spoken with the girl's mother, he accepted the father's explanation "that his child was sleeping with him and frequently made use of the chamber [pot], where he . . . dropped pieces of cotton on which he collected the pus flowing freely from his urethra."[29]

Nor did Sheffield voice any suspicions about incest in a second case, even after the patient's father admitted that he had initially lied about sharing a bed with his 4-year-old daughter. After Sheffield examined the girl and told her father that his daughter was infected with gonorrhea, the man became angry and refused to let Sheffield treat her. Three weeks later he returned with his daughter, who was complaining of abdominal pain and in "pitiable condition; the pus was literally pouring out of the vagina."[30] Sheffield offered no explanation for how she had become infected. Rather, he identified her as the locus of the disease in the home and stressed the importance of isolating her so that she would not spread the infection to her family, particularly through "the common use of privies, baths, beds, towels, etc."[31] Nor did Sheffield recommend that her father or other infected members of the household also be isolated, even though at least one of them had already spread the disease.

Sheffield's description of the girl's abdominal pain coincided with new reports about the severity of gonorrhea vulvovaginitis.[32] Men who practiced medicine in the 1890s had grown up believing that "the clap was a pesky infection, little more than a cold," or part of a college man's rite of passage.[33] But as doctors increasingly encountered girls suffering from the complications of untreated gonorrhea vulvovaginitis, they began to think of it as "one of the most dangerous microorganisms," not simply a rare or minor disease they might expect to see only in a medico-legal proceeding.[34] Doctors knew that the consequences of untreated gonorrhea in women include peritonitis (acute abdominal inflammation), salpingitis (pelvic inflammatory disease), and even death, but they were unnerved to discover the same sequelae in girls.[35] In November 1901, gynecologist Guy L. Hunner and bacteriologist Norman MacL. Harris told their colleagues at the Johns Hopkins Hospital Medical Society that although death was an infrequent complication of gonorrhea in females, the majority of females who died from acute gonorrheal peritonitis were girls.[36] The symptoms of acute gonorrheal peritonitis include abdominal distention, tenderness, and rigidity; vomiting; high fever; high blood pressure; pus filling the fallopian tubes; and pus filling the abdominal cavity and covering the appendix and intestines. Girls died slowly, after having endured heroic but futile surgical measures, including having their intestines manually wiped clean of pus, to save them. Hunner and Harris described the final hours of girls as young as 3 years of age as filled with "frightful pain, loud crying, failure of the pulse, vomiting and heart collapse," and they concluded that untreated gonorrhea in girls was an urgent medical problem.[37]

New questions about the severity of the disease were not the only problem. In 1903 Dr. Reuel B. Kimball, an attending physician at the Babies' Hospital, charged that "the medical world at large" had not realized "the extent to which gonorrhoeae occurs in children."[38] He accused medical journals—which published many more articles on gonorrhea ophthalmia in infants (a condition that affects the eyes and can cause blindness) than on genital infections in girls—of having contributed to the impression that gonorrhea vulvovaginitis was an insignificant problem. Kimball claimed that of the 600 children admitted to the public wards of the Babies' Hospital in 1902, only 1 became infected with gonorrhea ophthalmia, compared with 69 with gonorrhea vulvovaginitis.[39] The following year, Dr. Sara Welt-Kakels, who treated patients at the Children's Department of the Mount Sinai Hospital Dispensary in New York City, called gonorrhea the most pernicious and tenacious infection of early childhood.[40]

The new data suggesting that doctors had underestimated both the prevalence and seriousness of the disease did not, however, stimulate new research into its etiology. Instead it provoked new speculations about a causal link between infected family members and poor hygiene. Like Dr. John Morris in the prebacterial era, doctors rattled off lists of everyday objects that might be responsible for accidentally transmitting the bacteria from parents to their daughters. In 1898, for example, *JAMA* reported on a paper that pediatrician Isaac A. Abt had given at a meeting of the Chicago Medical Society, which included such a list.[41] Abt, who later became president of both the Chicago Pediatric Society and the American Pediatric Society, was on the staff at Chicago's Michael Reese Hospital, which had opened in 1881 to treat poor immigrants, and a professor of diseases of children and pediatrics at the Northwestern and Rush medical schools. Because he treated infected girls in both his private and hospital practices, Abt knew that gonorrhea vulvovaginitis was not confined to socially marginalized populations, and he warned that "none are exempt."[42]

Abt believed that rape was only "rarely" the source of infection. He claimed that girls (and not boys) became infected from everyday contacts with soiled sponges, bed linen, towels, thermometers, soap, bathwater, and unclean fingers.[43] Such a list assumed that at least one other member of the household was also infected. However, by displacing the source of infection from people to objects, Abt could account for a girl's infection without having to also confront the medico-legal implications that had traditionally been the focus of the diagnosis. Abt may have felt he had unlocked the etiology of the disease, but implying that the disease might be so easily spread that it could be endemic in girls offered no solution to the problem of curbing its spread.

Ward Epidemics

Because gynecological exams were not routine for girls, doctors often discovered girls' vaginal discharges only after they had been admitted to a hospital for treatment for another ailment. Doctors in the prebacterial era assumed such discharges were caused by uncleanliness and shrugged them off as leucorrhea (nongonorrheal vaginitis). But when doctors began to test girls' discharges bacteriologically in the late nineteenth century, they were astonished by the results: nearly every sample tested positive for gonorrhea.[44] Dr. Reuel B. Kimball had served as an officer on the board of directors of the Babies' Hospital since the 1890s, and he claimed that every physician who treated girls in a hospital had dealt with "ward epidemics" of gonorrhea vulvovaginitis.[45] Doctors assumed that girls acquired the infection during their stay, and so they turned their attention to ridding every vestige of *Neisseria gonorrhoeae* from their wards.

Doctors' suspicions about the ease with which infection could spread among girls in a hospital ward were well founded. In an era when hospitals had little money to spend on supplies, it was common for nurses to use the same instrument, such as a thermometer or washrag, successively for each child on the ward.[46] Without an effective antiseptic solution, a washrag used serially to clean girls' genitals or a thermometer soiled with infected pus and placed immediately into the rectum of one girl after another might have provided exactly the type of warm and moist environments in which *N. gonorrhoeae* could remain virulent. Doctors like Hunner and Harris, who had watched girls die of peritonitis, demanded that hospitals stop such careless practices and establish rigorous measures to prevent the spread of infection in girls' wards, regardless of cost.[47]

Dr. Henry Koplik, attending physician of the children's ward at New York City's Mount Sinai Hospital, agreed. In 1904 he expressed the frustration of many pediatricians when he called gonorrhea "one of the most annoying scourges of a children's hospital" because "the little sufferer . . . is a menace to others."[48] Koplik's strong words were justified. When he took charge of the children's ward in 1890 and tested the girls for gonorrhea, every test came back positive. A native New Yorker, Koplik graduated from the Columbia University College of Physicians and Surgeons, and after postgraduate medical training in Europe, he returned to Manhattan in 1887.[49] He first became attending physician and then director of the Good Samaritan Dispensary and then adjunct visiting physician to the Children's Service at Mount Sinai until his appointment as director in 1890.[50] With Abraham Jacobi, who established the first pediatrics service in the United States at Mount Sinai in the 1870s, Koplik co-founded the American Pediatric Society in 1888, of which he also served as president.[51] To eliminate gonorrhea from the girls' ward, Koplik devised strict new hygienic procedures that, because his colleagues described him as so severe that they "were literally afraid of him," we can assume they carefully implemented.[52] He was able to boast at the May 1903 meeting of the American Pediatric Society that his efforts had kept the ward free of infection for five months, and by March 1905 he claimed that the ward had not experienced an epidemic in two and a half years.[53]

Mount Sinai achieved greater success than most hospitals. Dr. L. Emmett Holt, attending physician at the Babies' Hospital, became so frustrated by his inability to eliminate the bacteria—which persisted even after the hospital erected a new building—that he likened the presence of an infected girl in a ward to Satan, a "serpent [who] came in the form of a child."[54] Holt's stature was such that he succeeded Abraham Jacobi, the "father of pediatrics," not only as attending physician at the Babies' Hospital but also as professor of diseases of children at the Columbia University College of Physicians and Surgeons. In addition to his pediatrics textbook, Holt wrote a popular guide for new mothers, and he was a founding member and two-term president of the American Pediatric Society.[55] At his memorial service at the New York Academy of Medicine in 1924, his colleagues eulogized Holt as the "foremost pediatric authority in America."[56]

Holt thought the task of eliminating gonorrhea vulvovaginitis so important that he implemented expensive and time-consuming sanitary improvements, even though most of the families the Babies' Hospital treated were too poor to pay for existing services. In 1903 the Babies' Hospital admitted 610 patients, collected $680 in fees, and spent only $1 a day per child.[57] Yet Holt insisted on costly improvements that ranged from giving each girl her own towel to fumigating the woodwork in the girls' ward. When none of these improvements worked, the hospital tried to screen out the bacteria by testing every girl before admission.[58] The hospital turned away any girl whose parents refused to permit her to be tested, or who tested positive for gonorrhea, except in a dire emergency. Still, the hospital sometimes discovered an infection only after a girl had already been admitted to the ward. In those cases the hospital either made the infected girl wear diapers to keep the discharge off the hospital linens, isolated her in a special detention ward, or tied a red warning ribbon to her bedpost.[59]

Holt was fully aware of the dollars and hours the hospital had wasted in its unsuccessful strategy to eliminate the disease. Yet he continued to insist that unsanitary hospital practices were the source of "ward epidemics" even though he also admitted that "no evidence whatever could be found to support this opinion."[60] At a loss for anything better to do, frustrated hospitals around the country followed Holt's lead and put similar measures into place.[61] But preadmission testing had disturbing ramifications for the health and well-being of girls. A positive preadmission test left a girl with few or no prospects of receiving inpatient care for whatever other illness had brought her to seek admission to the hospital in the first place.

Worse still, when doctors focused on ward epidemics, they ignored the greater problem of infection among the general population of noninstitutionalized girls, which continued to rise. Doctors knew that the disease had spread "even in children living under the best surroundings," but they avoided the more perplexing question of how girls who neither had been hospitalized nor reported a history of sexual contact had become infected.[62] Holt, for one, rejected the possibility of incest and was angered by what he perceived as the lay public's reluctance to discard the popular view that gonorrhea was sexually transmitted to girls. He believed that linking infection with sexual contact unfairly burdened "innocent" girls with a diagnosis that stigmatized them as sexually precocious and immoral. Holt advocated instead what he believed to be a modern and efficient solution, changing the name of the disease to disassociate it from any "venereal origins."[63] But rhetorical sleights of hand, even from someone as esteemed as Holt, could neither reduce the incidence of infection nor calm doctors' growing distress over it.

Reassessing the Meaning of Infection: The Infection at Home

As pediatricians battled infections among populations of institutionalized girls, medical examiners continued to diagnose gonorrhea among girls involved in criminal proceedings. After 1890, doctors for the New York Society for the Prevention of Cruelty to Children (NYSPCC) in Manhattan tested the vaginal discharges of every girl who claimed to have been sexually assaulted. That doctors could now confirm that girls were, in fact, infected with gonorrhea implied that girls' accusations about male sexual misconduct were more truthful than doctors had believed—and apparently more than they could accept. Just as they achieved the diagnostic certainty that should have resolved the concerns represented in the Jane Hampson case, medical examiners abruptly changed course.

Bacteriologic testing removed doctors' personal discretion in making a diagnosis. Physicians could no longer refuse to diagnose gonorrhea when it suited them, such as in those cases in which they wished to avoid insulting a girl's parents, questioned female motives, or deemed the alleged perpetrator an unlikely assailant. Medical examiners, who might have been expected to have most appreciated the new diagnostic tool, reacted to the loss of discretion—and the implications of the new data—by inserting new ambiguity into the infection's etiology. Surprisingly, they attacked the forensic utility of the diagnosis in cases of child sexual assault. After the NYSPCC was organized in 1875, it quickly began to influence every stage of the city's prosecutions for child sexual assault, and doctors who worked for the society examined victimized girls for corroborating physical evidence.[64] But sharply different views, even among doctors who worked for the NYSPCC, about the etiology of infection returned some measure of diagnostic control to doctors, particularly with regard to their willingness to acknowledge incest.

In 1894, NYSPCC medical examiners James Clifton Edgar and J. C. Johnston published a chapter about forensic evidence of rape in a medical jurisprudence textbook edited by R. A. Witthaus and Tracy C. Becker.[65] Little is known about Dr. Johnston, but Edgar was a prominent Manhattan obstetrician whose wedding plans the social pages of the *New York Times* followed closely. Edgar had graduated first in his class at the New York Medical College and completed postgraduate studies in Europe before returning to Manhattan as professor of obstetrics, first at New York University and then at Manhattan's Cornell University Medical Center. In addition to his work for the NYSPCC, Edgar had a private practice, served on the staffs of numerous hospitals, was a fellow of the American College of Surgeons and the New York Academy of Medicine, and was elected president of the New York Obstetrical Society.[66] He also translated German medical works, wrote a textbook on obstetrics, and in the 1920s co-wrote the introduction to a popular advice manual for new mothers.[67]

In their chapter in Witthaus and Becker's *Medical Jurisprudence*, Edgar and Johnston discussed their investigations into 176 allegations of child rape.[68] Edgar and Johnston had diagnosed some of the girls with gonorrhea, which they stated was "more often adduced as evidence of rape upon girls than perhaps any other symptom."[69] But they did not share their colleagues' view that the diagnosis was persuasive evidence of sexual contact. Edgar and Johnston believed instead that girls were susceptible to gonorrhea infection "in some way other than by impure intercourse," which, if true, undermined its forensic value.[70] In the decade since Dr. Jerome Walker wrote about his cases in Brooklyn, white middle-class New Yorkers had become increasingly alarmed about the "illiterate and ignorant swarms" crowding into tenement slums.[71] In 1890 Jacob Riis published his highly praised *How the Other Half Lives*, which documented the shocking conditions of the city's tenements and the hard-scrabble lives of the dark-skinned immigrants who crowded into them.[72] News of this urban underbelly both frightened and scandalized genteel white Americans. Stories in the *Times* condemned the greedy landlords and ineffective municipal government that had permitted and even encouraged the tenement-housing system, which they claimed had become the locus of the disease, immorality, and crime that had turned New York into the "costliest, wickedest, and deadliest" city in America.[73]

In step with their contemporaries, Edgar and Johnston argued that impoverished living conditions, particularly the "habits of herding and of uncleanliness, so common in the tenements . . . [,] are strikingly conducive to the spread of [gonorrhea], particularly among children."[74] Their only evidence was Dr. Burke Ryan's speculation. His views so perfectly supported their own social biases that, even as they declared their surprise that every tenement child had not become infected, they did not stop to question why, if gonorrhea was so easily acquired, this was so. Rather, they warned that infection spread to girls by "the use of common sponges, towels, even the seats of closets."[75] Girls who sought help and protection from fathers who victimized them bore the consequences of Edgar and Johnston's refusal to connect infection with assault. Their chapter offered no details that might explain the conclusions they reached about the girls they examined, which included nine cases involving father-daughter incest. In four of these cases, Edgar and Johnston refused to recommend that the defendant be prosecuted. Three fathers entered into plea agreements. Of the two fathers whose cases went to trial, only one was convicted.[76]

By claiming that poor hygiene was the source of girls' infections, Edgar and Johnston eliminated men, even poor and immigrant fathers, from suspicion of sexual misconduct. But the medical examiners were not simply exhibiting empathy for less fortunate members of their gender. Like the medical jurists, Edgar and Johnston insisted, as though it was a self-evident fact, that most accusations were false. Edgar and Johnston did not accuse the girls themselves of having ulterior motives but recounted Sir Astley Cooper's "famous 'warning'" about ignorant or conniving mothers who storm, threaten, and torment "the little one into accusing an innocent person, a man against whom she may have a grudge."[77] Edgar and Johnston went a step further, displacing all responsibility for girls' infections from men to poor sanitation, a shift that not only rendered every mother's suspicions baseless but promoted the view that women, who were supposed to keep their homes and children clean, were responsible both for their daughters' infections and for any subsequent miscarriage of justice.

Edgar and Johnston's colleague Dr. W. Travis Gibb viewed the matter differently.[78] Gibb was a well-known surgeon who, in addition to examining children for the NYSPCC, maintained his own Manhattan sanitarium and served with Dr. Prince A. Morrow (whom I discuss in the next chapter) on the staff of the New York City Home for the Aged and Infirm, which named a new surgery wing in his honor.[79] Gibb began examining girls in 1881, and he complained in 1894, the same year that Edgar and Johnston published their chapter on child rape, that doctors overlooked many instances of assault because they were too quick to attribute infections among poor girls to "filth or some constitutional disease [scrofulous]."[80] In "Indecent Assault upon Children," a chapter he wrote for a forensic medicine textbook, Gibb affirmed Alfred Taylor and Jerome Walker's view that gonorrhea infection was more convincing than genital injuries as forensic evidence of sexual assault. But Gibb's assumptions about poverty, nationality, and sexual practices limited his ability to detect child sexual assault and incest. Reminding readers that pedophilia was common in southern Europe, a notion he did not question, Gibb argued that most child sexual assaults occurred "among the poorest and most depraved classes of people."[81]

The disagreements between Gibbs and Edgar and Johnston made little difference. Whether a doctor attributed a tenement girl's infection to assault or to uncleanliness, these sources conformed to doctors' assumptions about "foreigners" and poverty. At the 1901 meeting of the AMA, Dr. Abraham L. Wolbarst, a specialist in venereal disease in males, argued that tenement housing encouraged exhibitionism, homosexuality, and incest.[82] Wolbarst was an assistant surgeon at the Good Samaritan Dispensary in New York City, a unique vantage point from which to think about the relationship between tenement life and disease. As a letter to the *Times* noted in 1889, "There are probably no other persons who see so much of the inside life of the poor and distressed of our city in their own homes as the visiting doctors of the various dispensaries."[83]

Good Samaritan, which had been called the Eastern Dispensary before 1895, was one of the city's oldest dispensaries. It was located in a four-story building on Manhattan's Lower East Side in what Jacob Riis called the "heart" of "Jewtown."[84] In 1885, the *Times* reported that the dispensary had spent an average of only twenty-three cents to treat each of the nearly 23,000 patients it had seen in the previous year, including the salaries of one house physician and three visiting doctors; sixteen additional physicians donated their services.[85] According to Riis, the managers of Good Samaritan "told the whole story when they said, 'The diseases these people suffer from are not due to intemperance or immorality, but to ignorance, want of suitable food, and the foul air in which they live and work.' "[86] But Dr. Wolbarst did not entirely share this view. He criticized the "cramped quarters" and "lack of proper sleeping accommodations" his patients endured, but he held the tenement dwellers responsible for the low morals and lack of self-restraint that led to the "promiscuous intermingling of the sexes."[87]

Whether they attributed gonorrhea vulvovaginitis to uncleanliness, overcrowding, or immorality, doctors were disgusted but not surprised by the incidence of gonorrhea vulvovaginitis among girls like those whom Riis photographed. But doctors' assumptions about degraded housing conditions and cultural differences did not explain how the infection had spread into the homes of "respectable" white Americans. Doctors found themselves caught in the midst of a public health epidemic that seemed to have no discernible origin and for which they were unprepared. For a generation so self-consciously modern, their response would be surprisingly backward looking.

A Progressive Era Anomaly

Doctors could not reach a consensus on the etiology of the disease even when they gathered at medical conferences to discuss it with the most renowned pediatricians in America. After Dr. Wolbarst presented a paper on gonorrhea in boys to the section on diseases of children at the AMA's 1901 meeting, *JAMA* published both the paper and the discussion that followed.[88] Wolbarst had developed a specialty treating gonorrhea in males, and he reported a spike in the incidence of infection among boys over the previous two years (he had treated twenty-two boys).[89] Doctors never speculated whether boys could acquire gonorrhea from nonsexual sources, and Wolbarst, who feared that the incidence of gonorrhea among both boys and girls was escalating, criticized a growing consensus among doctors that girls could acquire it from fomites. Like his colleagues, Wolbarst associated gonorrhea infection with poverty. But he believed that sexual contact, not poor hygiene, was the source of infection in children "in more cases than we are accustomed to believe." He blamed half of his cases on an infected parent, brother, or sister who slept with the child.[90] Wolbarst held female prostitutes responsible for the other half, claiming that they preyed on virginal boys because of a folk belief that sexual intercourse with an uninfected partner would pass on and so cure one's own infection.[91]

In the lively discussion that followed the presentation of his paper, no one challenged Wolbarst's remarks about boys. But many of his colleagues were distraught over his insistence that girls became infected in the same manner. One of the few to support Wolbarst was Dr. Alfred Cleveland Cotton, an authority on diseases of children and a future president of the American Pediatric Society.[92] Cotton told his colleagues how his views about child sexual assault had changed when he served as Chicago city physician from 1891 to 1894. Called to investigate girls who had been sexually assaulted, Cotton confessed to having been both "astonished" to discover the "large number" of assaults and "surprised at the large number of cases of gonorrhea" among them.[93] Cotton had assumed that most men who assaulted girls attempted intercourse, and so he had believed that gonorrhea infected only those girls who had been brutally attacked. But as city physician he had learned that men more often assaulted girls by "playing" with their genitals, which led him to conclude that even girls without genital injuries could be infected.

But Cotton never mentioned the possibility of incest, and the other physicians who shared his belief that girls became infected from sexual contacts proposed scenarios that explicitly avoided it. Dr. Edwin Rosenthal of Philadelphia, for instance, admitted to being "a little skeptical regarding . . . infection from dirty water-closets." He imagined instead that when multiple families were crowded into one tenement space, men assaulted one another's daughters.[94] Only Dr. John C. Cook raised the subject of incest overtly—and then only to reject the possibility out of hand. Cook specialized in diseases of children and was superintendent physician at Chicago's Jackson Park Sanitarium, known as La Rabida because it was housed in the renovated Spanish pavilion on the site of the 1893 Columbian Exposition. He formed the Chicago Pediatric Society in the 1890s and served as its president until 1899, when he turned the office over to Dr. Cotton.[95] Listening to his colleagues discuss the etiology of the disease, Cook lost patience with their inability to reach a consensus. He sharply criticized Wolbarst's paper as "lacking in scientific knowledge," and he argued that "the time has arrived when we should know whether it is possible to transmit the gonococcus from fabrics to human beings."[96]

Yet Cook did not follow his blunt assessment by offering either a more cogent explanation or one that might better reflect the dedication to empiricism and science that characterized the Progressive Era. Rather, he declared, "It is trying to our credulity to find a 4-year old daughter and a 35-year old father having gonorrhea at the same time with no other source of infection to the daughter other than the father, and yet I have observed this in a family of educated and refined people." Cook's relief was palpable when he added, "I am glad to hear it restated that it is possible to contract the disease in a water-closet."[97]

Cook's refusal to acknowledge the possibility of incest, and his eagerness to talk about sanitation instead, would become the hallmark of the Progressive Era position on the etiology of girls' infections. The logic of Dr. Sara Welt-Kakels, who diagnosed nearly 190 cases between 1894 and 1904, primarily in girls between the ages of 2 and 5, was typical of that position.[98] Welt-Kakels was attending physician at the Children's Department of the Mount Sinai Hospital Dispensary and one of the first women to receive hospital privileges in New York City.[99] Even though she had treated some of her patients for years and knew that some of their family members were also infected, Welt-Kakels admitted that it was often "very difficult, and even impossible, to obtain a history of the infection."[100] Yet she did not hesitate to claim that none of her patients had been sexually assaulted. She believed that the disease spread by "an indirect and accidental mode of infection," meaning that the girls infected themselves by "handling contaminated bedlinen, towels, or other utensils" and then "conveying the virus to the genitalia on their fingers."[101]

Welt-Kakels's experience with ward epidemics might have reasonably fueled such speculation, but she seemed blind to the possibility of incest—when she examined a 4-year-old girl suffering from a "very copious" discharge and vaginal bleeding whose parents had waited a year before taking her to see a doctor; when parents were reluctant to help her to identify the source of their daughter's infection; in response to the lack of evidence supporting her speculations about fomite transmission; and even in the presence of genital injuries.[102] Perhaps Welt-Kakels wanted to avoid judging her patients' families, mostly poor Russian Jews, too harshly.[103] But doctors were willing to accept any explanation for how girls became infected—other than incest. Manhattan physician W. D. Trenwith told the Section in Pediatrics of the New York Academy of Medicine in 1905 that none of the girls he had treated for gonorrhea, most of whom were between the ages of 4 and 6, had been assaulted.[104] Trenwith knew that many of his patients' fathers were infected and that these men shared beds with their daughters. But he explained that, after having made inquiries of "the most searching character," he had concluded that in 75 percent of his cases the gonorrhea resulted from "indirect infection by the father," by which he meant that the bed linens had, in a manner he did not explain, transmitted the infection. Trenwith held the girl's playmates and sisters responsible for the rest.

Reshaping a National Identity

In the 1880s, Dr. Jerome Walker had insisted that all sorts of men "outraged" children, including "apparently respectable men."[105] But new European research on sexuality reached America just as the revulsion white Americans felt toward dark immigrants—who they imagined were poised to overwhelm the increasingly chaotic urban centers in the North—reached its peak. Historian John Higham characterized nativism, the jingoistic reaction of white Americans to an increasingly diverse society, as "running at full tilt" in the 1890s, with white urban Americans viewing southern and eastern Europeans as a "particularly insidious representative of the whole foreign menace."[106]

Searching for ways to explain the behavior of people they assumed were so unlike themselves, American doctors embraced the work of Richard von Krafft-Ebing, a German-born psychiatrist and contemporary of Sigmund Freud in fin-de-siècle Vienna. An English translation of Krafft-Ebing's groundbreaking *Psychopathia Sexualis*, an "encyclopedia of sexual aberration," became available in America in 1893.[107] Many physicians in pre-Freudian America were deeply impressed with Krafft-Ebing's analysis of seemingly inexplicable or unnatural sexual behaviors.[108] They began to attribute behaviors like incest to psychopathologies that determined their occurrence, a premise that seemed to affirm a categorical difference between "ordinary" white American men and the immigrant and other lower-status men they imagined shared a proclivity to sexually assault children.

Perhaps unsurprisingly, considering the hostility with which Freud's peers reacted to his suggestion that father-daughter incest was widespread among the Viennese upper class, Krafft-Ebing found it "incomprehensible that an adult of full virility, and mentally sound, should indulge in sexual abuses with children." He declared men who molested children as not "ordinary."[109] *Psychopathia Sexualis* appeared in America only one year before Edgar and Johnston published their essay on rape, and they echoed Krafft-Ebing's views when they argued that the only men capable of sexually assaulting children were either a "broken-down debauchee or a very young man who, for some reason, has doubts about his virile powers."[110]

Genteel white Americans in the 1890s would never have considered a respectable "native-born" family man as having anything in common with such men. They conceptualized the "white father" as the exact opposite. The "white father," long the symbol of America's national strength, had mastered self-restraint and developed a refined moral sensibility.[111] Earlier in the century, Americans knew—from newspapers, trial testimony, and the revelation of incest in their own communities—that genteel white men not only were capable of father-daughter incest but committed it with appalling frequency. However, as Gail Bederman argues in *Manliness and Civilization*, turn-of-the-century discourses treated associations between civilization and whiteness on the one hand and primitiveness and blackness on the other as natural and evolutionary.[112]

As the social hierarchy convulsed from waves of immigration and the growing mobility of African Americans, sexuality and race became intrinsic elements in the construction of binary categories that described dominant and subordinate social identities, including white/black, ordinary/deviant, and civilized/savage. These binary categories served not only to represent but also to further marginalize people by race, gender, class, and sexuality. In more advanced societies, according to this logic, the incest taboo would have evolved into a deeply internalized barrier against primitive desires like father-daughter incest. In this sense, an "incest taboo" was evidence of the progress that left stark differences between civilized and savage societies. Such discourses were important elements of nation building, and incest was decidedly "un-American"; if incest occurred among "ignorant" southern Europeans, it could not occur among "civilized" white Americans.

Medical discourses about incest and gonorrhea helped shape this process. They lent scientific authority to the idea that respectable white fathers naturally protected their children from the sexual dangers that other types of men embodied, a paradigm that still left doctors unable to account for an epidemic of a sexually transmitted disease in respectable white families. As Amy L. Fairchild argued in her history of immigrant medical inspection, bacteriology "not only . . . failed to break the popular links between race, class, gender, and disease"; doctors who used the "ostensibly neutral language of science" to "objectify social fears" helped to forge those links.[113] When doctors responded to the new data on gonorrhea infection that bacteriology provided them, their willingness to revise their views about infection but not incest reflects their deep engagement with and commitment to turn-of-the-century ideologies of race, nation, and class.

The way Edgar and Johnston link normative African American male sexuality with the most repulsive sexual behavior illustrates the distance physicians placed between respectable, native-born, heterosexual, white males and "less civilized others." Edgar and Johnston placed all African American men, by virtue of their supposed intrinsic racial primitiveness, into the same category as the "derelict" (elderly men with syphilitic dementia) and "insecure" ("unmanly" or homosexual) men they believed were prone to assault girls. Just as urban northerners reacted nervously to dark-skinned immigrants, white southerners displaced their own fear and anger about a newly contested racial hierarchy onto a supposed horde of black male sexual predators holding white women and children under siege. Edgar and Johnston even charged that child sexual assault was so "prevalent" among "the negro [*sic*] race" that southern states with equal numbers of black and white residents had found it necessary to institute the death penalty to keep order.[114]

Associating incest with African American men would have resonated with popular white discourses in this period that described black males as "literally wild beasts, with uncontrollable sexual passions and criminal natures stamped by heredity."[115] In 1903, some doctors went so far as to identify "sexual madness" as a condition affecting African American males, based on a racialized physical stereotype, the supposed large size of the black penis.[116] That same year, the *Washington Post* published a story comparing the low morals of African American "animals" to those of a "grunting pig in the sty." And the author named "incest of the most complicated character" as one of the "blackest and most savage of crimes" that African American men regularly committed.[117] Such men were, in this schema, markedly different and so deservedly barred from entry into the sphere of manly respectability and power.[118]

Yet white doctors' racialized analysis left them without any way to account for father-daughter incest among respectable white Americans. Doctors attempted first to explain incest by acknowledging its occurrence where it affirmed their social biases: in the homes of immigrants, men of color, and white men from the "lower orders." But the ramifications of doing so did more than simply heap additional scorn on men who were already socially marginalized. Insisting that only certain types of men commit incest also made it more difficult for white Americans to see it where they did not expect to. Just as nineteenth-century ideologies endowed white "ladies" with certain moral qualities, including sexual propriety, morality, and goodness, the ideals that shaped the "white father" placed him above suspicion of incest. Conceptualizing sexual dangers to white women and children as located only outside the father's protective domestic sphere left white professionals unable or unwilling to believe that men such as themselves could engage in such heinous behavior.

When they refused to acknowledge that girls from respectable families acquired a sexually transmitted disease, professionals reinforced discourses that promoted the inherent superiority of white middle- and upper-class males. The frenzied lynching of African American males in this period illustrates the disastrous social effects of using race-based assumptions about sexuality rather than actual behavior to police sexuality and to organize society. In the same way that their characterization of black males as rapists papered over white males' own offenses against African American women and children, linking assumptions about incest with social position obscured white male sexual misconduct toward their own children.

The "Superstitious Cure"

To support their views about immigrants and people of color, doctors identified so-called foreign beliefs and practices that, they argued, actually encouraged men to sexually assault children, particularly their own daughters. Doctors ascribed father-daughter incest to the belief held by "uneducated, superstitious men" that a man infected with gonorrhea would be cured if he had sexual intercourse with an uninfected woman. To ensure they chose an uninfected partner, men sought out a virgin, usually girls, often their own daughters (whose virginity it was their duty to safeguard). Edgar and Johnston claimed that belief in the "wide-spread superstition, particularly in Europe, but imported into this country as well" was the second most common motive for child sexual assault.[119]

While at first glance their statement seems little more than an ethnic slur, the belief has a long history in western Europe, dating from at least the Italian Renaissance.[120] By the late eighteenth century, Londoners were so familiar with it that defendants raised it as a defense at the Old Bailey. When David Scott was charged in 1769 with the rape of an 11-year-old girl, the prosecution's case centered on the fact that both Scott and his victim were infected with gonorrhea.[121] In his opening statement the prosecutor tried to exploit the emotions of the jurors when he told them that "there is a fact . . . of a very important nature" that should deprive Scott of even a "ray of mercy," namely, "that too common idea of persons, having a certain disorder . . . which they foolishly think they can get rid of, by having connection with . . . a [virgin]." Later in the trial, Justice Rooke asked one of the medical experts for the prosecution whether the practice was effective.

> *Court. Q.* You say that you have had a great deal of experience in these venereal cases, can it be any relief to a person that has the gonorrhaea [*sic*] to be connected with a young child?
>
> *A.* Certainly not . . . Instead of being any service, it would . . . give more pain . . . because it encreases [*sic*] the irritation.
>
> *Court.* It cannot be too generally known, that it does harm, and not good.[122]

The jury, which found Scott guilty and sentenced him to death, may have shared the court's skepticism about both the efficacy of the practice itself and the sincerity of men who claimed it as a motive for raping girls.

Some doctors mentioned the superstition only to deplore its popularity. In 1853, Dr. William Wilde cited the prevalence of this "delusion" among Ireland's "lower orders" to explain why mothers had such a high index of suspicion about the possibility that someone had assaulted their daughters. He argued that widespread knowledge of the "cure" encouraged mothers to assume that daughters who displayed a vaginal discharge had been raped.[123] However, belief in the cure was not peculiar to the British Isles. Wilde had uncovered a reference to the superstition in a recent work on prostitution in Algiers. And in the 1890s physician and Sicilian folklorist Giuseppe Pitrè claimed that knowledge of the "virgin cure" was "infamous" among Sicilian peasants. Pitrè, who wrote with warmth and respect about Sicilian culture, had encountered the belief while researching his comprehensive ethnography of Sicilian folk medicine.[124] He expressed no doubts about the sincerity of those who claimed to believe in it but added that, because of the well-known ferocity with which Sicilian girls protected their virginity, he did not believe that Sicilian men acted on it. Even so, Pitrè did not consider how effectively a girl could protest if the diseased man was her father or if she was too young to understand or resist.

Still, the inclusion of the "cure" in Wilde's and Pitrè's accounts suggests that some men may have carried the belief in it with them to America. Before the 1890s, however, there is little evidence that it was commonly held in the United States, as medical jurists rarely mentioned it.[125] The first to do so was medical jurist and professor Robley Dunglison, who was born in England and trained in London, Edinburgh, Paris, and Erlangen (Germany).[126] In 1824, Thomas Jefferson recruited Dunglison to teach medicine at the University of Virginia, which opened the tenth medical school in the United States the following year.[127] Dunglison, whose patients included presidents Jefferson, Madison, Monroe, and Jackson, later became professor and dean of faculty first at the Jefferson Medical College in Philadelphia and then at the medical school at the University of Maryland.[128]

In his lectures at Virginia, published in 1827, Dunglison informed his students about "a common opinion amongst the lower classes . . . that . . . the best cure [for gonorrhoea] is intercourse with an uncontaminated female—hence gonorrhoea is frequently met with along with rape, especially in young children."[129] Dunglison did not indicate whether he was referring to the "lower classes" in Europe, but it is unlikely that his professional appointments and esteemed list of patients left him much time to converse with impoverished or enslaved Americans. The next reference to the superstition in the American medical literature appeared in 1855, in Wharton and Stillé's *Treatise on Medical Jurisprudence*. But they only cited the comments of European doctors, including Johann Ludwig Caspar and William Wilde, and did not mention the belief circulating in America.[130] In 1877 another American textbook on forensic medicine quoted the highly regarded Parisian forensic examiner Ambroise Tardieu, who instructed medical jurists that one of the most important questions to investigate was: "Is there any popular notion that explains the [sexual] attempts often made on young girls?"[131] Belief in the superstitious cure supplied an answer. Tardieu, who had studied more than six hundred cases of sexually assaulted females, most under the age of 15, had heard about the superstition and criticized those who believed in it.[132] By 1886 Dr. Jerome Walker had heard enough defendants in Brooklyn raise it that he admitted it was difficult to tell whether a man who claimed to believe in it was sincere or merely telling a story he knew a doctor would accept.[133]

Krafft-Ebing both popularized the idea and lent a patina of scientific fact to it when he cited it in *Psychopathia Sexualis* as a popular belief among peasants, to explain an increase in child sexual assaults in Europe.[134] Edgar and Johnston heard about the cure from men in Manhattan who were charged with child sexual assault, and after the 1890s references to the superstition became ubiquitous in the American medical literature on girls' infections. Elbridge Gerry reported it to the sophisticated reading public when he mentioned it in the *North American Review*, the country's oldest literary magazine. In an 1895 article entitled "Must We Have the Cat-o'-Nine Tails?" Gerry advocated "brutal violence" as the only meaningful punishment for immigrants who abused children. He supported his call for extreme measures by explaining that, "incredible as it may appear," some men will "inoculate" a child with an "ineradicable disease" because they believe that "the commission of the offence will rid the criminal of the results of vice with which he is infected."[135]

But why did American doctors take such claims seriously? After all, no one who had ever tried it had been cured. The frequency with which European doctors referred to it, and the cure's resilience even today as one way to account for acquired HIV infections among African children, suggest that the idea sprung neither from the professional imagination nor from any particular cultural ignorance but from desperation.[136] Men afflicted with a disease that interfered with their sexual functioning and caused significant physical discomfort may have explored all possibilities for treatment, however reluctantly. There was no cure for gonorrhea, and medical treatments that offered relief were so expensive, lengthy, and painful that few men from any class could endure the full course.[137] In addition, many immigrants, particularly southern Italians, were suspicious of American medicine, and some may have turned instead to folk beliefs they trusted.[138] However, the notion of infecting a virgin implies a one-time occurrence, whereas father-daughter incest often continued for years, too long a time to have sustained even the most uneducated man's belief in the efficacy of the "treatment."

Still, in those cases in which American doctors could not overlook the evidence of sexual assault, they found the superstition about the cure a useful way to explain why certain types of men committed incest and others did not. In 1905 New York social hygienist and venereal disease specialist Dr. Ferdinand C. Valentine told the Society of Medical Jurisprudence that belief in the superstition was "quite common" among poor families. He claimed that "a visit to any . . . large dispensar[y] will always show [girls] with bruised and torn genitals, infected with venereal diseases," and he readily identified their assailants as men from "the lowermost strata" who believed in the superstitious cure.[139] By characterizing men who assaulted children as displaying an exceptionally backward ignorance, Valentine placed respectable men, those whom Dr. John Cotton had described as "educated and refined people," categorically above suspicion.

Similarly, in 1906 Dr. Flora Pollack, attending physician at the Women's Venereal Department at the Johns Hopkins Hospital Dispensary, argued that the superstition was "the cause of rape by the colored man upon white women" and the source of infection in the girls she treated.[140] Pollack had graduated in 1891 from the Women's Medical College of Baltimore, where she lectured on embryology and physical diagnosis before going to Berlin for postgraduate study. When she returned to Baltimore, she worked at two city institutions before the Hopkins Dispensary appointed her attending physician in gynecology.[141] Pollack claimed that 8 percent of the female patients she treated for gonorrhea were girls, mostly between the ages of 2 and 4, and she did not doubt that some of them had been assaulted.[142] It is unclear whether Pollack meant that African American or white men were assaulting white children. But the rhetorical equation of the rape of white women by black men with the rape of children of any color was as incendiary a topic as any in the 1900s and called attention to the kind of "uncivilized" behavior that, unlike in the nineteenth century, no white person would have admitted possible from a respectable white male.

Even in those few instances when doctors acknowledged that men also assaulted children from white and upper-class homes, doctors omitted any mention of whether, like poor children, they had been assaulted by men of their own race and class. In 1908, NYSPCC medical examiner Dr. W. Travis Gibb published a second article on gonorrhea vulvovaginitis, and he did not shy away from revealing the extent to which men molested children, calling the numbers of child sexual assaults "enormous" and "much more prevalent . . . than many [doctors] are aware."[143] Gibb claimed that in fifteen years he had diagnosed gonorrhea in 9 percent of the more than 900 cases of child sexual assault he had investigated for the NYSPCC (he did not clarify whether this figure included boys). Gibb recognized that men also victimized children of the "well-to-do" but that the "better classes" avoided criminal prosecutions because they were "very loathe to endure the unpleasant notoriety."[144] But even if Gibb could acknowledge the occurrence of incest in a well-to-do American family, he could not explain it. He argued that immigrants who brought "their foreign ideas here" sexually abused children.[145]

When he emphasized the crowded living conditions and the "superstition and dense ignorance of our large foreign population, the brutality, perversion, and intemperance of many of the . . . perpetrators," Gibb discussed incest as a natural function of poverty and ethnicity.[146] He argued that immigrants were capable of committing incest because "childhood is not as sacred [for them] as with us," a fact he believed led immigrants to view incest with "indifference."[147] He claimed that belief in the curative power of intercourse with a virgin was a common motivation behind child sexual assault and incest "among certain classes, especially ignorant Italians, Chinese, and Negroes."[148] And he added that the majority of assailants were poor Italian, Chinese, and German Jewish immigrant men, of whom "frequently the worst . . . are the very ones who naturally ought to protect those who become their victims," namely, the girls' fathers.[149] Underscoring his division along national lines, Gibb did not mention a single native-born white American among the "perverts" he had encountered. He pointed instead to examples he may have believed his audience would find sensational but unsurprising, like the "big, lusty Italian" who said he raped his 8-month-old baby because God told him to. Gibb also sharply differentiated the types of girls likely to be attacked, citing the "precocity and low moral standard[s]" of victimized girls, a characterization that respectable white Americans would never have assumed included their own daughters.[150]

When doctors attributed child sexual assault to belief in the superstitious cure, they tried to skirt the troubling questions that the specter of father-daughter incest raised about white male sexuality in general and claims to white male superiority in particular. But as the epidemic of gonorrhea vulvovaginitis grew unabated, so did the number of people who claimed to believe in the cure. By 1909, Dr. Pollack claimed that belief in the cure was responsible for many of the infections among the nearly two hundred girls she had treated for gonorrhea vulvovaginitis.[151] In her second article on the subject, Pollack argued, "I . . . repeat with firmer conviction than ever, the cause of most of the outrages upon children is to be found in the superstition."[152] Unlike her colleagues who believed that the superstition helped explain why certain types of men committed incest and others did not, Pollack did not find the superstition useful because she believed that only socially marginalized men abused their daughters. She claimed that the superstition was common knowledge in Baltimore, "so deeply rooted . . . that were you to ask ten police officers, cab drivers, hucksters, etc. of the truth of it I think eight would affirm it as a '*fact*,' and all would know of its existence."[153]

So then why was Pollack interested in the cure? She found the "appalling number" of victimized girls, which she estimated to be at least 800 to 1,000 each year just in Baltimore, "too large to credit to perverts."[154] Pollack accepted Krafft-Ebing's distinction between "sadists," men who assaulted children to satisfy a "perverted sexual instinct," and "infectionists," men who acted in a premeditated, reasoned manner simply "to get rid of [their] infection."[155] Pollack may have been introduced to Krafft-Ebing's ideas during her medical training in Germany, but she accepted his theory about infectionists because, she noted, in America the idea that "sick people take the strength from the well, should they sleep together," is very popular among every class.[156] Pollack claimed to have overheard parents sitting in the waiting room sharing information about the cure. Some told Pollack, apparently without shame, that they had " 'gotten rid of their disease on a *baby*.' " One woman said that she and her husband had conceived expressly for the purpose of having a child they could use to rid themselves of the disease.[157]

In the end, it did not matter to Pollack, who viewed the duty of doctors in these cases differently than did the medical jurists, whether a man raped a girl because he was a pervert or infectionist. Nor did she consider gonorrhea a serious social issue because of the risk of complications from death or institutional epidemics, which she thought exaggerated.[158] Pollack thought the incidence of infection important because it demonstrated how often men sexually assaulted girls.[159] She did not mock the suspicions of frightened mothers but credited them for coming forward with the "first knowledge of the crime."[160] Pollack worked with parent organizations and community leaders to advocate more aggressive prosecution of assailants and she urged her colleagues to examine every girl diagnosed with gonorrhea for evidence of rape.[161]

Pollack was more committed to stopping child sexual assault than to protecting American manhood. She dismissed theories about nonsexual modes of infection as "a very useful shield for a guilty individual, [since such theories] impede justice, and make it extremely difficult to protect children from these assaults."[162] Instead, she declared, "Let it be known once and for all that gonorrhoea is thus acquired only most exceptionally; and this possibility . . . should only be accepted after the most exhaustive examination of all the attending circumstances."[163] Pollack argued that, until they could rule out the possibility, doctors should assume that an infected girl had been sexually assaulted, a standard that the American medical community would adopt ninety years later.[164] Until then, doctors repeatedly cited Pollack's statements about the superstitious cure but ignored her views about the sexual origins of infection.

As the number of infected girls continued to rise, doctors intensified their efforts to link incest and child sexual assault to nationality and class. In 1910, Manhattan physician Edward L. Keyes Jr., the son of an authority on venereal disease who took up his father's work and would become president of the American Social Hygiene Society, edited a textbook on sexually transmitted disease. The chapter on gonorrhea vulvovaginitis named poor hygiene, not rape, as the primary cause of infection.[165] Dr. Emily D. Barringer, the chapter's author, was a protégé of Dr. Mary Putnam Jacobi and began her career in 1902 at an emergency hospital on the Lower East Side where, she recounted almost fifty years later in her autobiography, child rape was "not infrequent."[166] Unlike most of her colleagues, Barringer examined infected girls for evidence of rape, but her assumptions about who committed incest influenced her ability to detect it. Barringer believed that most, if not all, of the men who infected girls were Italian tenement dwellers who believed in "the prevalent superstition."[167]

Chicagoans identified the superstition as rampant in their city, too. In 1912, a year after Barringer's chapter was published, Chicago assistant city physician Dr. Clara Seippel claimed that in only six months she had examined fifty-three girls who had been raped and infected with gonorrhea by "infectionists," adding that it was "common knowledge" that the superstition was deeply rooted among immigrants.[168] That same year the Vice Commission of Chicago concluded that "vicious and degenerate" men who sexually assaulted children defended themselves by claiming they had only been seeking a cure.[169] Jane Addams also published *A New Conscience and an Ancient Evil*, her condemnation of organized prostitution, that year, and she, too, complained that "a number of [girls] had been victims of that wretched tradition."[170] In 1915 Dr. William Healy, who had served on the Vice Commission, explained in his landmark book on juvenile delinquency that certain physical ailments so provoked some young men to crime that "even in modern America" they might "assault a pure little girl, in compliance with the old superstition."[171]

Such nativist sentiments had a direct bearing on how medical examiners viewed evidence of alleged child rape and incest. In 1912 Dr. Gurney Williams, the "Sometime District Police Surgeon of Philadelphia," boasted that he had performed "over fourteen hundred vaginal examinations" on girls who alleged they had been sexually assaulted. Gurney believed that only one hundred of the claims were true, and he shrugged off the rest by noting that girls lied "as easily as a morphine fiend."[172] It is difficult to imagine that Gurney found proof in twelve of every thirteen cases that a girl had fabricated an accusation for revenge or financial gain. But his beliefs about gender, race, and ethnicity influenced his diagnostic skills no less than they had those of his predecessors in the prebacterial era. Gurney attributed child sexual assault to the superstitious cure, which he claimed that only two types of men—Italians and African Americans—believed in.[173] Perhaps Gurney considered accusations made against men from other racial and ethnic backgrounds implausible. Similarly, in 1914 when Dr. Frederick Taussig, a social hygienist and gynecologist on the staff of the Washington University Hospital in St. Louis, tried to identify the source of infection in the girls he treated, he dismissed sexual assault. Referring to "German writers," Taussig noted that the "lower classes abroad" believed in the superstition, and he considered sexual assault as the possible source of infection only in those girls who "came from the ignorant, foreign-born population."[174]

So useful was the superstition as a way for doctors to explain seemingly incomprehensible male misbehavior that references to it appeared in the medical literature long after practitioners reported that their patients had ceased to offer it to explain their behavior.[175] In 1926 Harvard pediatrician Dr. John Lovett Morse, who had not mentioned the superstition in his 1894 article about five infected girls, claimed that it "is less prevalent now than it was a number of years ago."[176] Perhaps fewer men offered the superstition as an excuse, or perhaps practitioners noted it less often because European immigrants were becoming increasingly assimilated. Historian Matthew Frye Jacobson argues that in the 1920s white Americans reached the end of the process that consolidated European immigrants into the "white" race. And Amy Fairchild argues that the Immigration Restriction Act of 1924 "represented the culmination of American nativism."[177] These factors may have combined with new views among health care workers, who, as I discuss in the next chapter, had become convinced by the mid-1920s that poor hygiene was the primary cause of girl's infections. They may have simply stopped asking the types of questions about sexual conduct that would have prompted an anxious man to assert the superstitious cure. In any case, without a better theory to explain why men sexually assaulted girls, references to it continued.

By the 1930s it had taken on a life of its own. In 1934 the *Journal of Social Hygiene* cited the superstition as one of the "ignorant ideas" about gonorrhea infection that sex education could eliminate.[178] That same year the ninth edition of *Taylor's Principles and Practice of Medical Jurisprudence* referred to the "absurd belief" to differentiate between men who assaulted children out of "simple lust" and those who were too ignorant to know better.[179] In the 1936 printing of *Sex in Relation to Society*, the fourth volume of Havelock Ellis's influential *Studies in the Psychology of Sex*, Ellis referred to the cure as one motivation for sexual attacks on children, especially among "Italians, Chinese, Negroes, Etc.," and he cited Gibb's and Pollack's articles from the 1900s as authority.[180]

In 1937, Dr. John B. West, an African American physician with degrees from Howard and Harvard universities who was health director of the Central Harlem Health Center, felt it necessary to denounce the cure. When he wrote an article on gonorrhea for the *New York Amsterdam News*, he included an admonishment to readers, warning that the cure was "by no stretch of the imagination . . . true."[181] When social worker Phyllis H. Williams published a book in 1938 about southern Italian folkways, she, too, included the cure. Williams had translated Dr. Pitrè's book into English and claimed that her goal was to sensitize social workers and health care professionals to the social issues facing Italian immigrants.[182] As late as 1939, in a book on public health he coauthored, Dr. Nels A. Nelson, the chief venereal disease control officer in Massachusetts, called the superstition a "curious and disgusting tradition among some European males."[183] Nelson and his coauthor, public health nurse and instructor Gladys Crain, added, without elaboration, that "innumerable others are the result of sexual abuse for male pleasure only."[184]

Intellectual assumptions about universal and ahistorical truths that many now view as outmoded may have informed the work of some of these scholars, who did not recognize the diversity of Italian immigrants and Italian Americans and so did not question the likelihood of them subscribing to a folk belief that had been popular among Sicilian peasants, a group traditionally reviled by other Italians, forty or fifty years earlier.[185] But the tenacity with which doctors talked about the superstitious cure shows how the 1890s were a critical turning point, both in medical practice and in the doctors' eagerness to avoid the conclusion that American men were committing incest. American doctors discovered, but could not acknowledge, that native-born white American men took erotic pleasure in sexual contacts with children, particularly their own. The behavior that doctors treated as "taboo" was not incest itself but revealing publicly that men of their race and class were not only capable of but did, in fact, engage in such behavior.

To modern readers the medical community's aversion to rethinking its assumptions about associations between sexual behaviors and a man's race, class, or nationality may suggest a "conspiracy of silence." But a conspiracy would have first required doctors—male and female—to acknowledge the occurrence of incest among their peers. This they could not and did not do. They viewed sexual behavior more comfortably as a natural function of race, class, and citizenship, elements of personality that were immutable rather than socially constructed. The noisy silence with which doctors began to blanket the imminent discovery of incest left their social prejudices intact. But it also revealed the fragility of the ideological foundations on which their social vision and power rested.

CHAPTER FOUR

Protecting Fathers, Blaming Mothers

When social hygienists formed a national advocacy organization in 1914 with the ambitious goal of eradicating venereal disease from America, they affirmed, promoted, and institutionalized the culture of dissemblance about the etiology of girls' infections that doctors had begun. Rooted in the values of the Progressive Era, the American Social Hygiene Association (ASHA) shaped an interwar social hygiene movement marked by "scientific" investigation and the professionalization of "social hygiene" within nursing and social work. Absent their intervention, speculation about how girls acquired gonorrhea infection might have remained of interest to only a small circle of physicians and hospital administrators. But the same diagnostic and bacteriologic breakthroughs that led doctors to discover an epidemic of gonorrhea among girls also revived more general interest in venereal disease among doctors and reformers.

Although they trumpeted their devotion to science and rational inquiry as the sine qua non of a civilized, orderly, and just society, on the issue of infected girls Progressive Era professionals placed science, truth, and education at the service of obfuscation and denial. In stark contrast to their determination to dissuade doctors and the lay public from believing that adults could acquire venereal disease from nonsexual contacts, social hygienists argued—to parents, educators, social reformers, and health care professionals—that girls were an exception to the rule. But after two decades of speculating about the etiology of gonorrhea in girls, doctors had been unable to reach a consensus on how girls became infected or how they might avoid infection. As the rate of infection continued to rise unchecked in the 1920s, doctors, reformers, and social hygienists shifted tactics.

Social hygienists responded to the hedonism of the Roaring Twenties by redoubling their efforts to promote the male-headed nuclear family as the foundation of both social stability and personal fulfillment. At the beginning of the century doctors and reformers denounced men who visited prostitutes and brought venereal disease into their homes as "traitors" who "poisoned" their families. Now reformers changed their rhetoric. They stopped condemning philandering husbands and fathers and began to identify them instead as among the victims of the disease, which social hygienists recharacterized as a "family problem." But when reformers replaced critiques of male lust with paeans to the sexual satisfactions to be found in marriage, their schematic plan for marital and familial bliss among the newly emergent white middle class precluded the possibility of anything so threatening as father-daughter incest.

At the same time, social hygiene became increasingly professionalized within medicine, nursing, and social work, and by the late 1920s the medical literature shifted its focus away from the etiology of girls' infections. Professionals viewed the management of infected girls as a public health problem and turned their efforts toward devising procedures to prevent infected girls from spreading their infections to other members of the household or schoolmates. Health care professionals named poor sanitation and housekeeping as the source of disease and blamed mothers or other girls, not fathers or other men, for their daughters' illness. All told, as men began to disappear from discourses about infected girls, so did the connection between incest and infection.

Social Hygiene and Dr. Prince A. Morrow

Nineteenth-century purity reformers, who understood venereal infection as a moral issue with personal and social ramifications, first conceptualized the notion of "social hygiene." In his history of the purity movement, David J. Pivar identifies two strands of nineteenth-century social hygiene. Reformers who wanted to clean up and encourage a healthy and moral society by ensuring that its citizenry was virtuous and physically fit organized the first strand. The second was led by physicians whose primary interest was to control the spread of venereal disease by medical management of prostitution, a system that would permit only those women who passed a medical examination (proving they were free from infection) to legally ply their trade.[1] Purity reformers advocated the widely held belief that prostitutes were responsible for the spread of venereal disease and that infection demonstrated that the sufferer had engaged in "venereal" or "immoral" behavior, by which they meant sexual relations outside marriage. To achieve their goals, purity reformers, who included the Woman's Christian Temperance Union and Anthony Comstock, advocated age-of-consent laws and sex education; regulating or ending prostitution altogether; and eliminating the double standard of sexual morality. Proponents of a single standard of morality wanted men to join women in embracing the goal of sexual abstinence outside marriage.[2]

Ideas about how to achieve these goals shifted as twentieth-century social hygienists became enthralled by the medical, scientific, and technological breakthroughs of the period. Progressive Era professionals came to believe they could eliminate even the most intractable social problems through the use of science and reason. They argued that medical science—not religious or moral suasion—presented the most effective means with which to purify America: socially, morally, and medically.[3] And turn-of-the-century American physicians used the dramatic increase in "physiological knowledge" to vault medicine over the "church as the new authority on sex."[4] As they worked to cement their new authority, a loose coalition of health care professionals, social reformers, philanthropists, and clergy found common ground in an ambitious new goal: they devised a public health and social reform program to eradicate venereal disease from America. At a time when people worried about acquiring venereal disease from doorknobs and toilet seats, their plan was to educate the public about the link between sexual contact and infection by providing "frank" and scientifically accurate information. From the 1900s to the 1940s social hygienists lobbied local, state, and federal governmental agencies for measures that included ending prostitution; publishing sex education materials for adults, adolescents, and children; expanding the number of facilities that treated venereal disease; and improving the quality of medical care available to infected persons.[5]

Some of the era's most prominent figures, including John D. Rockefeller Jr. and Jane Addams, lent their money and prestige to the cause, but New York City physician Prince A. Morrow, a dermatologist, syphilologist, and professor of genito-urinary diseases at New York University, dominated the movement.[6] Morrow rose to prominence in 1901 when he published a report for the New York County Medical Society. He claimed that 80 percent of the men in New York City had at some point been infected with gonorrhea, a controversial estimate that could not be verified.[7] Three years later Morrow published his influential book *Social Diseases and Marriage*, in which he used his medical acumen to recast venereal disease from a moral problem to an urgent medical and social issue that deserved to be at the top of the Progressive agenda.[8] Morrow shared the Progressive Era view that medicine, coupled with professional and public education, could be the engine of social improvement.[9] As he put it, "there is no other department of knowledge of so much value to the human race as that of medicine" because the "genius of modern medicine is . . . in the . . . popularization of hygienic knowledge."[10]

Morrow identified venereal disease as important, not because it signified all that was loathsome about immorality but because the complications of untreated gonorrhea infection among native-born white Americans threatened their ability to reproduce.[11] Morrow treated female gonorrhea patients in Manhattan, and he disputed the popular view that the only women infected with venereal disease were prostitutes. He railed instead against family men who visited prostitutes and then "soiled and poisoned" their families.[12] With their newfound ability to confirm gonorrhea infection bacteriologically, doctors had been able to link the complications of untreated gonorrhea in women to a host of serious health and reproductive problems that Morrow recited unsparingly: pelvic inflammatory disease, miscarriage and infertility, and children who became blind or arthritic. Morrow considered the women and children he treated to be "innocent victims" because they had not been guilty of any sexual impropriety: "respectable married [white] women," children who became blind as a result of infections acquired during childbirth, and girls infected with vulvovaginitis.[13]

Morrow charged doctors with the duty to save civilization by ensuring the health and longevity of the white American family. Arguing against the stereotype that venereal disease was a disease of the lower classes, Morrow claimed that "venereal diseases respect no social position and recoil before no virtue; they ramify through every class and rank of society" and "approach with equal step the habitations of the poor and the palaces of the rich."[14] Morrow pointedly accused those "men who have presented a fair exterior of regular and correct living—often the men of good business and social position"—for having indulged in reckless, even traitorous, behavior for which their families, race, and country would pay the price.[15] Many doctors and reformers who feared that the growing population of immigrants would soon outnumber them considered the declining birth rate among native-born white Americans tantamount to "race suicide," and they were receptive to Morrow's call.[16] In February 1905 Morrow convened twenty-five physicians at the New York Academy of Medicine and organized the American Society for Sanitary and Moral Prophylaxis, whose goal was to educate the profession and the public about venereal disease. Within five years the group claimed seven hundred members, social hygienists in other cities formed affiliate groups, and when the influential American Public Health Association met in Havana in 1911, it endorsed the society's tenets.[17]

Morrow's emphasis on innocent infection was a defining facet of twentieth-century social hygiene and an effective rhetorical maneuver that he used to challenge the social stigma associated with an "immoral" disease. Doing so made possible his unprecedented public advocacy on behalf of venereal disease patients. But no one anticipated that so many innocent victims would turn out to be girls infected with genital gonorrhea—and Morrow's own willingness to pierce the veil of respectability went only so far. Morrow acknowledged that most infected girls shared a bed with an infected family member, and he lamented that "family epidemics are very frequent." But he refused to believe that respectable men infected their daughters in the same manner that they infected their wives.

Morrow had developed his views about the etiology of gonorrhea vulvovaginitis in the prebacterial era, and they reflected doctors' earlier anxieties about a misdiagnosis. After diagnostic technology changed, Morrow's views on girls' infections did not. He never stopped insisting that girls were "innocently" infected "by mediate contagion" from "fingers, thermometers, towels, sponges, etc."[18] He first wrote about the topic in 1885, when he cited Alfred Taylor's statement from the 1840s that "the existence of a purulent discharge from the vulva of children has often led to the unjust accusation and punishment of innocent persons for attempted violation."[19] Morrow repeated this statement even after bacteriologic testing had rendered nineteenth-century concerns about a misdiagnosis moot. In *Social Disease* he explained, "Vulvovaginitis . . . has . . . often led to the unjust accusation and punishment of innocent persons . . . The physician should . . . be exceedingly reserved in giving an opinion . . . [that] might lead to the gravest consequences. We now recognize that gonorrhoea in children is vastly more often due to accidental mediate transmission than to attempted intercourse."[20] Perhaps Morrow was influenced by his belief that "one knows the facility with which children are disposed to accuse and lie."[21] But nothing in the medical literature supported his unequivocal statement about the etiology of the disease. Just the opposite was true.

Doctors knew enough about *Neisseria gonorrhoeae* to doubt that it spread by nonsexual contact, which was one reason that their attempts to identify casual contacts as the source of girls' infections continued to trouble them. In a 1907 paper that Dr. J. Clifton Edgar presented to the American Medical Association (AMA) Section on Hygiene and Sanitary Science, he reiterated the uncontroversial view that "bacteriology proves the great difficulty of inducing the gonococcus to live outside of its regular habitat."[22] Doctors knew that the bacteria remain virulent only in a warm, moist environment and that as a vaginal discharge dries, the bacteria die rapidly. Still, even though they had no research to inform them how long or on what surfaces the bacteria might remain virulent, Edgar and others speculated that "the gonococcus can be spread by the handling of towels, washcloths, toys, etc."

One reason doctors thought that girls might be particularly susceptible to casual contacts was that the epithelial lining of a girl's genitals, which acts as a bacterial barrier, does not become thicker and more resistant to infection until puberty. Doctors wondered whether the epithelial layer might be thin enough in a prepubescent girl that *N. gonorrhoeae* could penetrate her "relatively more prominent and protuberant" genitals on casual contact.[23] Edgar reasoned that this developmental issue explained why "we find gonorrheal vulvovaginitis in young children a common condition, especially where lack of cleanliness exists, as in overcrowded institutions and where proper prophylactic measures are not enforced."[24] Edgar's views about the ease with which the bacteria spread to tenement children had not changed in the thirteen years since he had first written on the topic, even though in the intervening years doctors had discovered that the disease was not limited to tenement dwellers. Like so many others who wrote about girls' infections, Edgar speculated freely about what he supposed was the class-based etiology of the disease but remained curiously silent about how girls from clean and well-kept homes became infected.

When Progressive Era doctors repeated unsupported speculations about nonsexual transmission of the disease to girls, some of which, like Morrow's, voiced concerns that should have vanished with the prebacterial era, they created confusion where none had existed before. At a meeting of the AMA in 1901, Dr. Alfred Cotton admitted that during his tenure as city physician he had realized that his assumption that girls were only rarely sexually assaulted and infected with gonorrhea was mistaken. But five years later when he published his textbook *Medical Diseases of Infancy and Childhood*, Cotton equivocated. He wrote that "the usual mode of infection is by direct contact, although among infants and children there is ample reason to believe that it is frequently carried by intermediate agents."[25] However, medical science had not provided doctors with any such "ample reason." Nothing more than the sheer number of cases, particularly among respectable white Americans, had persuaded physicians to consider nonsexual modes of transmission the most plausible explanation. And, as might be expected from speculation that no one seemed eager to test, none of it had moved doctors any closer to identifying the origins of girls' infections.

Theories: Alice Hamilton

e Chicago Society of Social Hygiene issued a pamphlet in 1907 , "For the Protection of Wives and Children from Venereal Conation," it repeated Morrow's criticism of married men who visit stitutes and then infect their wives. But though the pamphlet also rned mothers about the danger of gonorrhea infection to girls, it did ot suggest that men infected their daughters in the same way.[26] Organized by the Chicago Medical Society, the Social Hygiene Society's membership included Jane Addams and some of the city's most renowned women physicians. Prominent among them were longtime Hull House resident Dr. Rachel Yarros, an obstetrician and head of the department of Social Hygiene at the Chicago campus of the University of Illinois. Yarros was part of a team of doctors, including Alfred Cotton and Clara Seippel, who lectured on social hygiene in Chicago, and she lectured nationally with Dr. Edith Spaulding on behalf of the AMA.[27] Chicago mothers would have reasonably expected that a pamphlet published by the society would contain authoritative information about a serious topic. But despite its warning about an "alarming" increase in gonorrhea among girls, the pamphlet neither explained how girls became infected nor suggested what steps a mother could take to protect her daughter from infection.[28]

Sanitation reformers filled in the gap. The early twentieth century was an era of heightened medical and public attention to cleanliness.[29] Progressive reformers who regarded improved sanitation as a remedy for many social problems readily included an epidemic of gonorrhea in girls among the problems they could solve. Although sanitary reforms included coercive measures aimed at "Americanizing" poor and immigrant families, cleaning up cities and educating the public about germs had markedly reduced disease and mortality.[30] But as applied to gonorrhea vulvovaginitis, the focus on sanitation was sadly misguided. In 1908 Hull House resident and public health pioneer Dr. Alice Hamilton published an influential review of the medical literature on gonorrhea vulvovaginitis. She scoffed at the notion that girls acquired gonorrhea primarily from sexual contacts, charging that such old-fashioned thinking had "contributed to retard the clearing-up of this subject."[31] Hamilton faulted poor hygiene for the infection's spread in institutions like hospitals, where, she warned, "all objects . . . which are damp and which come in contact with patient after patient in rapid succession should be regarded with the utmost suspicion," including the "seats of water closets, bath-tubs, towels, sponges, . . . thermometers, and, most important of all, the fingers of nurses."[32]

Hamilton's argument situated the disease within a popular and self-consciously modern reform movement with a record of accomplishment that the general public could understand.[33] She had established her authority on the relationship between sanitation and health five years earlier, when she made what her biographer calls "the most acclaimed discovery of her career."[34] The Memorial Institute for Infectious Diseases in Chicago was one of the first pathological and bacteriologic research centers in the United States. In 1902 Hamilton used its resources and those of Hull House to investigate the cause of a severe typhoid epidemic. Most public health officials believed water or milk was the source of the infection. Hamilton argued instead that flies transmitted infections to people from illegal, open privies, and she launched a successful political campaign to improve sanitary conditions in the tenements.[35] Her thesis turned out to be incorrect, but when the city covered up the actual problem (a broken pipe leaking sewage into the city's drinking water), her reputation remained intact.[36] Hamilton herself felt that she "gained more kudos" from this investigation than for any subsequent work.[37]

When Hamilton soon thereafter identified gonorrhea vulvovaginitis as a sanitation issue, she articulated what many doctors and reformers who wrote about the topic had been suggesting, and her reputation influenced all subsequent debate on the etiology of the disease. In his 1908 gynecological textbook, for instance, Dr. Palmer Findlay stated, "That gonorrheal infection is frequently conveyed by contaminated hands, instruments, and dressings is a fact too well established for comment."[38] Findley claimed there was "abundant proof" for the view that gonorrhea is transmitted from nonsexual contacts with unclean objects and that "examples are not wanting in which several members of a family have acquired the infection from the bathroom."[39] He instructed mothers to prevent the spread of infection by burning or disinfecting the clothes of their infected daughters.[40]

When Dr. Hamilton presented a second paper on the topic at the 1910 National Conference on Charities and Corrections, of which Jane Addams was president, she contextualized gonorrhea within these familiar discourses. Contacts with what Hamilton had characterized as unsanitary "dangerous things" were so numerous, and a mother's responsibility so broad, that Hamilton saw little reason to consider the possibility of infection from sexual contacts, including those with "other members of the family." She claimed that only a "small number of cases" arose from sexual violence, which she dismissed as "an insignificant minority."[41] Hamilton succinctly set out what she believed to be the modern view, that "the child victims of gonorrhea usually have been infected by their mothers," whom she blamed for transmitting the infection with "dangerous things" that had been soiled with the purulent discharge of gonorrhea infection, namely, "nurses or mother's fingers, towels, wash-cloths, sponges, bathtubs and the seats of closets," by which she meant toilet seats.[42]

Hamilton had no evidence to support her argument, but her logic neatly accounted for simultaneous infections in father and daughter in a way that exonerated the father not only from sexual misconduct but from having anything to do with the infection at all. In this scheme mothers, nurses, girls, or even household objects—anyone or anything except the father—transmitted the disease. As Hamilton explained, "Nothing is infectious about a girl with gonorrhea except the discharge itself or something that has recently come into contact with it," and she concluded that proper sanitary precautions would make it a "comparatively simple matter" to keep an infected girl from being "an object of terror."[43]

Progressive Era health care pioneers embraced Alice Hamilton's ideas about sanitation so thoroughly that they often excluded the possibility of sexual contact entirely from discussions of the etiology of girls' infections. Lavinia L. Dock's 1910 nursing textbook is a striking example.[44] Dock was the assistant superintendent of nurses at the Johns Hopkins Hospital, a resident of Lillian Wald's Henry Street Nursing Settlement in New York City, and one of the preeminent public health nurses and nursing educators in the United States in the first quarter of the century. She enjoyed an international reputation, and in 1909 she delivered a paper at the International Congress of Nurses in London on the "social significance of the venereal diseases and the crusade upon which women should enter in regard to them." A year later she expanded her thoughts into a nursing textbook, *Hygiene and Morality*.[45]

Dock's discussion of gonorrhea vulvovaginitis echoed Hamilton's views. She claimed that "infection from unclean water-closets is frequent, instances having been known where a whole series of cases of vulvovaginitis have arisen from this cause," although nothing in the American medical literature had identified and documented any such cases. Dock also promoted the notion that infected girls should not suffer from the stigma of venereal disease because "it is quite possible . . . to have contracted the disease from badly kept school closets [toilets], or from sleeping in the same bed and being contaminated by the sheets or clothing of diseased persons. Family towels . . . are sources of danger, and public baths and tubs are also known to have media of contagion."[46] Dock did not cite any examples of "water-closet epidemics," but she instructed nurses and mothers to adopt a lengthy list of measures she believed would prevent girls from becoming infected. These included prohibitions against masturbation, "alcoholic drinks," "rich stimulating food," "overwarm" clothing, and "idle luxurious living," but not a single recommendation that pertained to the possibility of sexual assault.

She warned instead about dangerous objects, especially the "seats of public water-closets," which she noted "may always be regarded as being more or less doubtful, and when used, may be covered with a clean piece of paper." And although there were no empirical data to support her view, she added that "many cases of gonorrhoea are caused by the dirty or ill-kept seats of public conveniences, especially in crowded places."[47] Dock recommended that mothers protect their daughters from infection by learning proper techniques for bathing and by having their children sleep alone, get fresh air, use rough towels, and have regular bowel movements. Some of her measures suggest that Dock believed the etiology of vulvovaginitis to be similar to that of tuberculosis: she admonished girls not to put money in their mouth and advised nurses to train infected girls to cover their nose or mouth when sneezing or coughing.[48]

Dock knew that child sexual assault was a problem, and she was a passionate supporter of age-of-consent legislation. But she believed that white slavery and prostitution, not incest, posed the greatest sexual threats to girls. Yet she chose to support her demand for stronger age-of-consent statutes by including a 1909 report from the *British Journal of Nursing* in an appendix to her book. The report, which Dock may have heard when she attended the 1909 nursing congress, investigated the disparity between the high frequency with which men assaulted girls and the relatively low rates at which girls prosecuted the crime. The nurse who conducted the investigation reported that when she asked why so few victims prosecute their assailant "the reply comes from all quarters that so often the culprit is the father, step-father, uncle, or brother of the victim."[49] Dock thought the report significant enough to include in her book, but nowhere in her discussion of gonorrhea vulvovaginitis did she suggest that American men might also be victimizing their daughters. She advised mothers how to protect their girls from infection from unsanitary contacts but mentioned nothing about the need to protect them from their fathers.

The Effect of Sanitation Theories on Treatment

Alice Hamilton believed not only that doctors diagnosed gonorrhea vulvovaginitis more frequently than in the prebacterial era but that it occurred more frequently because of the "close contact" of children in "modern city life." She reasoned that "an infected child who is kept at home will hardly spread the disease beyond her own family, but in a day nursery, or orphanage, or even a public school, she may be the cause of a widespread epidemic."[50] Because such institutions for children were expanding rapidly at the beginning of the century, Hamilton warned that epidemics of gonorrhea would become increasingly common. Yet even as she couched her argument in terms of modernity, she was repeating the assertions of Dr. William Wilde, who had sought to eviscerate the forensic meaning of gonorrhea infections more than a half century earlier when he noted the supposed proclivity of poor, dirty, scrofulous children to acquire gonorrhea infection.

Progressive Era reformers viewed public health and sanitation as directly connected to the quality of urban life, and they repeated claimed that "this disease is generally found in the children of the poorer classes, where overcrowding and unhygienic surroundings predispose to its transference."[51] To prove a presumed link between dirt and disease, doctors and social workers pointed to the number of tenement families in which most, if not all, of the members were infected. They did not consider that intrafamily "epidemics" might also be proof of incest. And although doctors limited their explanations about the etiology of the disease to the experiences of tenement families, the disease itself was not so constrained. Doctors occasionally acknowledged that "this infection is by no means confined to the children of the poor. Oftentimes most virulent and intractable cases have been found in a fashionable girls' school, or in the home of luxury."[52] But they were unwilling to consider incest as the source of infection in such cases, suggesting only that "the disease has occasionally been traced back to the erring nursemaid."[53] But what of those cases not traced back to a nursemaid?

Doctors were so eager to reject sexual contact as the source of girls' infections that they eschewed the traditional etiology of girls' infections without first identifying other means by which girls became infected. Physicians attributed the vast majority of infections to "source unknowable," a practice that continued through the 1930s, and did not investigate further. Some doctors may have simply found it more expedient to do so. But even as they failed, in case after case, to name a source of infection, doctors still insisted that few, if any, of their patients had been sexually assaulted. The reasoning of Dr. B. Wallace Hamilton, attending physician at the Vanderbilt Clinic, is typical. The Vanderbilt family underwrote the cost of the clinic, which opened in 1887 with a dedication ceremony that included speeches by many of the city's luminaries, including Cornelius Vanderbilt and Dr. Abraham Jacobi. One *New York Times* reporter who visited the clinic praised its lavishness, with its rooms "perfectly supplied with the latest appliances for surgical operations and clinical studies."[54] The clinic was just off Columbus Circle on the campus of the College of Physicians and Surgeons, which the Vanderbilts had also endowed and which would soon merge with Columbia University and move farther uptown.

That same year the Eastern Dispensary, which Jacob Riis had earlier noted was located in the "heart" of "Jewtown," also moved to new quarters on the Lower East Side and changed its name to the Good Samaritan Dispensary. When a writer for the *Medical News* toured both facilities in 1891, he praised Good Samaritan's "handsome new home" and its "elaborate appointments."[55] But the Vanderbilt Clinic amazed the reporter, who believed it too luxurious for the type of "wretched and densely ignorant" people treated at Good Samaritan to appreciate. Clinics were designed to offer free or low-cost services to the indigent, but the facilities at Vanderbilt were so impressive that the observer gushed that even "people in *comfortable* circumstances come from far and near" to be treated there. And although the practice of slumming for cheap medical care was a common, if disparaged, practice, the reporter claimed that well-to-do patients boasted about visiting Vanderbilt for treatment.[56] He did not exaggerate the clinic's popularity. Within four years the Vanderbilts planned a five-story addition, which would permit the clinic to double the number of patients it could treat.[57]

By 1910 the Vanderbilt Clinic had also become the nation's largest outpatient treatment center for girls infected with gonorrhea. Between 1907 and 1910, 334 girls, whose average age was 5, had visited the clinic. In a 1910 article in *JAMA*, attending physician Dr. B. Wallace Hamilton asserted that not one of these girls had been raped and that only one had been the victim of "attempted rape."[58] Hamilton's patients came from all classes, but when he discussed the etiology of gonorrhea vulvovaginitis, he spoke only of infection among that "class of patients" surrounded "by uncleanliness and poor hygiene, living in crowded tenements."[59] And though Hamilton could imagine innumerable scenarios in which an infected girl passed her infection on to her family and playmates, he did not address how the girl became infected in the first place or how her mother might have better protected her.

Hamilton's primary concern was to prevent an infected girl from spreading the disease to others by severely restricting her contacts with both objects and people. He gave each girl's mother a printed, detailed list of instructions, which he expected her to follow strictly, including: "The child should sleep alone. No one else should use any toilet articles, towels, napkins or wash-cloths used by the patient. All napkins, sheets, underclothing, towels and wash-cloths should be . . . boiled or immersed in a phenol solution before washing [as should] Bath-tubs, basins and everything else coming in contact with the patient . . . Parents are cautioned not to allow the child to mingle intimately with other little girls. The child should not attend school . . . lest other children become infected."[60] A mother reading this list, especially if she worked or lived in a crowded tenement, had more than one child, or was without household help, must have responded with despair. But that was not all. Hamilton also expected mothers to administer daily treatments at home, including cleaning their daughters' vagina at least four times every day with a watery solution of borax or boric acid crystals. And he ordered mothers to bring their daughters back to the clinic once a week—indefinitely—for an additional douche, a trip that would surely have extinguished any remaining pleasure mother and daughter experienced in each other's company.[61]

Other health care providers were more willing than Hamilton to acknowledge that some of their patients had been sexually assaulted, but they were no more eager to broach the subject of incest. From 1905 to 1908, the Social Service Department of the Massachusetts General Hospital (MGH) provided outpatient treatment for venereal disease, either at the hospital or at the Boston Dispensary, to ninety-eight girls and unmarried women.[62] Mrs. Jessie Donaldson Hodder, the director of a social work unit devoted to "sex problems," noted in the department's annual report for 1907–8 that "the source of gonorrheal infection makes a vast difference in the treatment appropriate to the case."[63] MGH attributed gonorrhea vulvovaginitis to assault, masturbation, sex play with other children, or "innocently, i.e. bad hygienic conditions in the home."[64] The hospital wanted to treat only those girls whose infections were the result of poor hygiene and whose mothers could meet all the demands of the treatment plan. If the social workers found a mother unreliable or uncooperative, MGH placed her daughter into the custody of the Massachusetts Society for Prevention of Cruelty to Children (MSPCC). Nor did the hospital treat girls whose infections they suspected were the result of child sexual assault. They, too, became the responsibility of the MSPCC.

Of the patients they did treat, twenty (one-fifth) were girls under the age of 10, and Hodder claimed that most were suffering from "innocent infection, by towels and the like."[65] Yet she confessed that these girls "form by far the most difficult group we have . . . Many subtleties . . . hinder successful social care of them, subtleties which explain . . . the reason the crusade against venereal diseases is not as active and above board as is the fight against tuberculosis."[66] But nothing in either her reports or her personal papers hints at the subtleties that she found so troubling. After Hodder left MGH in 1911 to become superintendent of the Massachusetts Reformatory for Women at Framingham, Dr. Richard Cabot, the "father" of medical social work in the United States, completed the department's next annual report. He reiterated Hodder's frustration and admitted that "her failure with girls of this class is [striking] . . . Because the diseased girl is often indifferent to her condition and wants no help or friendship, we have been obliged to confess our inability to accomplish anything . . . toward changing her life."[67]

The situation might have improved after MGH established a Children's Medical Department that assigned full-time social workers to visit patients' homes, including eight girls infected with gonorrhea vulvovaginitis.[68] To emphasize the importance the department placed on social work, Dr. Fritz B. Talbot, the department chief, placed a social worker's desk next to his own.[69] Yet in their report for 1912, Talbot and attending physician Richard M. Smith had no more to say about the issue than had Hodder or Cabot: "There is a big moral problem which comes to the attention of the physician through the cases of gonorrhoeal vaginitis in little girls; this should receive the especial attention of a paid worker. We have already gone far enough in this problem to feel that our chief efforts should be given to preventative work and to a close study of the condition."[70] But they neither identified nor elaborated on the "big moral problem."

An article Smith published in the *American Journal of Diseases of Children* a year later made clear his understanding of the problem. Parents who did not appreciate that gonorrhea vulvovaginitis was a highly contagious disease frustrated Smith, such as the mother who told social workers that because all her children were infected, she assumed that gonorrhea "runs in the family."[71] But most parents knew better. Smith recounted how surprised parents were when hospital personnel explained that their daughter's infection was "contagious by any other than the usual method of infection."[72] Yet however much this information may have eased some parents' minds, Smith's data did not support it. When he separated his patients into categories according to source of infection, Smith attributed infections among the largest group, about half (sixteen), to living with other infected family members, by which he meant the girls had become innocently infected from bed sheets or towels. Smith wrote off the next largest category (five) of infections as "undetermined": one girl's parents claimed she had used a "dirty railroad toilet," one was "feeble-minded," and in two other cases "no reasonable explanation could be found."[73] Playmates and assault, four each, constituted the next category. Smith listed infections acquired during a hospital stay last, with only a single case.[74]

Even when Smith acknowledged that some of the girls had some type of "sexual experience," he never mentioned the possibility of coercive male sexual contacts. Smith noticed sexual behaviors among some of his patients that he recognized as inappropriate for their age. Some, such as masturbation, were widely condemned for a variety of reasons. But Smith was particularly struck by how sexually precocious and unusually secretive the girls were and noted that the usual "embarrassment during examination seen in a child of six has entirely disappeared in many of these children."[75] Yet even though so many of the family members of his patients were also infected, Smith did not consider the possibility of incest. Rather, he identified other girls as the perpetrators—and the source of infection. When one infected 9-year-old claimed that an older girl had taught her to masturbate, Smith guessed that in the "learning process" the girls' hands had become soiled with discharge and the younger girl infected.

When he didn't blame other girls for spreading infection, Smith blamed their mothers, supposing that "as we learn more intimately the history of infected children, the less often we shall have to attribute the infection to toilets and other indefinite sources. The supervision given many children by their mothers is so inadequate that there are long periods of time when they are unprotected from the influences of evil companions."[76] Smith was particularly disgusted by the frequency of infection among "feeble-minded" girls, whom he criticized for not appreciating their responsibility to avoid spreading the infection. Yet he did not question why feeble-minded girls, who by definition could not protect themselves, might be disproportionately infected. Instead, Smith told parents to institutionalize their feeble-minded daughters—to protect the rest of the family from them.[77]

Similarly, Dr. Edith Rogers Spaulding, who ran the isolation clinic for infected girls at Boston's Children's Hospital, declared that none of her fifty-six patients had been raped, even though, like Smith, she could not identify a source of infection for almost half.[78] Most of Spaulding's patients were 5 years old, the youngest only 7 months. She attributed twenty-six infections to "history of discharge in one of parents," sixteen to recent hospital stays, and three to "history of contact [with other children]," all of whom she believed had become infected because of poor hygiene.[79] Because Spaulding did not believe the girls had been sexually assaulted, she did not view them as victims but as a threat to public health. Spaulding worried that, with infections that could last for years, every infected girl could be the source of an endless stream of new infections. Spaulding thought the school lavatory the most likely object of transmission between girls, a view that doctors would increasingly adopt over the next decade. To prevent the spread of infection, Spaulding recommend that the school nurse bar infected girls from using the toilet.[80]

The obvious unworkability of Spaulding's recommendation illustrates how characterizing gonorrhea vulvovaginitis as a sanitary disease frustrated rather than improved doctors' ability to prevent or control the spread of the infection. Doctors talked about gonorrhea vulvovaginitis as an uncontrollable, nearly random phenomenon disconnected from any specific factors they could identify, let alone change. Viewing the infected girls as the problem to be controlled, doctors advocated isolating or placing them in "quarantine" until—in the era before there was a cure—the girls tested negative. Dr. Spaulding praised the Frances Juvenile Home, a small residential quarantine facility in Chicago, as the ideal environment for infected girls.

In 1909 philanthropists founded the Frances Home so that girls infected with gonorrhea (and children with hereditary syphilis) could be placed in quarantine without disrupting their medical care and schooling. The board of education provided a schoolroom and teacher, and the older children received "practical instruction in domestic science," suggesting that the girls were not from the white middle and upper classes.[81] Dr. Clara Seippel, a gynecologist just beginning her practice, was attending physician at the Cook County Hospital children's venereal disease ward. She served without compensation as attending physician at the Frances Home and as president of the association that managed it. Seippel also held the post of assistant city physician for twelve years and was an active member of the Chicago Social Hygiene Society and various women's and civic organizations, including the Medical Women's Club, which elected her secretary and then president.[82] When she left Chicago in the 1920s to become dean of women and "medical advisor to the women students" at the University of Arizona, her references for the position included the president of the Chicago Medical Society and Jane Addams.[83]

The Frances Home, which moved twice to larger quarters before it closed in 1941, filled a perceived medical and social need for isolating infected girls.[84] As Seippel put it in a 1911 fundraising plea, "We keep them until they are cured and can be allowed to . . . mingle with other children without danger of infecting healthy children."[85] Seippel claimed that though the Frances Home admitted twenty-one children in 1910, it turned away "dozens" more.[86] Demand for services elsewhere confirms Seippel's implication that infection was widespread. That same year, Cook County Hospital treated 166 girls, of which 111 were under the age of 6.[87] By 1913 Chicago's "high grade hospitals" were so overwhelmed by ward epidemics, even though they had adopted preadmission screening for infection, that they refused to treat infected girls altogether. The health department responded by ordering that all future cases be transferred to the county hospital, which set aside fifty beds.[88]

Seippel's superior in the city's medical office was Dr. Alfred Cotton, and she shared his awareness of the frequency with which girls were sexually assaulted and infected with gonorrhea. In an article published in 1912, Seippel claimed to have personally investigated fifty-three cases of child sexual assault in just six months. Thirteen of the girls were younger than 10 and all but one infected with gonorrhea. And Seippel noted that these cases were not a representative sample. Parents who "shrink from the notoriety" or do not wish to put their daughter through an "ordeal [that] is one of the most revolting experiences one can possibly imagine" had declined to prosecute "hundreds" of other cases.[89]

Parents' responsibility to protect their daughters' reputation was a heavy one, and their wish to avoid a trial and publicity is understandable. Some families, however, may have been more concerned with protecting a girl's father or other male member of the household by trying to control the types of circumstances in which the incest secret might erupt out of the family circle, such as the extended conversations they might imagine occurred between health care workers and hospitalized girls. Seippel complained about parents who suddenly took their daughters out of the hospital, without any explanation, after only a few days or weeks of treatment. Perhaps some of these parents feared that suspicious doctors would ply their daughters with questions about how they had become infected. But Seippel did not ponder these questions in her article. Like Dr. Flora Pollack, she believed that the superstitious cure motivated many men to assault girls. But unlike Pollack, Seippel claimed that belief in it was limited to the "foreign classes."[90] It may be that her assumptions about the inability of foreign-born men to behave properly so heavily influenced her reasoning that it distracted her from the possibility of incest among white, native-born families. Seippel knew that "the little daughters of the rich" were also infected with gonorrhea, but she did not discuss the possibility that they, too, had been sexually assaulted. She could only lament that "one is occasionally utterly unable to trace the source of infection in a child surrounded by every protection and comfort money can procure."[91]

Yet Seippel did not call for increasing scrutiny of the origins of these infections, just as she did not question whether incest had occurred. Rather, she told her colleagues at a joint meeting of the Chicago Medical and Social Hygiene societies that they had a civic duty to protect "hundreds of children exposed to infection" by raising "standards of hygiene."[92] To illustrate the danger of soiled bed linens, Seippel told the story of one family in which both parents and all four children (three girls, one boy) were infected with gonorrhea. After a fifth baby acquired *ophthalmia neonatorum* during childbirth and became blind, the father "lost interest" and deserted the family. By the time the mother saw a doctor, her infection had advanced to the point where doctors had to remove her fallopian tubes, and they placed her daughters in the Frances Home.[93] Doctors published similar accounts of case after case in which an entire family was infected, a circumstance that contradicts the assumption that a girl or a male other than the head of the household brought the disease into the home. Yet at no point in this harrowing story did Seippel suggest that the father had infected his children, even though a man who deserted his family was hardly respectable or deserving of the privileges and protection that status implied. Seippel worked heroically on behalf of her patients and sexually assaulted girls in Chicago, but, like so many other dedicated health care professionals and reformers, she seemed blind to the possibility of father-daughter incest in homes where she imagined it did not occur.

Incest, Infection, and Class

Progressive Era reformers would acknowledge histories of father-daughter incest only among one group of infected girls: poor, immigrant, and working-class girls incarcerated in penal institutions. Progressive Era reformers wanted to right the course of wayward lives, and they felt confident that if they could identify the root causes of female sexual delinquency they could keep girls on the right track. Early twentieth-century reformers identified girls who were or who they feared would become sexually active outside marriage as "delinquent." And, as happened repeatedly in the Progressive Era, reformers who studied these girls' histories discovered incest. Jane Addams thought the connection important enough that she noted it in her presidential address at the 1910 National Conference on Charities and Corrections. Addams named father-daughter incest a major cause of female juvenile delinquency, saying, "Out of the total number of 500 girls in the [Geneva] Illinois Industrial School committed for their first sexual immorality, forty-six had become involved with members of their own families, nineteen with their fathers, the rest with brothers and uncles."[94]

Whether delinquent girls acquired gonorrhea from their first "sexual irregularity" or from subsequent promiscuous behavior, beginning in the 1890s researchers consistently noted the frequency with which girls in reformatories, nearly all of whom were from poor and working-class families, were infected with gonorrhea and reported a history of incest.[95] The Chicago Vice Commission gathered information for its 1911 report from interviews with more than two thousand women and girls it suspected of engaging in prostitution. It concluded that "a large proportion" had their first "experience in sexual irregularity" with a family member, in most cases their fathers.[96] The following year Dr. Edith R. Spaulding coauthored an article in *JAMA* in which she claimed that as many as 70 percent of female delinquents were infected with gonorrhea, including 30 percent who had a history of incest.[97] In 1912 Jane Addams again mentioned a link between incest and delinquency. In her book on prostitution, *A New Conscience and an Ancient Evil*, Addams argued that girls often became delinquent as the consequence of early "wrongdoing," including assaults by some of "their own relatives."[98]

Progressives trumpeted data about gonorrhea infection and incest among socially marginalized families because it supported what they already believed, not because it demonstrated something new and surprising. The *Journal of Social Hygiene*, published by the American Social Hygiene Association, rarely mentioned incest, and then only as a behavior found among the poor and working classes. In 1920 it reprinted the text of a speech that Arthur W. Towne, superintendent of the Brooklyn Society for the Prevention of Cruelty to Children, gave to the American Prison Association. Towne was concerned about "motherless girls who are left to live with fathers, uncles, and older brothers, of questionable character." He believed that "the amount of incest in every community is much greater than ordinarily supposed" and warned that girls were especially vulnerable to "early acquaintance with depravity."[99]

Reformers often repeated their findings about the supposed relationship between class and ethnicity or race on the one hand and incest and infection on the other, despite having done no research into the behavior of men who lived in white middle- and upper-class homes. The Society for Prevention of Cruelty to Children was no more concerned with white middle- or upper-class families in the early twentieth century than it had been in the nineteenth. Like other reformers who worked with socially marginalized groups, Towne based his conclusions about incest on the experiences of only a narrow, unrepresentative sample of society. As a result, by the 1920s influential studies such as *The Unadjusted Girl*, by sociologist and Hull House associate W. I. Thomas, mentioned the high incidence of infection and histories of incest among delinquent girls as unsurprising and even expected.[100]

Doctors also insisted that a causal connection existed between class status and disease. In what seems an attempt to reinsert prebacterial-era ambiguity into the diagnosis, and with it the ability to avoid having to make a diagnosis of incest, a majority of pediatricians who answered a survey in 1915 from the Committee of the American Pediatric Society on Vaginitis in Childhood said that they believed that "true," or venereal, gonorrhea affected only indigent and low-income patients.[101] Because doctors believed that uncleanliness was the primary source of infection, they claimed that the incidence of infection was rising at a higher rate among ward and dispensary patients than among the better-off patients they treated in their private practices.[102] The following year, the Health Bureau of Rochester, New York, distributed a handsomely printed pamphlet about gonorrhea infection claiming that although men who were "members of exclusive clubs as well as the habitués of saloons" were infected, infection was "more common among the idle, wicked, and vicious."[103] But it also warned fathers and "other relatives" not to "leave about cloths, towels, etc." because their daughters, "using these articles, and wiping their genital organs with them also contract gonorrhea."[104] An incestuous father reading such pamphlets could have easily deduced how medical science could shield him from blame.

Studies of poverty, delinquency, and infection affirmed Progressive Era reformers' belief in a natural link between class, ethnicity, race, and sexual behaviors. Progressive Era reformers asserted their claim to social authority, in part, on the presumed objectivity of their social scientific methods. But they used highly selective empirical data, a method that turned their social biases about qualities they supposed were natural to foreign and less civilized men—or the poor housekeepers they married—into social scientific fact. Their unwillingness to broaden the scope of their investigations, particularly in response to an epidemic of gonorrhea that they could neither explain nor control, is a glaring omission among a generation of professionals who deified empirical research. In contrast to the tenacity with which they attacked other issues, such as their repeated marches through tenement neighborhoods to count and inspect privies, they ignored or mislabeled evidence that might have led them to identify father-daughter incest as a social problem and threat to the purity, health, and well-being of girls. And they continued to do so even when none of the measures they proposed to stanch the spread of the disease proved successful and the rate of infection continued to rise.[105]

No one attempted to study the incidence with which the infection arose within the white middle and upper classes, just as no one documented the occurrence of father-daughter incest among the "better sorts." Doctors and public health officials were still discussing gonorrhea vulvovaginitis as a class-based disease into the 1930s. In 1929 a physician leading a well-funded and highly publicized study of gonorrhea vulvovaginitis in New York City claimed that from 1 to 5 percent of the girls in "poorer" sections of large cities were infected. And though he had no data on which to base his estimate, he dismissed as negligible the numbers of girls infected in the "best residential districts." Dr. Walter Brunet assumed that differences in rates of infection could be explained by the "general care as to personal hygiene and the provision of separate sleeping arrangements among the population of the better sections."[106] When public health nurse Gladys Crain published an article in both *Public Health Nursing* and *Commonhealth*, the official bulletin of the Massachusetts Department of Health, that same year, she, too, identified poverty as the cause of gonorrheal vulvovaginitis.[107] Citing Brunet's speculations about differing rates of infection, she concluded, "The reason for this difference lies in bad housing, crowded sleeping quarters and poor hygiene."[108] Unable to explain how so many men like themselves could be abusing their daughters, reformers and health care professionals displayed a blithe indifference about the source of girls' infections. And although they decried incest among the poor and working classes, professionals' unwillingness to acknowledge father-daughter incest among respectable white families left those girls unmercifully on their own.

The National Social Hygiene Campaign and Family Values

After Prince Morrow died in 1913, anti-vice and anti–venereal disease groups joined forces as the American Social Hygiene Association, an umbrella organization that would provide a national platform from which they could address their interlocking concerns: venereal disease and prostitution.[109] Each organization had a stellar cast of founders representing many of the major public health and social reform movements of the period, including David Starr Jordan, president of Stanford University, vice reformer, and eugenicist; philanthropist Grace Dodge, known for her work with factory women; Chicago reformers Jane Addams of Hull House and Louise deKoven Bowen, philanthropist and director of the Juvenile Protective Association; New York City health commissioner Haven Emerson; Katharine Bement Davis, director of the New York Bureau of Social Hygiene, where she conducted research into female sexuality; Dr. Edward L. Keyes Jr., a gonorrhea specialist and friend of Dr. Morrow's; and California public health official Dr. William Snow, who spearheaded a successful campaign in California in 1911 to require physicians to report venereal disease infections.[110]

The breadth of the ASHA board illustrates its position as a nexus between concerns about venereal disease and broader reform issues. John D.

Rockefeller Jr., who had long been active in a variety of anti-vice and anti–venereal disease endeavors, provided the major funding for the new association and recruited its leadership. He convinced both Dr. Snow and former Harvard president and social hygienist Charles W. Eliot to reject prestigious other offers in favor of positions with ASHA as executive director and president, respectively.[111] Some of the key members of the old boards also joined ASHA, including Jordan, Addams, Emerson, and New York's Catholic cardinal James Gibbons. New board members included Dr. Hugh Cabot, chief of urology at Massachusetts General, active social hygienist, and Dr. Richard Cabot's brother; adolescent "protective" worker Martha Falconer, who worked with delinquent girls; and influential women's rights advocate Anna Garlin Spencer. The ambitious organization continued to attract high-profile board members through the 1930s, including a U.S. surgeon general; two New York City health commissioners; Hull House associate Grace Abbott, director of the Children's Bureau; Harvard Law School dean Roscoe Pound; and the country's most prominent venereal disease researchers, including Keyes, Dr. John H. Stokes, Dr. Percy S. Pelouze, and Yale public health expert Charles Emory Avery Winslow.[112]

The inclusion of reformers, philanthropists, and educators with such a diversity of interests ensured that ASHA's influence would spread far beyond the cadre of health care professionals and reformers concerned primarily about venereal disease or prostitution. For many social hygienists, educating the public about venereal disease was only a means to an end: convincing white Americans that sexual intimacy should occur only within marriage, if for no other reason, because their own health and the lives of their families depended on it. In a 1913 speech that ASHA would often reprint, Charles Eliot argued that male lust was too powerful a force for even the combined energies of "Christianity, democracy, and humanitarianism" to control. Sharing Morrow's view of social hygiene as a nativist or eugenics issue, Eliot called venereal diseases the "worst foes of sound [white] family life, and thence of civilization."[113] Two years later Snow reiterated that "continence outside of marriage . . . is the greatest factor in the prophylaxis of these diseases."[114]

However, before World War I, few Americans seemed interested in ASHA's message. The war briefly provided the organization with the patriotic mission of reducing the incidence of venereal disease in the military. But after the war ASHA's focus on health and morality fell out of step with the eroticized consumer society of the 1920s. As young adults hurried to distance themselves from the sexually conservative mores they associated with their parents, ASHA renewed its commitment to the moral high ground. The *Journal of Social Hygiene* continued to publish articles recommending sexual abstinence rather than condoms to prevent disease, and it denigrated plays and motion pictures for sexually suggestive content just as the popularity of the movies was exploding among the white middle class.[115] As late as 1926 the *Journal* republished Eliot's critical appraisal of male lust.[116] Such views must have sounded painfully out of fashion to a generation of hedonists steeped in postwar nihilism. Many local groups—such as the Massachusetts Society for Social Health, which had been founded by Eliot, Hugh Cabot, and other upper-class Bostonians to promote "the highest standards of private and public morality"—lapsed into inactivity.[117]

By the mid-1920s some hygienists began to explore new directions that would both appeal to young adults and convince them to harness their rampant sexual desires. Whereas Morrow and Eliot had held promiscuous husbands responsible for the degree to which venereal disease threatened the white American family, in the 1920s social hygienists made a profound shift in emphasis and direction. They stopped condemning individual men for their sexual misbehavior and role in spreading infection to their families and laid the blame instead on the laws and mores of "respectable society." Hygienists argued that polite society's ban on "frank" and public discussions of sexual matters had curtailed access to reliable information about sexuality that, as a result, had left men without the knowledge necessary to properly manage their sexual desires.[118] Twentieth-century social hygienists had always conceived of their mission as brave and modern because it attacked the cultural mores codified in the Comstock laws, which broadly restricted the publication and sale of items pertaining to sexuality. They complained bitterly about laws that treated all discussions and materials about sexuality equally—from science-based sex education for young adults, to contraceptive information for married couples, to the risqué of burlesque, to the crudest pornography.[119]

Social hygienists in the 1920s still bemoaned the newly commercialized culture of sexual behavior, but they identified ignorance, rather than male lust, as the problem and education, not condemnation, as the remedy.[120] Jazz Age social hygienists believed that if sexual education could improve marital sexual satisfaction, fewer men would seek extramarital contacts and spread disease. To affirm the potential for sexual satisfaction within marriage, ASHA developed a strategy for making married life seem attractive to adolescents and married couples: first, by delivering the practical instruction necessary for couples to achieve sexual fulfillment within the marital bedroom; and second, by promoting an ideology that characterized marital happiness and parenthood as the crowning achievement of adulthood.[121] Education in preparation for marriage and parenthood might have offered benefits to people from all classes, but social hygienists directed their plan of "scientific education" to the middle class.[122]

The new pro-family rhetoric also appealed to respectable African Americans, a growing part of a burgeoning middle class, some of whom joined ASHA as active participants.[123] African American newspapers had started printing educational material on venereal disease after 1910.[124] By the 1920s, prominent African Americans like Mary Church Terrell were also using the new data about venereal disease infection among white Americans to rebut the assumption popular among white Americans that they maintained higher standards of morality than did African Americans.[125] In 1927 the *New York Amsterdam News* attacked assumptions about "Negro immorality" by reminding readers that two white men were recently charged with father-daughter incest.[126] And when a white author published a novel in 1929 that linked African Americans with "illiteracy, superstition, incest, syphilis, and final ruin," a black reviewer derided it as "the sort of thing we usually see when a white author writes about Negroes."[127]

Social hygienists hoped that a middle-class audience learning to consume would find their message about the idealized family attractive. ASHA's new tone may have been simply a conservative response to anxieties raised by woman's suffrage, women's rising visibility in the workplace and in consumer culture, and the sexual license of the 1920s, issues that competed in the larger cultural conversation with the anti-vice faction's earlier focus on prostitution as the major threat to marriage and the family. But it was also practical. In 1929, a researcher from the Russell Sage Foundation criticized ASHA's advocacy of the single standard of morals. Its study concluded that people who attempted to achieve the " 'highest standards of private and public morality' " usually became "dismayed" when their efforts fell short.[128] By shifting from a negative critique of male sexuality to a positive presentation of sexual fulfillment in marriage, health care workers and sex educators struck a more pragmatic note. It worked. Middle-class men and women responded to this message and began to revive moribund social hygiene chapters.[129]

Speakers at the 1924 National Social Hygiene Convention no longer inveighed against the immorality that led to venereal disease infection. Social hygienists chose instead to affirm the primacy of the nuclear family, which they framed positively in the convention's first resolution: "To preserve and strengthen the family as the basic social unit."[130] The roster of invited speakers featured not only key supporters of social hygiene but also some of the country's most prominent female protective workers. Protective workers sought to discourage poor, working-class, and immigrant adolescent girls and women from engaging in extramarital sexual activity because it led to other social problems, including delinquency and venereal disease infection. Protective workers at the conference included Hull House associates Grace Abbott, now director of the federal Children's Bureau, and Jessie F. Binford, director of Chicago's Juvenile Protective Association. But speakers also included Katharine Bement Davis, whose "Study of the Sex Life of the Normal Married Woman" the *Journal of Social Hygiene* had published serially between April 1922 and March 1923.[131] Reformers no longer dwelled on the dangerous aspects of sexuality to the exclusion of the healthy or positive.

In the most dramatic turnabout, the *Journal*, which still characterized prostitutes as irresponsible carriers of disease, began publishing articles that referred to men who visited prostitutes and became infected as naive rather than immoral, and deserving of empathy rather than condemnation.[132] Where Dr. Morrow had castigated such errant husbands and fathers, social hygienists in the 1920s recuperated them. And as social hygiene became increasingly professionalized, health care workers reconceptualized venereal disease from a punishment for individual indiscretion to a "family problem."[133] Physicians subsumed their interest in the individual patient within a new commitment to minister to the "venereal disease family." Rather than condemn a philandering husband for lack of restraint, health care workers scrutinized the material and affective quality of the patient's home life for clues that might explain (and so provide a path to fixing) a husband's infidelity.[134] In 1923 New York City health commissioner Dr. Haven Emerson suggested that the term "family service ward" replace "venereal disease ward" to describe the place where "sympathetic unraveling and scientific as well as social treatment of the tangled lives blighted or threatened by venereal disease" occurs.[135]

Social hygienists and health care workers now expressed little concern over how the disease was introduced into the family. They stressed instead a dispassionate, scientific approach to managing the disease once it had entered the home. Social hygienists were not interested in taking a stance that might have encouraged marital discord, possibly ending in the dismemberment of the family in the divorce or criminal courts.[136] But to recuperate the social standing of errant husbands and fathers, social hygienists blamed the spread of infection exclusively on the women within a man's own home: his wife, whom they presumed was sexually withholding and unhygienic, and her daughters. Social hygienists now reversed Morrow's analysis of venereal disease as a family problem. They no longer viewed mothers and daughters as the innocent victims of husbands and fathers who poisoned their families. Rather, social hygienists now identified mothers and daughters, not prostitutes, as the bearers of disease and husbands and fathers their victims. As a result, when a girl became ill, nurses and social workers excoriated her mother for poor housekeeping; they ignored her father.[137]

"Make the Mother Accept Responsibility": Displacing the Source of Infection

After 1920, the public health movement shifted its focus away from the environmental sources of disease and toward identifying and treating sick individuals. Public health officials now considered disease control and prevention more pressing than personal hygiene and civic sanitation. "The housewife's contribution to the control of infectious disease [through cleanliness] diminished accordingly," historian Nancy Tomes noted, "her duties came to consist primarily of obeying the health department's quarantine regulations and obtaining immunizations for her children."[138] But, as was the case with almost everything about this disease, doctors viewed gonorrhea vulvovaginitis as an exception. Doctors and social hygienists in the 1920s and 1930s intensified their class-based condemnation of the mothers of infected girls, whom they blamed for the "unbelievably dirty environment" that had resulted in their daughters' infections. Caseworkers judged mothers harshly and offered them little sympathy.[139] Articles in the medical, social work, and social hygiene literature invariably describe the mothers of infected girls as poor, ignorant, or incompetent. Even though they could not explain how, health care providers assumed that careless or slovenly mothers transmitted the disease to their daughters, and they held them, not their husbands, responsible for placing "a blot on the white page of girlish innocence."[140] If social workers or health care providers found infected girls living in clean homes or the well-kept homes of the wealthy, they did not mention them.

When they focused on maternal behavior and sanitation, social hygienists narrowed the number of possible sources of infection. In 1916, a social worker employed by the Boston Dispensary and Massachusetts Society for Sex Education could not determine which of the many possible sources of infection was responsible for the infected girls with whom she worked.[141] Bertha C. Lovell believed that women acquired gonorrhea only from sexual contacts, but she was less certain about the source of infection for girls, in part because the possibilities were so numerous: "Was it an accidental infection from a dirty toilet seat, or an infection, as accidentally incurred in the daily exigencies of life in a crowded household where the mother or father or perhaps an older sister had the disease? Or was it one of the rare cases of rape . . . or did it open up a quagmire of immoral practices among a group of children; or was it a case of precocious sexual development?"[142] Lovell's bewilderment characterized the first two decades of the twentieth century. Had she been writing in the late 1920s, however, Lovell would not have mentioned the father at all, either as a possible carrier of disease into the home or as his daughter's assailant. Social hygienists had erased the father's behavior, and with it the possibility of incest, from view.

In the 1920s, Edith M. Baker, director of social services at the St. Louis Hospital and president of the American Association of Hospital Social Workers, worked to professionalize medical social work by setting standard protocols. In cases of adults infected with gonorrhea, Baker required social workers to take extensive patient histories that included interviews with the patient's family, relatives, friends, and employers. But when the patient was a girl, Baker cared little about how she had become infected. Baker considered it more important for caseworkers to determine how many people shared towels with the infected girl than to learn whether a girl's father might also be infected.[143] In the case of one infected 12-year-old girl, Baker ordered the staff to isolate her; to explain to her mother the "precautions" she needed to use regarding the toilet seat, bath, laundry, and the sleeping arrangements; and to test the mother and any other daughters for infection. Only after the mother tested positive did Baker recommend that the rest of the family, including the father, be tested. But the purpose was not to investigate the possibility of incest or even to determine who was responsible for bringing the infection into the household. Baker wanted the information so that she could impress upon the mother the full extent to which the disease had spread to her family and, presumably, the urgency with which she needed to adhere to Baker's instructions.[144]

Baker's focus on the high standards she expected mothers of infected girls to attain would not have surprised anyone in the 1920s. Nineteenth-century Americans had considered maternal love a naturally "benevolent force" that provided mothers with the moral and emotive basis for nurturing children. But, as historian Rebecca Jo Plant has observed, as a growing class of professionals increasingly sought to rationalize every aspect of Americans' lives, they rejected the presumed natural ability of mothers to raise their children properly and advocated "scientific motherhood" instead.[145] Mothers could no longer enjoy the luxury of basking in a social status grounded in sentiment; they now had to earn any status associated with motherhood by demonstrating their facility with modernity.

Changes in technology and labor in the same period, which seemed likely to simplify housekeeping, often meant instead that mothers had to adapt and become accomplished at new types of physical labor. The most obvious change was that African American women increasingly left domestic service for better-paying jobs in industry, leaving white middle-class women to do housekeeping tasks themselves. But changes in housekeeping affected everyone. Before World War I, most poor families had no indoor plumbing, and even for those that did, the cost of a washing machine—fifteen dollars—was beyond their reach. Even if they did obtain a washing machine, mothers then had to spend hours hand-cranking it. Bathing one's family entailed repeatedly filling a wooden tub with water, a task so onerous that most family members bathed fully only once a week.[146] With large families cramped into small, often poorly maintained quarters without closets or storage space of any kind, and which a coal oven might continually cover with soot, cleaning up was, as historian Ruth Schwartz Cowan observed, a "Herculean endeavor."[147]

Changes in the availability of consumer goods and the increasing reluctance of African American women to work as domestic servants dramatically increased the amount of work a middle-class mother had to do personally to keep her home clean. Whereas well-to-do households in the nineteenth century had employed more than one servant, by the 1920s mothers were expected to perform all the housework themselves.[148] Middle-class families obtained luxuries like indoor plumbing, electricity, and household machines decades before their poor rural and urban counterparts. Yet these new goods did not always offer the respite that many women imagined might follow. The installation of a modern bathroom, for instance, also meant that mothers were newly charged with keeping the bathroom spotlessly clean, lest their families become ill.[149] And although most American homes acquired electricity by the end of the 1920s, machines like vacuum cleaners and washing machines failed to reduce the time mothers spent doing housework, as new ideals about cleanliness required women to launder their families' bed linens, underwear, and clothes more frequently than in the past.[150]

Nor did the new appliances decrease the time a mother devoted to caring for her children, the demands of which became increasingly complex as modern women learned to pay attention to their children's diet, education, and clothing in new ways—and without the help that had been available earlier in the century from domestic servants and nursemaids.[151] New ideas about children's physical, emotional, and developmental fragility and a growing chorus of experts also emphasized the mother's role in actively protecting and nurturing children, whom they now characterized as vulnerable and in need of constant, detailed attention.[152] Popular acceptance of the germ theory, coupled with a barrage of articles in magazines by doctors and advertisers selling products that promised to keep one's family safe from disease, raised standards of cleanliness and health care and intensified the anxieties and duties of mothers.[153] Even accidents in the home became a mother's responsibility, for "who else can keep a child from falling from a window, from pulling over a vessel of boiling water?"[154] As Peter N. Stearns concluded in his study of parental anxiety in this period, "professionalism replaced sentimentality."[155]

Not coincidentally, the steady exit of female domestic employees—Irish immigrants in the nineteenth century and African Americans in the first decades of the twentieth—from white middle- and upper-class homes reduced the possibility that a servant could be blamed for having brought venereal disease into the home and spreading it to white children.[156] The white mother now shouldered the entire burden herself. Consistent with new social pressure on mothers to personally keep their homes clean, and with the guilt they were expected to experience if their efforts fell short—so widely shared that even advertisements for household products from Campbell's to GE tapped into it—the central goal of social hygienists and health care providers treating infected girls was to "make the mother accept . . . responsibility."[157] They not only blamed mothers for the conditions in which they lived and the health of their daughters but also held them responsible exclusively for preventing other family members from becoming "contaminated." As one doctor explained, "We look to the mother as the sine qua non in carrying out home treatment and principles of prophylaxis."[158] In 1915, Dr. Frederick J. Taussig recommended that nurses teach the parents of infected girls about the "use of separate towels and wash-cloths . . . [,] a separate bed [for the infected girl], care as to a thorough cleansing of any contaminated clothing, and special precautions in the use of the lavatory. It would of course be best for the child to use its own vessel instead of the common lavatory."[159] Even if a mother could isolate her child from all social contact, items such as a separate bed for the infected child (an idea growing popular for all children in this period) and extra clothing and linens would have been beyond the financial and material means of many households, especially since often more than one child in the home was infected.[160]

Doctors also expected mothers to bring their daughters to the clinic for treatment at least once a week for a minimum of one year. When she began working at the Boston Dispensary, Bertha Lovell discovered one 8-year-old girl who had been receiving treatment since she was 2.[161] Because there were so few clinics and most treated only certain groups of patients at specified times, many mothers and daughters had to travel across town to reach a clinic during the one or two hours a week it reserved for treating girls. Clinic visits, which entailed pelvic examinations and vaginal douches, could be emotionally draining. Many girls became hysterical; sobbed and shook; tried to hide; became rigid, sweaty, and clammy; clenched their teeth; and acted as if they "were being half killed while the smear was being taken." Some girls were so terrified that it took three nurses to hold them down. Perhaps the girls were frightened because they had little experience with doctors and they felt particularly vulnerable in the physical position required for a gynecological exam. Others may have experienced the focus on their genitals, as contemporary pediatricians who examine girls for sexual assault warn, as a repeat of the trauma of having been raped.[162] But whatever the reason, the girls' pleas for sympathy fell on deaf ears. One social worker sniffed in the 1930s that the "agitation and terror" of girls who screamed "over and over again that they [the doctors] were going to hurt her" was "all out of proportion to the immediate problem."[163]

Health care professionals interpreted their inability to check the rising number of infections as proof that the disease was highly contagious, not that their emphasis on sanitation was misplaced. Their demands on mothers became increasingly labyrinthine. In 1927 Dr. B. Wallace Hamilton updated his printed instructions, which reflected his belief that "gonococcus vulvovaginitis is a highly contagious," not a sexually transmitted, disease.[164] If complied with, Hamilton's instructions would have left girls almost completely isolated from any social contact and their mothers exhausted. Still, the U.S. Public Health Service, the New York City Health Department, and ASHA endorsed them.[165] That same year, when Dr. Stephen Yesko identified gonorrhea as the second most common childhood disease, he also advised parents to completely isolate their infected daughters. If literal quarantine was impossible, he suggested the same measures that Wallace Hamilton did, to which he added: "The child . . . should be made to wear a vulvar pad at all times . . . [and should not] mix freely with playmates, but . . . [be] restricted to the sick room. After each use of the toilet, the child's hands must be washed and cleaned with bath alcohol. This . . . is a tedious task, especially in crowded surroundings. The mother is cautioned that clothing and the infected child's bed linen must be washed in a separate laundry . . . [and] first be saturated in a bichloride of mercury solution . . . followed by boiling from one-half to one hour . . . The seat of the toilet . . . should be protected by a towel, followed by washing with . . . creolin solution."[166] By 1934, doctors also expected mothers to "see that the patient sits in a basin or tub of warm water for fifteen minutes twice a day," wash her genitals with a boric acid solution and insert mercurochrome into her daughter's vagina four times each day, and change her daughter's sanitary pad each time.[167] By the late 1930s social workers in New Orleans had expanded the zone of contamination to include all kinds of recreational activities, including skating, playing ball, or swimming in public pools. But still they were unsatisfied. To avoid the "chance that through childish carelessness another little girl may be exposed," they urged responsible mothers to forbid their daughters to go outside and not to let friends visit them at home.[168]

Some girls were as frightened and upset by the home treatments as they were at the clinic, and it is difficult to imagine any mother who could sustain the necessary patience or exert such a high level of control over a child for long. It is easier to imagine how the incessant pleas and misbehavior of frustrated girls, who must have felt unjustly punished, and the task of having to meet such an exacting regimen—for days, months, and years—would have worn down even the most diligent mothers until they became so exasperated they gave up. However, to social workers and nurses, the fact of infection was enough to prove a woman's inadequacy as a mother. After professionals felt compelled to step in to protect her family where she had failed, it was too late for a mother to question their authority, let alone attempt to argue that she possessed any measure of superior knowledge. If a mother failed—or refused—to meet each requirement, her daughter suffered the consequences.

Nurses and social workers removed girls from mothers who didn't meet their standards and placed them in a reformatory or quarantine facility. Edith Baker praised the swift action of a caseworker who put a girl in the Convent of the Good Shepherd, a reformatory, as soon as she determined that her mother was "careless and self-centered."[169] Social workers might have removed even more girls if not for their insistence on the "catchiness" of the disease, which made juvenile institutions and foster homes unwilling to accept them. The rhetoric that identified girls as "highly contagious" may have shielded some fathers from detection, but it also delayed or even prevented social workers from moving victimized girls to safer environments. Social workers noted that some girls "improved rapidly" once they were taken out of their homes, and they expressed their regret that, because of the girls' infections, they could not be placed into an institution or foster home much sooner.[170]

Social workers intent on laying down the rules for improved sanitation rebuffed mothers who tried to report incest. Some mothers may have viewed a social worker's threat to remove her daughter from the home and place her in an institution as an opportunity, however desperate, to protect her. But social workers attributed a mother's resistance to accepting the diagnosis, following up with treatment, and isolating her child to ignorance about the "nature and dangers of the infection."[171] Public health nurses complained about mothers who too often had "a highly emotional attitude toward the situation." Gladys Crain, assistant director of the National Organization for Public Health Nursing, advised nurses that "only by a sane interpretation of facts can the nurse help her to a wholesome understanding of the problem."[172] What was the "problem" that mothers of infected girls couldn't understand? When Edna Pearson Wagner, a caseworker at the New Orleans Children's Bureau in the late 1930s, reviewed social workers' files on infected girls, she found no references to incest, rape, or sexual assault. What she found instead were notes documenting that most mothers refused to believe that their daughters had "gotten such a disease." Many permitted doctors to treat their daughters only after a social worker had convinced them that the disease was common in children and easily acquired.[173]

Mothers of infected girls often expressed anger and hostility at caseworkers and doctors, whose attitudes they found "insulting."[174] In cases in which social workers asked a girl's mother but not her father to be tested, many mothers refused, saying that they saw "no reason at all for this" or that the request insinuated that, in the words of one mother, she "had been a slut." Others blamed shared toilets, a response that social workers accepted.[175] Were mothers surprised that their daughters were infected or offended at the implications about their housekeeping? Or were they angry because the diagnosis had confirmed their fears? Some may have worried that the incest would be detected and the girl's father or another male family member arrested. Others may not have expected the diagnosis, which they learned only after the disease was detected by a doctor treating their daughter for another condition. Some mothers described feeling overwhelmed by "the whole problem" and despaired at being able to care for their sick child, while others responded to the diagnosis by arguing that they kept their children too clean to have become infected.[176] In any case, when mothers expressed guilt, social workers did not note the reason why.[177]

Nineteenth-century medical jurists like Sir Astley Cooper, who had mocked suspicious mothers, still recorded a woman's fears that her daughter had been sexually assaulted. Twentieth-century doctors, social workers, and social hygienists did not. Their records and writings infrequently noted a mother's presence, let alone what she or her daughter might have had to say about the girl's history, diagnosis, or treatment. Were health care workers simply not recording mothers' suspicions, had mothers stopped sharing their fears, or had mothers come to accept that poor housekeeping was to blame? Knowing that they would be blamed for their daughters' infection no doubt intimidated many women into silence—whether its source was incest or uncleanliness, the fact that her daughter was infected discredited a woman's value as a wife and mother.

Like the nineteenth-century medical jurists in England and America, Progressive Era Americans and interwar social hygienists fancied themselves as selflessly protecting the nation. Progressive women did not hesitate to demonstrate the evils of prostitution by criticizing philandering husbands for acquiring gonorrhea from contacts with prostitutes and then spreading the disease to their "innocent" wives.[178] And they readily detected incest and acknowledged its connection with infection among socially marginalized families. But accusing "respectable" white men of incest was another matter. Even when they obtained information about an "epidemic" of gonorrhea vulvovaginitis, Progressive reformers, especially women, did not use it to expose sordid male sexual behavior.

Progressive Era reformers concentrated instead on public dangers to girls, even though their own research had shown that incest could be a pivotal event in a girl's life.[179] Early twentieth-century female reformers promoted marriage and the family because it provided the environment in which women could maximize their social power through the exercise of their innate morality, not only by nurturing moral children but by taming men and male sexuality.[180] Jane Addams, for one, promoted the nuclear family as a corrective to social chaos, and concerns about the viability of the family may have discouraged both Progressive Era and interwar reformers from identifying its more sinister features.[181] How could white middle- and upper-class fathers be the lynchpin of America and the promise of modern civilization if they were also inhuman perverts of the most beastly kind, no better than dark-skinned immigrants or African American men?

Unable to answer that question, reformers and health care professionals embraced speculations about sanitation and maternal incompetence that erased from view the role of respectable white men in causing their daughters' infections. The reluctance of Progressive Era doctors and reformers, and of their protégés in the interwar years who professionalized social hygiene, to do more than speculate about the etiology of girls' infections, their uncharacteristic refusal to ask hard questions about the empirical data they collected on gonorrhea infection in girls, and their unwillingness to articulate incest as a social issue may have left American ideologies intact. But like fairy tales by the Brothers Grimm, they had inadvertently exposed the harsh dangers and realities underpinning their dreams.

CHAPTER FIVE

Incest Disappears from View

IN MAY 1931 the *Journal of Social Hygiene* published a letter from Paul Popenoe to the American Social Hygiene Association (ASHA), his former employer:

> The question has frequently been raised as to precisely what we should tell the public about innocent infection with venereal diseases . . . We have . . . been rather vague . . . in stating just what the extent of this innocent infection is [and] just how [it] occurs . . . Are there actually enough gonorrheal infections from toilet seats to justify us in urging that paper seats be regularly used in public places?
>
> In other words, is all this talk about innocent infection largely a bugaboo, and can we say that the matter is of little or no importance to those who live under normal civilized conditions? My impression is that virtually all of our present literature is equivocal on this point and that . . . we should clear it up thoroughly.[1]

At the time he wrote this letter, Popenoe had been employed as a social hygienist for more than fifteen years. He began in 1913 as editor of the *Journal of Heredity*, which was published by the American Genetic Association. During World War I, Popenoe served as a sanitation (venereal disease) officer, and after the war he became ASHA's executive secretary.

In the 1920s Popenoe moved to Pasadena to work for the Human Betterment Association, a eugenics organization whose founders included some of the most influential names in interwar California, including University of Southern California president Rufus B. von KleinSmid, Stanford University president and chancellor David Starr Jordan, and Harry Chandler, the publisher of the *Los Angeles Times*. Popenoe's first task was to advocate on behalf of California's pioneering eugenic sterilization program, and the *Journal of Social Hygiene* serialized portions of articles he wrote that praised the program's goals and fair-minded administration.[2]

In 1929 the association organized the Institute for Family Relations in Los Angeles, which it promoted as the first marriage counseling service in the United States, and it named Popenoe its director. Although the majority of the institute's clients were couples preparing for marriage, the association had not opened the institute to quell the anxieties of newlyweds. It had designed the institute to collect data on heredity and eugenics, which it accumulated by requiring clients to answer painstakingly crafted questionnaires, the answers to which the association used to support its political agenda.[3]

But however prying young couples may have found the questionnaires, the institute remained popular. Because printed, medically accurate information about sex and especially venereal disease was difficult to locate, couples raised questions on these topics with Popenoe and his staff.[4] Popenoe used the title "Dr.," but he had neither graduated from college nor been trained as a counselor of any kind. Still, he and the institute seemed to meet a need among couples, and by the 1950s Popenoe had become a nationally recognized expert on marriage. He wrote an advice column for Los Angeles newspapers, the *Ladies' Home Journal* carried his column for more than twenty years, and he dispensed advice on daytime television's popular *Art Linkletter's House Party*.[5]

Dr. Walter Clarke, ASHA's director of medical measures, replied to Popenoe's letter.[6] There may have been no person better suited to the task. Clarke began studying social hygiene and settlement work while in graduate school in the second decade of the twentieth century, and he joined ASHA in 1914 as its "field secretary." His first post was Hull House, where he resided for three years until joining the military as a sanitation officer during World War I. After the war Clarke became director of the Bureau of Social Hygiene for the League of Red Cross Societies (in Geneva) and then an investigator for the League of Nations' Commission to Investigate the Traffic in Women and Girls. In 1928 he completed his medical degree, and ASHA named him its medical director. For the next ten years he visited and wrote about venereal disease treatment facilities in Europe and America before rising to the post of executive director in 1938, a position he held until he retired in 1952.[7]

Yet despite his experience and training, Clarke did not provide the unequivocal answer that Popenoe had requested. "My own impression with regard to gonorrhea . . . is that, except in the case of little girls . . . infections by means other than sexual contact are rare . . . The so-called 'accidental' infections from bath tubs, toilet seats, etc., are so rare in adults that every doctor treats alleged cases with great scepticism."[8] Clarke conceded that opinion was divided as to whether gonorrhea in "little girls, for example in school children, may be contracted from toilet seats," but he added that "recent investigations seem to prove that it is entirely possible."[9] Clarke did not cite these investigations; there were none.

Clarke's inability to provide a succinct response to Popenoe's letter, two decades into an epidemic, is striking when compared with the proliferation of new research and public education materials that ASHA had prepared and disseminated in the same two decades about gonorrhea in boys and adults. After blaming girls, mothers, and ordinary household objects for the epidemic of infection among girls, health care professionals and social hygienists in the 1920s placed the blame on toilet seats, displacing the act of transmission from a person to an object. Even then, they removed the guilty object from the home entirely and settled on girls' lavatories in public spaces, particularly schools, as the primary danger to girls.

Public Dangers, Intimate Infections

In her history of sanitary reforms, Nancy Tomes noted that late nineteenth-century reformers urged mothers to protect their family's health by keeping their homes clean. As doctors and reformers began to accept that bacteria, and not odors or miasmas, were the cause of many illnesses, sanitarians quickly singled out bathrooms and toilets as hazards to family health. Tomes cited an 1887 sanitary tract stating that women who had "lost a child, a husband, or other relative" to disease should consider "whether the source of trouble may not be in the water-closet."[10] No one familiar with the conditions of urban privies would have doubted it. In 1912, the president of the National Housing Association credited a 1901 tenement law with having reduced the number of "school sinks"—open privies in tenement house yards that were used by all the building's residents and which were rarely flushed with water—from 9,000 to 375. Most were replaced by indoor toilets to be shared by two families.[11]

By the early twentieth century the duty of mothers to protect their families from death and disease by keeping their homes clean had been firmly established, and home economists pressed mothers to become "active agents in the pursuit of the safe toilet."[12] But learning how to use the new technology and to keep toilets and bathrooms clean enough to keep serious diseases at bay was hardly "light work."[13] Reformers began to instruct tenement mothers to flush their toilets regularly; schooled farm mothers on "sanitation privies"; and explained to mothers everywhere the "exacting daily maintenance" necessary to keep their toilet spotless—and their families free from disease.[14]

The intensity with which Americans feared the invisible germs they now learned could attack their families, and the disgust with and demonization of outmoded, filthy facilities that accompanied the cultural shift from outdoor privies and outhouses to indoor plumbing, also attracted the attention of doctors eager to identify a nonsexual source of gonorrhea vulvovaginitis. In October 1914 Dr. Frederick Taussig, a gynecologist and member of the executive committee of the St. Louis Society for Social Hygiene, published an article on the etiology of gonorrhea vulvovaginitis in which he concluded, "The most frequent source of infection is from child to child, and the most common manner of its transmission is through the school lavatory."[15] Taussig treated girls at Washington University Hospital, and he assumed that the same factors that contributed to "ward epidemics" also spread the disease to girls in close contact in similar settings. In a paper he delivered to the St. Louis Medical Society in May 1914, Taussig echoed Alice Hamilton, warning that "wherever children congregate in considerable numbers, in schools, playgrounds, or tenements, and use the same lavatories or towels, infections of this sort are likely to spread."[16] His speculations, like so many others, were incorporated quickly as fact into the literature on vulvovaginitis. ASHA included a slightly revised version of Taussig's article in the second issue of the *Journal of Social Hygiene*.[17]

However, Taussig was not as comfortable with this explanation as his unequivocal declaration seemed to suggest. In his detailed discussion about various possible sources of infection, Taussig's doubt about a nonsexual source of infection that was both medically and logically consistent with the empirical data about infection is palpable. He relied on data from fourteen patients, girls between the ages of 5 and 7, who came from different parts of the city and different social backgrounds. Taussig knew that gonorrhea was a sexually transmitted disease, and he acknowledged the possibility of rape in four cases. But he discounted it as the general source of infection for two reasons. The first was that none of his patients exhibited genital injuries, but the second, Taussig's assumptions about ethnicity, proved more persuasive. Referring to "German writers," Taussig noted that the "lower classes abroad" believed in the superstitious cure. And because "only a few girls" in his study "came from the ignorant, foreign-born population," he brushed aside the possibility that sexual assault could be the source of infection.[18] Taussig does not seem to have considered the possibility that "infectionists" or "perverts" might also include native-born white American men among their ranks.

Neither could Taussig imagine that the infection arose from poor sanitation, as his patients were from the middle and upper as well as the lower social classes.[19] He was inclined to believe that mothers transmitted the infections, but he admitted that the data did not support such a conclusion.[20] Taussig reasoned that because of the intensity of mother-baby physical contact, such as changing diapers and cleaning the genital areas, if mothers' hands were to blame, the majority of infections transmitted from mother to daughter should be among infants. But the ages of infected girls, most of whom were not infants, failed to bear out this idea. Nor did Taussig believe that the bacteria were transmitted on shared towels or linens, which he believed dried too quickly to transmit the disease.[21]

Taussig thought the toilet seat the ideal vehicle for spreading infection. Straining to identify some common factor in each of the girls' lives, he noted that most were school age and that one 5-year-old girl became infected shortly after she began attending school. Taussig rejected the possibility that she had been infected at home because no one else in her family was infected and, in any case, her "home lavatory was immaculate and used only by the family."[22] But when he considered possible sources of infection outside the girl's home, he described the mode of transmission in a way that skirts the border between the medical and the pornographic imaginations: "Lavatories, even in grammar schools, are as a rule so high that the smaller children in using them are forced to have their genitals and clothing rub over a considerable portion of the seat. The greater the number of persons using the same lavatory, the less interval of time is apt to elapse between its use, and hence the greater likelihood of carrying infection." Taussig concluded that "the lavatories in tenements, playgrounds, and public schools are . . . a source of considerable danger."[23]

Having deduced the source of infection, Taussig devised a plan to reduce the likelihood of transmission. Taussig first recommended that a visiting nurse inspect tenement toilets and report any she found "in an unclean condition."[24] To decrease the risk of infection among all girls, Taussig recommended that public toilet bowls be lowered, that U-shaped toilet seats replace the old-style closed or O-shaped seats, and that school lavatories provide paper seat covers. In the interim, he advised that placing "an attendant in the school lavatories" could "aid materially in the discovery of girls having a discharge, in the proper use of the paper covers, and in the general cleanliness of the lavatory."[25] Such an intrusive procedure did not seem unwarranted to Taussig. He was already in the habit of voluntarily reporting the names of infected girls to the school board, and he required his patients to stay home from school for at least the first two months of treatment, until the telltale discharge disappeared. Even after Taussig permitted a girl to return to school, he expected a teacher to make certain that she did not use the school toilet.[26]

Enforcing such measures would have increased surveillance over girls and created or reinforced any shame they felt about the disease, their genitals, and their bodily functions. In some classrooms, a circle of empty desks cordoned off the desks of infected girls who had been barred from school, a measure that had no effect on containing the spread of infection but spoke volumes about the stigmatization that infected girls could expect.[27] What the attention to girls and toilets seats did succeed in containing was any suspicion that men and fathers were the source of girls' infections. Where Dr. Alice Hamilton emphasized the role of mothers in transmitting infection, Taussig blamed girls. The results were the same. When doctors and nurses scrutinized the intimate functions of girls, they overlooked the intimate activities of the girls' fathers.

Many early twentieth-century health care and social workers shared Taussig's concerns about toilet seats in general and school lavatories in particular. In Chicago, Dr. Clara Seippel warned that "a drop of gonorrheal pus on the toilet seat in a public school can start [an] epidemic."[28] In Boston, social worker Bertha Lovell, fearing that her patients might infect their "susceptible little classmates by means of [spreading discharge on] the closed [O-shaped] toilet seat," forbade them to attend school.[29] The notion of a school epidemic arose from the nineteenth-century German medical literature on gonorrhea vulvovaginitis, but doctors had never documented a "school epidemic" having occurred in the United States. Yet the need to improve the condition of schools, particularly the lavatories, was an issue with which doctors and reformers were already engaged. Most health care professionals and reformers readily endorsed the notion that public toilets were the source of the epidemic.[30]

Since the 1860s, doctors had considered schools to be a particularly fertile environment for communicable disease. The physical plants of many public schools, especially in rapidly growing and overcrowded cities, were run-down and dirty.[31] In the 1890s, as doctors began to formalize public health projects in large municipalities, physicians, public health activists, sanitarians, and social reformers intensified their efforts to improve sanitary conditions in the schools.[32] When Seattle's health officer discovered in 1891 that the water closets of the South (Primary) School were in such bad condition that the smell of "secretions" circulated throughout the two-story building, he recommended that they be removed and replaced with outhouses before someone became sick.[33] Beginning around 1910, governmental and philanthropic groups that had been organized to unravel the problems of tenement housing also tried to pressure municipalities into cleaning up and repairing broken-down public school buildings and toilet facilities.[34] When the New York Association for Improving the Condition of the Poor studied the physical condition of schools in 1916, its published report included photographs of dilapidated buildings and broken-down bathrooms that, in muckraking style, were calculated to incite public furor and force political action.[35] And in the rural South, Progressive Era reformers linked medical inspection of children and improved school buildings with children's ability to learn.[36] Numerous observers described sanitation in rural southern schools as "too vile for description" and "germ breeding . . . places."[37] Yet as late as the second decade of the twentieth century, fewer than half of rural southern schools had a privy of any kind, and not every state required them.[38]

But cities and school boards already strained by rapid growth and corrupt governance lacked the funds to keep lavatories clean, let alone to remodel and staff them, and school boards at first paid little attention to the possibility of outbreaks of gonorrhea. In December 1915 the New York

City Board of Education complained only that odors from backed-up school toilets were "one of the greatest annoyances and a menace to health."[39] Within a year, however, the political winds had changed. The board recommended that infected girls be suspended from school because they "constitute a real menace" to other children. Yet the board also noted that gonorrhea in children was "not a common disease," an assertion that contradicted the reason given to justify excluding infected girls from school. If infection spread so easily, schoolwide epidemics of gonorrhea vulvovaginitis, like those of measles, should have been a major problem.[40] In any case, the board's recommendation was more a public relations gesture than an acknowledgment of a threat to child health, as there was no bureaucracy in place that could identify infected girls and enforce their exclusion from school, and the board did not create one.

In 1921 the push to ban infected girls intensified. In an action that seemed to herald a new level of municipal attention to the issue, New York City passed an ordinance prohibiting children infected with syphilis and gonorrhea from attending school.[41] Once again, however, the ordinance lacked muscle: five years later, not only had the city failed to implement any mechanism to enforce the ban, but it had not yet even begun to collect the girls' names. Another law required physicians to report infections to the city health department's Bureau of Social Hygiene, which was then to notify the girl's school so that she could be suspended. But few doctors complied with the law, and even if they had, the bureau did not develop any record-keeping system.[42]

Some doctors, including those at the Vanderbilt Clinic, contacted their patients' schools directly, apparently without regard to confidentiality or parental consent, or perhaps even as a rebuke to those parents they judged unable or unwilling to responsibly control their daughters themselves.[43] At the Bellevue Hospital clinic, doctors recommended that parents keep their daughters home for twelve weeks but made no reports to either the health department or school board.[44] In the late 1920s the Bellevue-Yorkville Vaginitis Clinic reported only those girls they considered "a menace" to other students.[45] Whatever the degree of actual monitoring, the emphasis on regulating transmission between girls using school lavatories shifted the locus of infection from the home to public, all-female spaces, excluding family members in general, and fathers in particular, from the chain of contacts leading to infection.

New Investigations

Social hygienists, public health officials, philanthropists, and reformers in the interwar years formed "blue ribbon" committees to investigate the extent of venereal disease in general and gonorrhea vulvovaginitis in particular. But they displayed no more interest in investigating the part played by sexual contacts in its etiology than had their predecessors. In contrast to the increasingly rationalized workplace of the era, many treatment facilities kept only nominal records about infected girls, if any. When the Charity Organization Society of New York City (COS) tried to study the problem, it discovered that even a rudimentary count of the numbers of girls infected or treated was impossible.[46] Troubled by the number of venereal disease patients among its workers' caseloads, and worried that venereal disease was a sign of "sociological disorder," the COS in 1922 formed the Committee on Venereal Disease to investigate the extent to which the families with which it worked were infected.[47] The blue ribbon committee met in various forms over six years, and in addition to COS personnel, its members included the director of the New York City Department of Health's Venereal Clinic, Dr. Louis Chargin; Louis I. Harris, chief of venereal diseases of the New York City Health Department and then health commissioner; Harry Hopkins, director of the influential New York Tuberculosis and Health Association (and later administrator of the Works Progress Administration [WPA]); and Dr. Walter Brunet, secretary of the Tuberculosis and Health Association's Social Hygiene Committee. Two board members from ASHA also served on the committee, including adolescent protective worker Martha P. Falconer.[48]

In November 1925 the committee formally joined forces with ASHA to study the incidence and treatment of venereal disease in New York City generally and to develop lectures and educational materials to distribute to the public.[49] Some committee members had been impressed by an article in the January 1926 issue of *Venereal Disease Information*, the bulletin of the U.S. Public Health Service, on environmental factors that might contribute to infection. Among other measures, the author had recommended that U-shaped toilet seats be installed in public lavatories to reduce the risk of infection to girls. As the *American Journal of Public Health* had recently noted about soiled, O-shaped seats in the Philadelphia schools, girls had "to climb onto the seat, thus facilitating the transmission of infection from the seats to the genitals."[50] Some committee members wanted the COS to pressure public schools and tenement owners to replace O-shaped seats with new U-shaped seats.[51] But Dr. Louis Chargin strongly objected. In 1915 he had written, "Children or other innocent persons may catch gonorrhea from persons who have carelessly dirtied towels or other toilet seats."[52] But now Dr. Chargin argued that the trouble and expense could not be justified because gonorrhea was "seldom contracted" from toilet seats and that, in any case, gonorrhea vulvovaginitis was a rare disease.

Considering the reluctance of New Yorkers to collect data on venereal disease infection, Chargin could not have been so certain.[53] When the Venereal Disease Committee attempted to count how many girls were being treated in New York City, it discovered that the incidence of infection went virtually unrecorded: clinics lacked the interest and resources to maintain thorough records; doctors almost unanimously refused to report venereal disease infections, which they considered a breach of patient-client confidentiality; and the city's health department did not systematically record the random reports it did receive. The few reports and statistical compilations that did exist were incomplete. They omitted information such as the age of the patient or even whether the patient was an adult or child.[54] Frustrated by the lack of reliable data, in 1926 the committee decided to undertake its own research.[55]

Committee members began by combing through COS case files. They were surprised to discover that caseworkers had documented gonorrhea infections among sixty girls under the age of 14, from thirty-three different families.[56] Many of the girls had been charged with sex delinquency, and all of them were from the impoverished population that the COS served. The committee used these data to argue that overcrowding stimulated adolescent sexual activity and was the primary cause of gonorrhea infection among girls. The COS had no empirical data with which it could compare the incidence of infection between social classes. Yet in its final report, which ASHA published in 1927, the committee concluded, "While it is true that gonorrhea vaginitis in children is not limited to tenement conditions, it must be admitted that in their crowded and unspeakably offensive living quarters, there is a wide distribution of the disease. The sleeping accommodations . . . offer one dirty bed for not less than three occupants . . . [and] present a particular medium of infection."[57]

Their statement suggests that committee members believed infection to be more common in tenements than in respectable homes, but it leaves open the question of whether girls became infected from dirty sheets or sexual contacts. From the patient histories, the committee identified a list of possible sources of infection, which they called "recurring social factors":

11—sex delinquency (six involved incest)

2—rape

8—infected in institutions other than hospitals

3—infected while in hospitals

2—toilet seats

14—parental exposure (girl slept with infected mother)

20—unknown[58]

The COS did not discuss the six cases of incest it documented, even though these cases represented 10 percent of the total. Nor did it investigate whether incest or sexual assault was the cause of any of the cases categorized as "source unknown." Still, the sample was enough to convince the COS that vulvovaginitis was a significant health problem.

Determined to learn more, the COS formed another committee in June 1926, this time including socially connected philanthropists and public health activists. The COS charged the new Committee on Vaginitis with studying the incidence of vulvovaginitis throughout the city, the sources of infection, and the type of medical services provided to infected girls.[59] The committee devised an ambitious plan and directed social worker Kathleen Wehrbein to examine patient records and speak directly with clinic personnel in every hospital, clinic, and dispensary that treated infected girls.[60] Yet this effort, too, would prove fruitless. Wehrbein discovered that many clinics did not maintain any patient records, and those that did made only cursory notes.[61]

But Wehrbein was tenacious, and she was able to collect enough information to convince her that the number of infected girls in the city exceeded the 544 cases reported to the health department in 1925 and 1926.[62] At Bellevue Hospital's clinic, Wehrbein found records for 105 girls under the age of 12 who had received treatment in 1925.[63] However, at the Vanderbilt Clinic she found the most dramatic information about the extent of infection: in just one month, September 1926, the clinic had seen 213 new patients.[64] Because most hospitals still refused to admit infected girls, the only places they could receive inpatient treatment were Metropolitan and Bellevue, the city's public hospitals. The Bellevue vaginitis ward, which had been set up as an isolation unit for girls whose infections had not been detected in preadmission screening, consisted of ten beds. Metropolitan had a seventy-two-bed "vaginitis ward" that was always filled to capacity, in some cases with girls who had been in the ward "for years," a fact the committee deleted from her draft when it published Wehrbein's final report.[65]

Wehrbein became even more dispirited about her ability to complete her task when she interviewed the doctors who treated infected girls. She found doctors' answers to her questions so confusing that she concluded that there was little consensus among them regarding the etiology and treatment of gonorrhea vulvovaginitis. Even though doctors admitted they kept few records about gonorrheal vaginitis among the "better classes," they insisted that most infected girls were from the "poor classes."[66] Many told Wehrbein they believed that a direct correlation existed between a girl's standard of living and her risk of infection: the higher the standard of living, the lower the incidence of gonorrhea vulvovaginitis.[67] Yet physicians seemed hesitant to identify incest as the source of infection, even among the poor. Wehrbein knew that the COS had documented incest in its prior investigations, but she did not press doctors on the issue. Nor did she press them on the practice of dismissing cases as "source unknown." Might not some of those patients and their families been unwilling rather than unable to identify a probable source of infection? If she considered the possibility, Wehrbein never put it on paper.

Instead, she focused on tracking down reports of contaminated toilets. In most cases Wehrbein was unable to confirm even the existence of the accused toilet, let alone verify that a girl had become infected after using it, though she traveled dutifully around the city to check toilets that parents had blamed for their daughters' infections. At one address she found only an empty lot, the building having been torn down years earlier. But even the fact that parents might have intentionally sent her on a fruitless search seemed not to have raised any doubts in Wehrbein's mind about their credibility. Even after she had been unable to find any evidence to support the notion, she still reiterated that "all medical literature includes the toilet seat as a possible medium of infection." And she added, again without having found any evidence, that "direct [sexual] contact is responsible for a minimum incidence of infection."[68]

Nothing in the committee's minutes suggests that it was any more interested than Wehrbein in pursing the possibility of incest. In its final report the Committee on Vaginitis argued that a direct causal relationship existed between infection and economic status: "In all 435 case histories . . . the highest incidence and distribution of vulvo-vaginitis is in homes where filth and poor living conditions coexist."[69] The report criticized poor housekeeping in "tenement households, where filth, neglect, and general living conditions are bad," as well as the morality of parents who were also infected, presumably because of their promiscuous sexual activities. Even though the COS had never investigated the homes of any infected girls from middle- and upper-class families, the committee concluded that "a specific relationship between living standards and vulvo-vaginitis, can no longer be ignored."[70]

The *Archives of Pediatrics* published the committee's final report in 1927 and reprinted it as a pamphlet. Yet the efforts of two blue ribbon committees had failed to produce a single new insight into the epidemic. Worse, reiterating long-standing assumptions about a correlation between dirt and disease proved an ineffective response to the epidemic, as the number of infected girls continued to increase.[71] However, the Committee on Vaginitis was so worried that the disease was becoming an urgent public health problem that it again proposed more research.[72] To find the financial and institutional support for a new study, the committee contacted every major funding source in New York active in public health: the Rockefeller Institute, the Bureau of Social Hygiene, the Committee on Maternal Health (a Rockefeller enterprise), the Commonwealth Fund, the New York Academy of Medicine, the American Social Hygiene Association, and the Welfare Council of New York City. All of them turned the COS down.[73]

Bellevue-Yorkville Health Demonstration

Any further investigation into girls' infections might have stopped there had it not been for the success of the Vaginitis and Venereal Disease committees in convincing health commissioner Dr. Shirley Wynne to add a "vaginitis clinic" to the newly opened Bellevue-Yorkville Health Demonstration. A joint venture between the health department and the Milbank Foundation, the health demonstration was opened to assess existing public health resources and to educate the public about health and disease prevention.[74] A board of New York's most prominent public health workers organized and managed the demonstration, including public health nurse Lillian Wald, who operated the nearby Henry Street Settlement.[75] The clinic's goal was ambitious: to gather reliable data about the incidence and pathology of gonorrhea infections in girls under age 12, which public health officials could then use to develop procedures to treat and control the disease.[76] Dr. Walter M. Brunet, secretary of the Social Hygiene Committee of the New York Tuberculosis and Health Association and a member of the COS Committee on Venereal Disease, argued that the clinic could provide important research because there was "no worth while data . . . available" about a topic over which medical opinions differed sharply.[77]

On July 14, 1927, the Bellevue-Yorkville Health Demonstration Vaginitis Clinic began its three-year-long study, the first clinical study of gonorrhea infections in girls.[78] As many as two hundred thousand people, mostly native-born white families, lived in the Bellevue-Yorkville neighborhood, which was bounded by Fourteenth and Sixty-third streets on Manhattan's East Side.[79] The public health officials and social workers who oversaw the clinic considered vulvovaginitis a two-pronged problem and set up advisory committees on medicine and social science.[80] The clinic's focus was on working-class families, and it assigned a social worker to conduct an in-depth interview with every family member.[81] If clinic personnel lacked confidence in the ability of a girl's mother to comply with their protocols, they sent the girl to Willard Parker, a contagious-disease hospital where the health department opened a fifteen-bed inpatient ward for girls in the study. Although the offer seemed generous, mothers would have understood it as a profound threat. In 1925 a visitor to Parker had compared the isolation rooms to "cells for the incarceration of criminals," and by 1929 public distrust of the hospital was so high that it became an issue in the mayoral campaign. Socialist candidate Norman Thomas captured the bitterness of New Yorkers when he complained that the staff seemed intent on running the hospital in a way calculated "to provoke the ill-will of almost all the victims who suffer within its walls."[82]

Yet even as the clinic emphasized mothers' responsibility to protect their families, social workers discouraged them from talking with one another about their daughters' infections.[83] One of the clinic's primary functions was to reeducate mothers who still believed that girls became infected from sexual contacts. But its educational materials were both less accurate and more ambiguous than the common sense or folk beliefs they sought to disparage. The clinic distributed pamphlets produced jointly by the city health department, the Tuberculosis and Health Association, and ASHA, but they included only general information about how girls became infected. English and Italian versions of pamphlets with titles such as *Information for Women* advised mothers that "the usual way for one person to acquire the disease from another is through sexual relations. It is possible, however, for little girls to become infected from a drop of the discharge left on a toilet seat, or on a towel, clothing, or bed clothes."[84]

The clinic's unwillingness to investigate the possibility of incest affected its ability to achieve its ambitious goals. Over the three years in which it was in operation, the Bellevue-Yorkville Vaginitis Clinic examined 322 girls from 248 families. More than half were younger than 6 years of age, and nearly all (97 percent) tested positive for gonorrhea. Clinic personnel boasted that they had gathered complete patient histories for 241 girls. But when the three-year study ended in 1930, the clinic's final report had nothing to say that Dr. Taussig hadn't considered, and rejected, nearly twenty years earlier. Perhaps for this reason, when Brunet and his team prepared their report about the most comprehensive investigation into vulvovaginitis to date, no medical journal would publish it. Board members blamed the report's length, no doubt a strong factor during the depression. Yet even after the board sharply edited the report down to ninety-eight pages, it still could not find a journal willing to print it. The project's board members were upset by the rejections, but their records do not provide any explanation. Board members finally used their influence to convince *Hospital Social Service Magazine*, neither a prestigious nor an especially appropriate choice, to publish the report as a supplement in March 1933.[85] And, as happened repeatedly with the medical literature on the topic, even though the report offered little new information and no definite conclusions about the etiology of the disease, doctors who wrote about gonorrhea vulvovaginitis cited it as the most authoritative word on the topic.

How could such a prestigious and well-funded project have failed so completely? The structure of the study all but guaranteed that clinic personnel would not discover incest. The final report claimed, "Every attempt is made to have full information on each case in order that any study which is desired may be possible."[86] To achieve that goal, the clinic had specifically hired a nurse who had also been trained as a social worker to gather information about the "modes of infection and examination of contacts" of the infected girls, and clinic staff developed special questionnaires to ensure that this process would be thorough. But none of the forms contained any questions pertaining to the possibility of sexual assault. The Medical Sheet, for instance, included broad questions about "source of infection," "when infected," and "condition on treatment" but none that queried patients specifically about genital injuries or a history of sexual assault. Questionnaires focused instead on identifying the names, occupations, incomes, place of birth, and religion of everyone who lived in the girl's household. A questionnaire prepared by the clinic's Health Department seemed to have been designed to confirm links between poor sanitation and infection. It asked social workers to rate the patient's "housing conditions" according to amount of rent paid; "sleeping arrangements for patient"; and the number of rooms, beds, and "toilet facilities," which were categorized as either "private," "hall," or "other."[87]

Perhaps the flawed questionnaires were simply an oversight. But every aspect of the Bellevue-Yorkville Health Demonstration treated sexual assault and incest as having already been ruled out. For instance, even though researchers claimed they wanted to test every family member of an infected girl, fewer than half of the families agreed to be tested, and then only mothers and a few siblings cooperated. Of those who were tested, nearly all—92 percent—tested positive.[88] Fewer than one-third (76) of the mothers agreed to be tested, but more than half of them (46) tested positive for gonorrhea.[89] Researchers were perplexed that so many mothers refused to be tested and by the hostility with which they responded to the request: "Either from fear of the detection of a possible infection, or a firm conviction that there was no possible basis for suspecting an infection, the mothers . . . resisted, and sometimes resented, the suggestion . . . Many painstaking explanations were necessary . . . Several . . . resentfully left the clinic . . . and could not be induced to return."[90] Yet even though the high rates of infection among the girls' families and their mothers suggested that a girl's father might also be infected, researchers considered his status so unimportant that "no provisions were made to examine the fathers in the clinic. They were urged, however, to be examined elsewhere, but satisfactory reports on their examinations could not always be obtained."[91] Of 248 families, only 12 fathers agreed to be tested; 9 were infected, a rate of 75 percent.[92] The authors of the report made no comment on this fact.

The health demonstration's final report did not explore the implications of even the few data it did collect about fathers. It simply asserted that "no definite data concerning incest were secured," implying that researchers had investigated the possibility.[93] Dr. Brunet acknowledged that the possibility of sexual assault had been raised in at least eight cases but that it had not been investigated because none of the girls exhibited genital injuries.[94] Even so, Brunet did not hesitate to declare that of the clinic's three hundred cases "not a single case of rape occurred in this series," proving, he argued, that "cervico-vaginitis was rarely due to sexual contact or violation." How else, then, had the girls become infected? Brunet had no answer. With the clinic having investigated every possibility except sexual assault or incest, its final report only repeated what doctors had been saying for years: "In these patients the exact manner of transference of the infectious material from the ill to the well is unknown."[95]

Clinic personnel had not even been able to substantiate the supposed link between infection and unsanitary contacts. Brunet reported that parents had suggested a long list of possible sources of infections, including "soiled diapers, playing in sand or dirt, hospital admissions, accidental contamination from infected towels, thermometers, wash cloths, bath tubs, toilet seats, especially in schools, etc."[96] Yet he thought it more likely that girls became infected from "close, intimate personal contact usually within the home," by which he meant contacts with their mothers. In the end he concluded, with less confidence than Alice Hamilton had displayed when she first made the speculation, that mothers were the "presumptive source."[97]

The Bellevue-Yorkville Health Demonstration Vaginitis Clinic had intended to bring clarity to an elusive social and medical issue. But instead of moving medical research forward, the project lent new credibility to old speculations about the susceptibility of girls to casual contacts. And any success the project may have achieved in reeducating mothers may have also dissuaded them from acting on their worst suspicions. One of the clinic's pamphlets for mothers, *Important Information: Special Instructions for Vaginitis Cases,* was printed on pink paper and emphasized the pivotal role of mothers in preventing infection: "Your child is suffering from a contagious (catching) disease. You must be careful to prevent its spread to others . . . Care in this cause means CLEANLINESS. Your child's hands should be constantly kept clean."[98] For lack of any better information, the city, with funding from the WPA, continued to distribute the pamphlet, without change, as late as 1937.

Toilets Redux—School Exclusions

As doctors continued to debate the etiology of the disease, some states implemented new reporting and data collection procedures to determine the extent of infection. By the early 1930s, numbers collected by the Massachusetts Department of Public Health consistently showed that girls under the age of 14, three-quarters of whom were younger than 10, accounted for more than 10 percent of all female gonorrhea infections in the state.[99] Frightened by these numbers and by reports about the supposed danger to girls from public toilets, parents demanded action. School boards and public health departments around the country debated or put into practice programs to exclude infected girls from public schools. In 1923 Dr. John H. Stokes, chief of dermatology and syphilology at the Mayo Clinic, decried gonorrhea vulvovaginitis infection because it "spells the end of the child's education."[100] Stokes's prediction nearly came true. Only the inefficiency of poorly staffed and funded municipal and school bureaucracies, which could not keep track of the girls, averted the wholesale disruption of the education of two generations of girls.[101]

The vague and confusing medical literature available to health care workers, public health officials, school boards, and parents hampered their ability to reach a consensus on whether excluding girls would stop the spread of infection. Dr. Nels A. Nelson, assistant director of the Division of Communicable Diseases of the Massachusetts Department of Public Health, was one of the most vigorous opponents of "school exclusion," policies that barred symptomatic girls from school. Dr. Nelson was so pressed on the issue that he included it in the first of a series of radio talks about venereal disease—aired on Christmas Eve, 1930. Nelson introduced the subject by telling listeners that hundreds of girls infected with vulvovaginitis were "doomed to months and sometimes to years of treatment, suffering, and exclusion from school."[102] But he also reassured parents, saying that the physical complications of gonorrhea vulvovaginitis were not, in and of themselves, serious enough to merit attention as a public health problem. In his view, gonorrhea vulvovaginitis had become a public health problem only because of the practice of excluding girls from school.[103]

But Nelson's speech did not calm frightened parents, and medical uncertainty about the etiology of the disease provided support for experts with conflicting views. When the Neisserian Medical Society of Massachusetts, the state's professional association of physicians and researchers working in the field of gonorrhea, held a symposium on gonorrhea vulvovaginitis in November 1931, the most contentious issue was whether to ban infected girls from school.[104] As he had on many occasions, Nelson argued that there was no proof that a girl had ever initiated an outbreak of gonorrhea at a school and that doctors had never documented a school-based epidemic.[105] Nor was it likely that they would, unless someone inspected toilet seats for gonorrhea discharge after each and every girl left the lavatory or examined every girl for symptoms of the disease, an event unlikely to occur, as school boards did not permit school medical inspectors to examine girls below the waist.[106]

Proponents of exclusion, however, persisted. Dr. S. H. Rubin, a school physician and assistant professor of pediatrics at Tufts Medical School, argued that the fact that Boston had not suffered any "school epidemics" since implementing a suspension policy proved that such bans worked.[107] And he added that there was no evidence proving that gonorrhea vulvovaginitis was not contagious, especially in school lavatories. But Rubin's arguments had more to do with politics than medicine. He warned "those who protest against the exclusion of pupils" to "take into consideration the prevailing public sentiment in this matter, which would undoubtedly turn upon us the moment we waived our present policy of exclusion."[108] I could not find any evidence with which to determine whether Rubin overstated the level of parental anxiety. But any relief that parents may have found in an exclusion policy was illusory.

At around the same time Rubin was pressing his case, ASHA and other public health professionals evaluated the school exclusion policy in Washington, D.C., and concluded it had failed. District physicians reported that 10 percent of the more than 3,000 cases of syphilis and gonorrhea diagnosed in 1929 were children under 16, mostly girls being treated at hospitals.[109] Using these numbers as an estimate, at least 300 girls (10 percent per year) should have been suspended. Two years later, when ASHA sent Dr. Walter Clarke to see whether the district's policy worked, he found the system of record keeping so poor that he could not ascertain how many children should be excluded, let alone whether they were, in fact, absent from school.[110] Even worse, those records that did exist showed that some girls had been absent for two to three years, leading Clarke to wonder whether some suspended girls simply never returned. When the Children's Bureau and the Washington, D.C., Social Hygiene Society made a second study of the district's policy in 1932, which they published in the *American Journal of Public Health*, they concluded that more than 300 children had been ordered to stay home from school but that most ignored the order.[111]

Rubin's faith in school exclusion to solve the problem was misplaced, and his boast that no school epidemics had occurred in Boston lent support to the opposite argument: that infected girls regularly populate American schools with no adverse consequences to other students.[112] In the twenty years since Taussig had speculated that girls transmitted the infection to one another through the school toilet, not a single "school epidemic" had been reported in the medical literature. But the absence of data did not dissuade health care professionals and reformers from their focus on the toilet. In 1933 Texas gynecologist Paul Stalnaker noted that most girls infected with gonorrhea were between the ages of 5 and 9, the same ages at which girls begin going to school, where they "attend crowded assemblies and movies, and use community toilets." The problem, as Stalnaker saw it, was that "little girls are too small to seat themselves easily and they wabble, so to speak, on the toilets, getting their vulvas in contact and smeared as they place themselves on the seat."[113] The following year, Philadelphia banned infected girls from school, and the Philadelphia General Hospital vaginitis ward started to hold classes so that their patients could keep up with their education, a program the San Francisco General Hospital had begun two years earlier.[114]

By the time he gave his eighth radio talk on venereal disease in 1935, Nelson had become the top public health official in Massachusetts in the area of venereal disease. In a talk entitled "The Darkness of Ignorance and the Fog of Prudery," he reported that more than 30,000 women in Massachusetts became infected with gonorrhea each year. Of these, about 3,000, or 100 of every 1,000 females, were girls under 14 years old, including babies.[115] These numbers were shocking, and Nelson tried to reassure the citizens of Massachusetts by telling them that while some girls were infected by "attendants" (he did not say how), most became infected by sleeping with a mother or sister who had the disease.[116] His message did not assuage parents' fears.

In a March 1935 article published in both *Commonhealth* and the *Bulletin of the Massachusetts Society for Social Hygiene*, Nelson decried the "misunderstanding and resultant confusion, almost to the point of hysteria, over gonorrhea and syphilis in the public schools." He declared, "It is time that those who have to do with the schools and school children familiarized themselves with the facts," and he reiterated his view that "gonorrhea is acquired by adult females and by males, only through . . . sexual practice . . . Girls under the age of puberty are susceptible to infection with gonorrhea through intimate contacts not necessarily of a sexual nature. These contacts . . . do not occur in school. There is no evidence that school toilets have been vehicles of transmission of the infection."[117] Yet as strongly and clearly as he worded his statement, Nelson had not rejected the possibility of nonsexual transmission, and he did not say exactly how girls became infected. Apparently giving into political pressure at last, Nelson recommended that infected girls be excluded from school until their vaginal discharge cleared up, usually a period of six to eight weeks.[118]

With Nelson having fallen into line, school boards across the country also moved to substitute political caution for medical science. In 1935 the U.S. Public Health Service recommended that infected girls stay home from school, and the Los Angeles County School Board followed suit with a policy suspending infected girls.[119] But Nelson continued to debate the issue. In December 1936 nearly a thousand health care workers and social hygienists attended a federal Conference on Venereal Disease Control Work.[120] Dr. Percy Pelouze, an ASHA board member and an expert on gonorrhea, joined Nelson in a report titled "The Control of Gonorrhea," in which they characterized gonorrhea vulvovaginitis as more of an "annoyance" than a serious health problem. Nelson and Pelouze again argued that the disease had become a public health problem because of its social, not medical, ramifications, and they cited "its disturbance of the family and of the child psychologically and socially, its interference with education through exclusion from school, and its cost in months of treatment."[121] When Nelson and Pelouze prioritized a list of twenty "problems awaiting solution" for gonorrhea researchers and clinicians, they ranked "the collection of evidence as to whether or not there is any foundation in the traditional notion that gonorrhea may be spread by toilet seats" dead last.[122]

Still, Nelson had little success persuading either his peers or the public that toilet seats were not dangerous because he could not—or would not—suggest an alternative source of infection. In a 1936 public education pamphlet he wrote for the Massachusetts Society for Social Hygiene, of which Nelson was a board member, Nelson advised parents that it was possible for children to become infected from casual contact with objects, even though "nobody knows" exactly how it happens.[123] Social hygienists and health care professionals had drawn a bleak and confusing picture in which girls seemed exposed to infection at every turn. ASHA's educational materials often did not distinguish between means of transmission in adults and children. A 1935 pamphlet, "Gonorrhea: How Does a Person Catch Gonorrhea?" simply stated that "the most common way [of acquiring infection] is through sexual relations. Other possible ways are by using a towel or a toilet that has just been soiled by some of the discharge."[124] Some of the ambiguity might have been intentional and even helpful toward meeting AHSA's goal of increasing the numbers of Americans tested and treated for venereal disease. Social workers noted approvingly, for instance, that some fathers who initially refused to allow their daughters to be treated for gonorrhea relented when told that girls could become infected at school.[125] A father's reluctance to permit his prepubertal daughter to be treated for what he believes to be a sexually transmitted disease seems counterintuitive. A father who learns his daughter is pregnant, and is surprised by the news, would certainly want to know who had impregnated her. Similarly, if a father thought that a gonorrhea infection connoted that someone had sexual contact with his young daughter, it is difficult to imagine he would not have been upset—unless he had something to hide.

Nearly forty years after doctors declared gonorrhea vulvovaginitis epidemic among girls, no one could explain how the bacteria were transmitted or how mothers could protect their daughters from infection. Valuable information might have been gleaned from patient records that doctors had amassed after having spent years responding to an "epidemic"; the number of clinics treating girls had increased, and new state laws required physicians to report infections. But record keeping remained—almost insistently—scattershot. In 1936, for instance, when the Social Hygiene Committee of the New York Tuberculosis and Health Association, whose members included doctors Walter Clarke, Louis Chargin, and William Snow, investigated gonorrhea clinics that treated women and children in New York City, only eight of the thirty clinics the committee visited maintained reliable patient records. At twenty-three clinics committee members could not even estimate the numbers of patients being treated, let alone glean any facts about the source of infection.[126]

Vague statements about the etiology of the disease became the rule rather than the exception. When Dr. Clarke spoke at a meeting of the Section on Pediatrics of the New York Academy of Medicine in 1936, he said that one thousand new cases of vulvovaginitis were reported in New York City each year, a number he assumed to be far below the actual incidence of infection. And he added that he had no idea how the girls had become infected.[127] Yet a nursing textbook on communicable diseases published in 1937 instructed students that gonorrhea vulvovaginitis was very contagious and that it was "usually obtained from an infected public toilet seat" or from sleeping with other infected children or infected servants.[128] And that same year the head of the Central Harlem Health Center warned the African American readers of the *New York Amsterdam News* that parents should teach their daughters not to sit on O-shaped toilet seats without first placing paper on the seat because of the danger that an "infected toilet seat" posed to girls.[129]

In 1938, Reuel A. Benson and Arthur Steer, attending physicians at the busy vaginitis ward at New York City's Metropolitan Hospital, claimed to have no idea how the bacteria spread, but they still insisted that most infected girls were from the "poorer classes."[130] But if, as advocates of the dangers of toilets had reasoned, school lavatories provided the means by which infections in girls spread, then infection would not have been limited to poor girls. And if doctors had indeed felt certain that infection was contained within the "poorer classes," they might have been less interested in the dangers of school toilets and more interested in identifying incest. But of the 195 girls they had treated, Benson and Steer attributed nearly half of the infections to "source unknown."[131] Yet the meagerness of their data did not prevent them from claiming that "the infection is usually contracted from adults, but young sisters and playmates also frequently spread the disease . . . The part played by toilet seats and other infected objects can only be guessed."[132] Benson and Steer acknowledged that patient histories provided by parents might be compromised and that it was impossible to verify the accuracy of patient histories when "things such as toilets seats, linens and bathing water were accused."[133] But, like their colleagues who also wrote about girls' infections, they still dismissed incest as "an infrequent source of infection."[134]

In 1938, shortly before he left his post to teach at the Johns Hopkins School of Hygiene and Public Health, Dr. Nels A. Nelson was still arguing about toilets. In a public health textbook he co-wrote with nurse Gladys Crain, Nelson became the first doctor since Flora Pollack to state that there was no evidence to support speculations about girls' susceptibility to infection from casual contacts: "How the gonococcus finds its way to the child's vulva, if not by direct sexual contact, has never been determined. Toilet seats, bed sheets, an infected mother's hands and various other objects with which the vulva might come into contact have been blamed but not proven to be vehicles of transmission." Nelson and Crain admitted that "it is possible that a freshly and grossly contaminated old-fashioned toilet seat, so high that the child must drag herself onto it, might be a vehicle for infection" but that "a great deal has been said, but little if any evidence has been presented, against the public school toilet seat as a factor in the spread of gonorrhea among girls. There is this evidence of its innocence, that although a whole school may use the toilets during a brief recess, no *epidemic* . . . in a public school has ever been recorded."[135]

Nelson had always refused to believe that sexual assault was the source of so many infections, but now he implied that parents and health care workers blamed toilet seats to avoid the truth: "It is less embarrassing to accuse a toilet seat than to seek for sexual sources or to request the examination of other members of the family. And it requires much less tact and no labor . . . When all the old-fashioned seats are finally done away with, they can no longer serve either as an alibi or a source of infection."[136] But however much Nelson had crusaded against the "fog of prudery," victimized girls paid the cost. In 1932, Dr. Nelson reported in the *New England Journal of Medicine* that for the years between 1919 and 1930 doctors in Massachusetts had reported 13,000 cases of gonorrhea infection in females, of which more than 10 percent (1,382) were girls under 14 years of age, more than three-quarters younger than 10.[137] Nelson added, however, that the true prevalence of infection was probably eight times greater.[138] In 1939, the Public Health Service estimated that over a million Americans became infected with gonorrhea each year.[139] If half of the infections were among females, 10 percent of whom were under age 14, some 50,000 girls may have been infected annually.

Yet in 1939 the only preventative measure that even the most experienced physicians could recommend was to instruct mothers in the proper use of the toilet. Benson and Steer praised one children's clinic for having placed a toilet in the waiting room so that social workers could instruct mothers on "the precautions to be observed" when their daughters used the toilet.[140] Well-funded research, clinics, and the increased availability of educational materials had not had the intended effect of clearing up "thoroughly" the etiology of infections in girls. Even as federal monies in the late 1930s poured into educating Americans about the link between sexual contact and venereal disease, on the eve of World War II doctors remained unable to recommend a single measure that could effectively prevent or reduce the risk of infection among girls. The best efforts of Progressive reform and modern medicine had succeeded only in hiding the occurrence of father-daughter incest from view.

CHAPTER SIX

Incest in the Twentieth Century

IN EARLY APRIL 1926, Melba McAfee's father asked her and her 6-year-old brother to take a ride with him to a rabbit farm near their San Diego home. Melba was 11 years old. That November, when she testified at her father Robert's trial for incest, she would recount that after they set out for the farm, her father said the car had a flat tire and then stopped on a deserted road. After he took out the tire pump and placed it next to the fender, Robert McAfee told Melba to walk down the road with him, where he sat down on a clump of grass and told Melba to come and sit by papa. Once seated, Robert kissed and "loved" his daughter a bit. Melba testified that her father's affection surprised her because whenever she had previously tried to kiss him he had told her to "get out of his way." But Melba's pleasure with Robert's attentions was short lived; by the time she returned to the car, Melba felt so sick that she lay down on the back seat.[1]

What had transpired at the side of the road? Melba testified that after sitting down, her father had ordered her to spread her legs. When she resisted, he told her that if she didn't go along, he would force her. Still, Melba started to run away, only to have her father drag her back. Robert forced his daughter down on the ground, took off her bloomers, and inserted a finger into her vagina. Melba screamed, kept saying no, and cried because it "hurt so bad." Her father, perhaps aroused by his daughter's fear and his power over her, took out "that thing between his legs," pulled a jar of Vaseline from his pocket, and forced his penis into Melba's vagina. The pain was so intense that Melba tried to get up, but her father forced her down and got on top of her, telling her it was nothing and that it would stop in a few minutes. Melba then described how her father's body made an up and down motion for a few minutes until he stopped and said, "That's what is called the maiden's head." Then he penetrated his daughter again. By this time blood was pouring down Melba's legs, which she tried to wipe off with her bloomers after her father let her up. As they returned to the car, Robert told Melba not to tell anyone what had happened, explaining that if he was found out he would have to go to jail. When Melba replied that she wouldn't tell, perhaps in an unconvincing voice, her father added that he would kill her and her mother, Nancy Morgan, his former wife, if she reported what had occurred. Back home, Robert told Melba to quickly burn her bloomers before her stepmother came home. Melba hid them under the house.[2]

Two months later, Robert McAfee assaulted his daughter again, saying, "I am not only going to do it this time, but I am going to do it always to you, do you savvy?" Melba again bled profusely, her limbs felt numb, and she "hurt a lot." When they returned home, Melba's stepmother, Mildred, saw the blood on the bottom of Melba's petticoat; after Robert went out, Melba told Mildred what had happened and that something "yellow" had come out of her vagina. "I could not go to the toilet hardly, it just hurt me so bad." Nothing happened until a week later, when Robert sent Melba to live with her mother. Melba still felt sick and kept crying but, heeding her father's threats, was afraid to say anything. But when Nancy pressed her, Melba confided in her mother, who took Melba to the police that same day, as soon as she had heard the story. When the police learned that Melba had a vaginal discharge, they asked a physician to examine her. Dr. Thomas Sidney Whitelock prescribed pills for her stomach and medicine for douching, which she was still using five months later.[3]

At trial, Dr. Whitelock testified that he had microscopically examined a smear and diagnosed Melba with gonorrhea. When the prosecuting attorney asked him how girls acquire gonorrhea, Dr. Whitelock, who had graduated from medical school in 1899, answered, "It must be contracted by someone that has it." The prosecutor continued to question Whitelock:

Q. In other words, from a sexual contact?

A. Well, not always. It may possibly be accidental; but it is most generally supposed that it is from some other individual.

Q. Sexual contact?

A. Yes.

The defense did not dispute that gonorrhea vulvovaginitis was sexually transmitted. Instead of debating whether Melba might have acquired the disease from a casual contact, McAfee turned the diagnosis into a weapon against her. McAfee admitted that he had been infected seven years earlier but claimed that he was not now infected, as he had no symptoms, and so could not have transmitted the disease to his daughter.[4] The defense suggested (but offered no evidence to prove) that Melba had become infected from intercourse with boys her own age, thereby questioning both her veracity and her innocence. And they accused Nancy of instigating the incest story because of lingering hostility toward her former husband. After five days of testimony the jury found McAfee guilty of incest and lewd and lascivious conduct. On appeal, the court upheld the convictions for incest.[5]

Melba's testimony about her father's assaults and Robert's strategy at trial are as sordid and depressing as they are typical of courtroom narratives in cases of father-daughter incest in the first half of the twentieth century. No longer focusing on the shocking dissonance of a man revealed to be a beast in human form, discourses about father-daughter incest turned instead on the character of the victim. Early twentieth-century parents, social workers, and defense attorneys embraced the misogyny of nineteenth-century medical jurists who warned of the proclivity of girls to lie, updating it only to deem girls fully capable of lying on their own, without the prodding of a mother or other vengeful woman. In trial transcript after trial transcript, fathers characterized their daughters as sexualized harpies who turned child's play into a costly travesty that victimized and ruined innocent men.[6]

Defense attorneys, attuned to how jurors measure courtroom testimony against what they accept as common sense, incorporated such narratives into their trial strategies. In McAfee's case, it would have been difficult for him to have argued that no sexual contact had occurred. Numerous witnesses testified that Melba was both infected with gonorrhea and displayed physical marks consistent with sexual intercourse. Under these circumstances, the defendant had little to gain by trying to convince the jury that Melba had been infected by a toilet seat. The defense tried instead to exploit the physical evidence to its advantage. The state could not prove that Melba had become infected on the dates of the alleged assaults, which left open the possibility for the defense to suggest that Melba had been infected by someone else. They tried to convince the jury that a sexually promiscuous 11-year-old was not a morally innocent victim whose word against her father could be trusted.

The defense also attacked the motives of the only other witness to testify on Melba's behalf, her mother, Nancy. (When Mildred testified, she denied Melba's claims.) To augment the trial narrative that sexually aggressive girls lie, the defense attacked Nancy by claiming that she had been the first woman to victimize Robert McAfee: she had been unfaithful during their marriage, becoming infected with gonorrhea, which she passed on to her husband, who then divorced her on the grounds of adultery. When the defense tried to destroy Nancy's credibility by arguing that she was motivated to initiate a false claim against her former husband, it did not need to also explain why, perhaps assuming that jurors would draw the link themselves—the mother of a precocious girl eager to turn on her father must herself be an untrustworthy harpy.

In the early twentieth century defense attorneys sometimes raised the possibility of infection from casual contacts, but they did not stake their case on jurors finding such an argument persuasive. Defense attorneys who represented men accused of sexually assaulting a girl focused instead on questions that had absorbed defense attorneys for a century: whether the defendant was infected at the same time as the victim, and whether the victim's infection preceded the date of the alleged assault.[7] In 1943 Ralph H. Manchester pushed the forensic implications of gonorrhea infection even further when he turned the fact that his 10-year-old daughter was *not* infected with gonorrhea into an argument supporting his claim of innocence. After being convicted of incest and sentenced to Folsom prison, Manchester filed a writ of habeas corpus claiming that he could not have been the man who raped his daughter because she never tested positive for gonorrhea, while he suffered from "chronic" infection.[8]

What is strikingly absent in defendants' arguments is any implication that their daughters, though deceitful and preternaturally sexually aggressive, had initiated sexual contact with their father. Among the California trial and appellate records I reviewed, only in the late 1940s did psychiatrists begin to testify at incest trials, and then it was to support the medico-legal designation of the father as a "sexual psychopath," a new label that permitted the court to sentence him to a mental hospital rather than a prison.[9]

A defense lawyer would have been keen to avoid the pitfalls of trying to blame sexual contact between father and daughter on the daughter. If a defendant in a murder case claims self-defense, he or she implicitly admits to having killed the victim. Doing so is risky because the defendant must concede that he or she killed the victim, usually the most important fact the state would otherwise be forced to prove. Similarly, if McAfee had claimed that Melba seduced him, he would have had to admit to having had sexual contact with her. Such an admission would have bolstered Melba's credibility on at least that fact and narrowed substantially the elements of the prosecution's case about which the defense could assert reasonable doubt. Even more important, by the twentieth century consent was irrelevant in most prosecutions between father and daughter. Except in those states that interpreted the crime of incest as applying only to consensual sexual relations between adult kin, a prosecutor could choose to file under an incest or rape law. Under most incest and all statutory rape laws, the daughter's consent (implicit in the notion of her being the aggressor) is immaterial.[10] No testimony would be allowed to support a defense that a girl seduced her father, however willing a participant the defense might wish to portray her.

Still, men like McAfee found themselves convicted of a felony and, though prison sentences were generally shorter and less severe than those handed out in the nineteenth century, facing hard time.[11] Just as social hygienists in the 1920s and 1930s were in the process of reeducating Americans about the sources of girls' infections, so too were views about the inherent innocence of girls in the process of change, partly in response to the increase in sexual assault claims filed on behalf of girls against adult males after states began to enact and enforce age-of-consent laws in the 1880s. But the revelation of a new set of underage female victims did not result in a reevaluation of male prerogative and sexual behavior, just as the revelation of an epidemic of gonorrhea among prepubescent girls had not resulted in a reevaluation of father-daughter incest. The fact that the girls had been sexually assaulted was turned instead into a sign of their sexual precocity, and fathers tried to cast themselves as the victims not only of vicious women but of their rapacious female offspring as well.[12] This characterization of girls became so prevalent that in 1944 a California appellate court reversed Harry Rankins's conviction for incest because the trial court had refused to instruct the jury to "view the testimony of [his daughter] with caution . . . since in this class of cases . . . ample opportunity exists for the free play of malice and private revenge."[13]

Such a negative view of girls and women could not have achieved widespread resonance among early twentieth-century Americans without professional and popular insistence that father-daughter incest occurred only among the poor, the working class, and people of color. Progressive Era views about protecting a precocious girl from becoming a wayward woman rose out of the belief that socially marginalized parents were intrinsically unable to provide the moral and material environment necessary to nurture innocence. Tenement dwellers endangered their children's welfare by raising them in a setting that exposed rather than protected their children from urban vice, whether in the form of male adult boarders with whom a girl shared a bed, prostitutes with whom she shared the streets, or fathers who returned from the saloon only to take out their frustrations on their families.

Many social hygienists also shared the increasingly popular eugenics notion that bad behavior was the product of bad genes.[14] Reformer and philanthropist Ethel Sturges Dummer had organized and funded Chicago's Juvenile Protective Association (JPA) in 1909, and her response to hearing about father-daughter incest was typical. In a speech on feminism in 1916, Dummer noted that she became interested in eugenics after having become depressed while listening to a report at the JPA about several cases of incest, "which custom I had imagined to have passed with the Old Testament period."[15] But whether such bad behavior was due to nature or nurture, the idea that a deleterious connection existed between an impoverished home life and immorality seemed justified by the statistics—the vast majority of girls adjudicated "delinquent" had been accused of some sort of sexual impropriety, and virtually all of them were from the poor and working classes.[16]

As the notion that girls lie about sexual assaults began spreading throughout American society, judges and prosecutors found themselves having to address juries on the issue of whether nightmarish accounts about paternal behavior like Melba McAfee's were so farfetched as to seem improbable.[17] Or, as Harry Hobday argued in 1933 after he had been convicted of having raped his 14-year-old daughter, "the outrage of a child by a parent is a thing in itself so inherently improbable and revolting to the normal human that every intendment should be against the verdict."[18] Hobday, a 35-year-old baker, had raped his 14-year-old daughter while his wife was in the hospital. When he told his daughter that she would have to fill her mother's role, she thought her father was talking about housework until he explained that her duties included having sex and talking about sex. In his closing argument, the prosecuting attorney told the jury that Hobday knew how difficult it was to believe that incest really occurs but that he had showed both that it is possible and that it does. But even the trial judge shared the jurors' skepticism. He admitted that he had found it so difficult to believe that such a level of depravity was possible that he conducted his own "independent investigation" before concluding that Hobday was not only guilty but unworthy of any mercy in sentencing.[19]

The details of McAfee's and Hobday's victimization of their daughters were incompatible with anyone's vision of the idealized American home, but that is not the type of home the McAfees or the Hobdays—or countless other Americans—actually inhabited. As the appellate court noted in *Hobday*, "It has become something of a commonplace to urge that charges of this character [incest] are difficult to disprove. Yet current or past history records few miscarriages of justice. The truth is that these charges are often too difficult to prove and justice is aborted."[20] However, soon the line of improbability would shift from the likelihood of behavior as repugnant as incest to the likelihood that a middle-class white man was capable of committing such an act. Even though the testimony of multiple witnesses had led to Harold W. Roberts's conviction in 1941 for having raped a 9-year-old girl, Roberts argued on appeal that the judgment should be reversed "because of the . . . inherent improbability that he committed the offenses charged, in view of his long and prominent association with outstanding musical and civic organizations in this community, as well as the normality of his family life."[21]

Beginning in the 1890s, white Americans embraced racialized notions of masculinity. Their discussions of father-daughter incest enlarged these ideas by assuming that a set of moral and behavioral qualities, including the ability to commit incest, were inherent in a man's racial and class status. Acknowledging incest only among those immigrants, African Americans, and poor people who materially and symbolically stood outside the American dream reaffirmed the fiction—or fantasy—that incest occurred only where native-born white Americans expected it, not in the homes of the white middle- and upper-class Americans who symbolized the idealized domesticity that America promised. Resistance to acknowledging incest in any home stiffened as the value of forensic medical evidence was diluted, prosecutions focused primarily on socially marginalized families, and misogynist views about women spread to include girls.

Girls Lie

In the early twentieth century, the premise of an innate childhood sexuality advanced by Sigmund Freud replaced Victorian beliefs about the inherent modesty of little girls. Professionals from the fields of psychiatry, law, and medicine now advocated the notion that women—and especially girls—lie about sexual assaults. Some psychiatrists and law enforcement officials even began to characterize girls as accomplices in, rather than victims of, the crime of incest, regardless of the evidence.[22] Progressive reformer Dr. William Healy was a Chicago neurologist and member of the Chicago Vice Commission. In 1906 he served with Ethel Sturges Dummer on a committee investigating recidivism among delinquents. When the committee developed a plan three years later for the Juvenile Psychopathic Institute, Dummer funded it and named Healy its director. The institute's mandate was to provide juvenile court judges with medical and psychological examinations of the children whose cases they were deciding.[23]

In 1915 Healy began to publish the results of his examinations, and he included the story of 9-year-old Bessie both in his groundbreaking work, *The Individual Delinquent*, which he coauthored with his wife, Mary Tenney Healy, and in a monograph titled *Pathological Lying, Accusation, and Swindling*. Healy had examined Bessie after she accused her father and brother of incest.[24] Healy interviewed all three parties and examined the physical evidence, noting that the girl's symptoms included a vaginal discharge and genitals so swollen that she could not be examined. But despite repeated bacteriologic tests that ruled out gonorrhea and his characterization of the father and brother as "decent," Healy concluded from the physical evidence that sexual contact had occurred.

Yet Healy's assessment of Bessie's demeanor, how poised and articulate she had been when describing the assaults, and her claim that she had been sexually active—with boys, girls, men, and objects—since age 5 (a claim Healy seemed not to question) led him to conclude that she was lying about having been coerced. Like doctors in the prebacterial era, Healy used his assessment of the victim's character to conclude that she had been the "instigator" of the sexual contacts, and he did not criticize the father and brother for having failed to rebuff Bessie's "advances."[25] The charges against them were dropped, and Bessie was placed in an institutional school "for her own protection." Once she was confined, Healy concluded that Bessie's mind was "so continually upon sex subjects" that he recommended she remain "long under the quietest conditions and closest supervision."[26]

The Individual Delinquent had helped to launch the field of "child guidance," which shifted psychiatry's focus in the 1920s from studying the pathological histories of "delinquent" children to identifying instead the developmental and behavioral issues of ordinary children from the white middle class.[27] In 1917 William Healy moved to Boston to direct the newly formed Judge Baker Clinic, which became the twentieth-century model for integrating psychiatry, psychology, and social work to shape child behavior.[28] Healy's attention to a child's individual psychology, rather than heredity or environment, to understand and change behavior influenced contemporaries from various fields, some of whom included his findings in their own, equally eminent works.[29] In the 1926 supplement to his *Treatise on the System of Evidence in Trials at Common Law*, law professor John Henry Wigmore, the authority on American rules of legal evidence, included Healy's "proof" that girls lied about incest to support his argument that girls frequently made false accusations of sexual assault.[30] Wigmore believed that girls fantasized so often about being raped that a responsible judge should require every girl who claimed to have been sexually victimized to submit to a psychiatric examination, a step that prominent child psychiatrist Dr. Karl A. Menninger endorsed heartily and Wigmore's *Treatise* repeated in subsequent editions through 1940.[31]

Forensic examiners also registered a high degree of skepticism about the credibility of girls. In his 1930 textbook on legal medicine, Ralph Webster, a medical examiner in the Chicago coroner's office and professor of medical jurisprudence at the University of Chicago and Rush Medical College, stated that parents often pressured their daughters so intensely to name their assailant that "terrified" girls often accused "perfectly innocent individuals."[32] Webster may have observed such interactions between frantic parents and their distressed daughters. But he did not consider the possibility that a girl may have been so frightened, and perhaps deceitful, because her assailant was also one of the people pressuring her, her father. Yet the weight of professional opinion was on Webster's side. The ninth edition of *Taylor's Medical Jurisprudence*, published in 1934, continued to remind physicians of their duty to be skeptical of a girl's accusation because "a child may, either through mistake or design, accuse an innocent person." By this time, *Taylor's* should have had plenty of medico-legal evidence to support such a claim. But it simply referred to the eighteenth-century case of Jane Hampson, repeating the adage that "some cases are reported by which it would appear that men have thus narrowly escaped conviction for a crime which had not been perpetuated."[33]

Medical examiners and other professionals might have been less skeptical of girls' accusations of sexual assault if they believed, as Dr. Jerome Walker had acknowledged in the 1880s, that all kinds of men regularly assault girls, including their own daughters. In the nineteenth century, a casual reader living in New York City or on a ranch outside Albuquerque could randomly open a local newspaper or national scandal sheet and find a report of incest. Reports of fathers—from illiterate laborers to refined clergymen—who sexually assaulted their daughters were available in nineteenth-century print culture, in the active response of a community to an unmarried girl's pregnancy, and in courtrooms across the country. After 1919, however, even the most informed reader might have concluded that American society had indeed evolved past incest. Even though the number of prosecutions had increased, the revelation of incest no longer rippled through the public sphere. By the 1920s such disclosures had disappeared from public view as individual anonymity intensified in a newly urbanized America; newspapers stopped reporting incest cases; judges routinely granted defense attorneys' motions to exclude the public from incest trials; and professional child protection workers—who never policed the domestic affairs of respectable families—replaced community intervention.

As noted in the introduction, a search through the same archives and electronic databases in which I found more than 500 different incidents of father-daughter incest published in more than 900 newspaper reports over the nineteenth century turned up only 136 cases in newspapers between 1900 and 1940. Multiple stories about the same accusation raised the total number of stories to 221 (again, not counting stories of adult-child sexual assault in which the relationship between the parties is unspecified, or accounts of incest between other family members). Between 1900 and 1909 newspapers published stories about 73 cases of father-daughter incest, more than the combined total for the next three decades:

44 cases between 1910 and 1919; 7 in the 1920s; and 13 in the 1930s. I could not locate even one report for many of the years between 1920 and 1939, and none of those years produced more than two cases. Many stories appeared only in the small local paper in the town or county in which the crime occurred, and few included the details—salacious or otherwise—that had so enthralled nineteenth-century readers.

A search through a partial database of historically black newspapers yielded another ten cases from the 1920s and 1930s. The court dismissed charges against three of the African American defendants, and the outcome of three others, including one man shot by his 18-year-old stepdaughter, is unknown.[34] The papers published reports of three white fathers accused of incest, including the "common law husband" of an African American woman.[35] And a 1939 article in the *New York Amsterdam News* complained about "Georgia Justice" after the governor commuted the death sentence of a white man convicted of incest with his daughter and the murder of their baby but let six African American men convicted of murdering four white men go to the electric chair.[36]

Certain narratives popular in the nineteenth century seldom appeared in news accounts after the turn of the century, particularly after 1919, including links between alcohol and incest; wayward clergymen; and race-based lynchings. Other themes continued: the absence of mothers; pregnancies that raised suspicions about incest; fathers' use of violence to coerce and control their families, sometimes ending in murder; men who chose suicide over dishonor; friends of the accused who proclaimed his innocence but refused to post bond; and the lengthy period of time over which the incest continued, sometimes serially among daughters.[37] New themes also emerged, including increased confidence in a more professionalized court system; prosecutions against men who had procured abortions for their daughters; the expanded role of child welfare organizations with "semi-official standing" in discovering incest and bringing prosecutions; courtrooms routinely closed to the public, most often at the insistence of attorneys for defendants; the more visible role of women, including mothers, aunts, sisters, and neighbors, both in bringing the incest to light and in obtaining public intervention to stop it and punish the father; and gonorrhea infection as either a woman's first clue or corroborating evidence of the crime.[38]

At the beginning of the century, accused men still attempted to explain or defend their actions by asserting a temporary loss of manly volition. In 1900, a father described only as a 45-year-old "white man" accounted for having raped his 16-year-old daughter by coolly stating, "Whisky did it."[39] But men soon made new and less familiar claims to explain behavior that they themselves seemed unable to understand. In 1909 the *Los Angeles Times* reported that Sherman Caudell sobbed as his 11-year-old daughter testified that she had endured hundreds of rapes at her father's hands over the past three or four years and for as long as she could remember. On trial during the same year in which Sigmund Freud made his celebrated visit to the United States, Caudell ceded all volitional ability not to intemperance but to his "mind," an interior monster over which he seemingly had no awareness or control: "There must have been something wrong with me to make me do it. My mind must have been wrong." Perhaps agreeing that Caudell had lost all control, the judge set his bond at an unusually high ten thousand dollars.[40]

But few men were as willing as Caudell to admit they had committed incest—under any circumstances. Most defendants denied that any sexual contact had occurred and fell back on the traditional tactic of maligning the motives of females who accused men of sexual assault. Newspaper accounts and court records show the consistency with which men defended themselves from incest allegations by impugning the motives and loyalty of their wives, children, relatives, and neighbors, all of whom they accused of having—often with little or no proof—"trumped up" the charges as part of a scheme to get rid of the father or to obtain his wealth and property by putting him in jail.[41] In these narratives, familiar from nineteenth-century medical jurists, fathers portrayed themselves as the innocent victims of vengeful females. But in the twentieth century, simply accusing a man's wife and children of being greedy liars was not enough.

Fathers also accused their daughters of maliciously making false claims in order to evade punishment for their own sexual misconduct or to hide a romance with a boy or man of whom the father disapproved.[42] In other words, fathers tried to defend their actions by claiming that their daughters were so committed to sexual misadventure that they would do or say anything to circumvent their father's wise exercise of paternal authority. Some fathers claimed their daughters made a false claim out of spite, usually in response to a punishment or a rule they deemed unfair, an appropriate line of defense to establish a motive to lie.[43] Such claims were occasionally true, and some daughters, mostly teens and young women, confessed to having filed false claims because they wanted to escape fathers they found too strict or domineering.[44]

However, the viciousness with which fathers publicly savaged their daughters' characters, while sometimes an effective courtroom strategy, belies the credibility of their claims about the selflessness with which they guarded their daughters' reputations. In 1929 Claude N. Hewitt's 15-year-old daughter testified that her father raped her on three occasions, twice with a gun at hand and once while holding a wrench. Hewitt defended himself by claiming, without tendering any evidence, that his daughter had "concocted" the story because of a "sex complex" that caused her to want to be with boys at any cost and to resent her father for not permitting her to do so.[45] In 1944, when Harry C. T. Rankins testified at his trial for statutory rape and incest, he denied his 16-year-old daughter's claim that he had assaulted her for five years, asserting, "I wouldn't ruin my reputation and stoop that low, with the reputation I have at church and all, to commit a crime of that kind. I wouldn't be that low. I wouldn't have the audacity to sit on this stand in front of this jury or anyone else and be doing such a thing. I wouldn't be that low."[46] But Rankins wasn't above implying that his daughter was keeping company around town with and had acquired gonorrhea from "Big George," an African American man who worked at a barbeque stand that sold beer and where, with her father's approval, Mamie Rankins had worked late nights for the past two years. And just to be sure the jury understood how low Mamie had sunk, the defendant put a newspaper photo of dark-skinned Big George into evidence.[47]

Even if a father could not escape a guilty verdict by attacking his daughter's reputation and morality, the courts and community would not punish too severely a man whose daughter they believed to be sexually delinquent. When James B. Cowles Sr., a church sexton, pleaded guilty in 1925 to raping his 16-year-old daughter, who became pregnant, the judge sentenced him to only nine months in prison because his daughter had already "gone wrong."[48] Other communities achieved a sort of parity by imprisoning not only guilty fathers but also the daughters who complained about them. When R. P. J. Munroe, a house painter, admitted in 1902 to having raped his "feebleminded" 27-year-old daughter, the prosecutor recommended mercy, and Munroe was sentenced to only eight years in prison. His daughter was committed to the Home of the Feeble-

Minded, presumably for life.[49] In 1929, when the police arrested Henry Lester and two other men for the rape of Lester's daughter, they also charged the 12-year-old victim with danger of falling into vice.[50] Similarly, when the daughter of Edward M. Cook became pregnant in 1937 after having been gang-raped by her father and some boys, Cook was arrested for incest, and his daughter was arrested for lascivious carriage and placed in the House of the Good Shepherd pending trial, after which she was convicted and sent to the State Farm for Women.[51]

Siblings who revealed the incest or tried to protect a sister from it might also be penalized for their efforts. When John Shield tried to sexually assault his 16-year-old daughter, his 15-year-old son shot him. The front page of the *Placerville (CA) Republican* reported in 1937 that, as a consequence, both children and another sister, a 13-year-old girl, were sent to institutions where they were to remain until they turned twenty-one.[52] The progeny of incest, the visible, enduring reminder of what had occurred, perhaps fared the worst. After Thorten Stadstad was arrested for incest with his daughter in 1911, the *Grand Forks (ND) Daily Herald* commented that their union had produced a 7-year-old "idiot child" who, the paper churlishly noted, "is one of the lowest forms of human life imaginable and will undoubtedly be taken to some feeble-minded institute where it can be cared for."[53]

Unlike in the nineteenth century, when evidence of father-daughter incest was more abundant in newspapers than in many other forums, in the twentieth century a variety of professionals also began writing about father-daughter incest. Their published writings and records suggest that girls continued to disclose their fathers' assaults but that the noisy silence surrounding the topic among the white middle and upper classes had spread out so thoroughly that it drowned out their voices. Doctors and social hygienists insisted that only poverty or ignorance produced incest or gonorrhea infections in girls, a position that their data did not support but that professional white Americans, eager to affirm the American dream, embraced.

Containing Incest

Looking at the men imprisoned for sex crimes in general and for incest in particular, celebrated trial lawyer Clarence Darrow wrote in 1922 that "incest . . . is peculiarly the crime of the weak, the wretched and the poor."[54] Darrow was correct that poor men were most often prosecuted and convicted of incest, in part because child protection agencies rarely intruded into respectable homes. But there are no data to support the inference that the class of men imprisoned for incest fairly represents the class of men who commit incest. Prewar researchers who wrote about father-daughter incest characterized it as exclusively a crime of the poor, and they ignored evidence that suggested otherwise. In 1935 Dr. Jacob A. Goldberg, who succeeded Dr. Walter Brunet as the secretary of the Social Hygiene Committee of the New York Tuberculosis and Health Association, and his wife, Rosamond W. Goldberg, published *Girls on City Streets*. The Goldbergs analyzed 1,400 criminal prosecutions and convictions for child sexual assault and incest, of which nearly 20 percent (280) involved girls who had been assaulted in their own homes, most often by their fathers.[55] As the Goldbergs realized, "The fact remains that many young girls have been and continue to be subjected to violation by those upon whom has rested the highest degree of responsibility for their protection and care."[56]

The Goldbergs argued that incest was a crime of opportunity, a conclusion that fit their social biases but not their data. They warned of the dangerous father who drinks "alcohol and seizes on his young daughter as a readily accessible female to vent overstimulated senses." Or the man who, because his wife is absent (whether at work, ill, or dead), "seeks to channel normal urges to an abnormal channel."[57] Or, in the most pornographic scenario, the stepfather who casts "longing eyes upon the adolescent girl . . . over whom he has acquired parental responsibility and control."[58] Yet only a few men in their study conformed to these patterns. The Goldbergs' data provided abundant evidence to show that father-daughter incest was only rarely a crime of passion or a momentary loss of judgment or control. They documented instead the details of the careful, deliberate, and methodical planning that preceded the first assault and the ensuing complications that fathers attempted to master as they struggled to conceal repeated assaults over a period of days, weeks, months, and years.[59]

Because most of the fathers in their study were from the poor and working classes, the Goldbergs interpreted their data as support for their belief that men who lack self-restraint commit incest.[60] Ignoring the business and professional men in their sample, the Goldbergs singled out overcrowding as the major cause of incest, and they identified certain traits, including "low mentality and its concomitants in the guise of superstition, cruelty, lack of social sense, etc.," that led men to commit incest.[61] When the Goldbergs acknowledged that middle- and upper-class men also experienced sexual desire for their daughters, however, they also dismissed the notion that such men acted on their unseemly desires. They argued that middle- and upper-class men turned to psychoanalysis to successfully divert their incestuous desires.[62] Their remark not only expresses a great deal of confidence in psychoanalysis but also fails to explain how the majority of men, who had not been psychoanalyzed, managed their desires. But it mattered little to the Goldbergs, who blamed an "attitude of silence" about incest among the "lower classes" for girls' vulnerability to their fathers' attacks.

The Goldbergs' distorted use of their data helped to erase father-daughter incest in middle- and upper-class homes from the professional literature on the subject. Jacob Goldberg was also a social hygienist (in 1938 he coedited a textbook on social hygiene), and his social hygiene colleagues followed suit with corollary statements about the supposed relation between gonorrhea and poverty, even though they, too, had gathered empirical data that implied otherwise. Social hygienists had worked hard to decrease the stigma associated with venereal disease, and health care professionals were afraid of asking patients any questions—including their names, let alone any financial information—that might diminish the likelihood that they would visit a clinic for testing or return for prolonged treatments.[63]

But investigations, such as the *Survey of Female Gonorrhea Clinics in New York City* done under Goldberg's direction at the New York Tuberculosis and Health Association, make it clear that by the 1930s people from all social classes sought treatment for infected daughters at venereal disease clinics.[64] Clinics and dispensaries had been traditionally dedicated to serving those people who could pay little or nothing for services. However, venereal disease clinics had discovered that charging even nominal fees—particularly during the depression—"removed the taint of charity" that had discouraged some patients from seeking medical services.[65] And because treatment could be both expensive and shameful, some patients who could afford to see a private physician for other illnesses went to clinics and dispensaries because of the anonymity and reduced rates they offered.[66] One clue to the economic diversity among patients at New York City's vaginitis clinics is the disparity in fees. Various clinics charged the parents of infected girls as little as twenty-five cents or as much as two dollars per treatment plus the cost of drugs. The average cost of the complete course of thirty treatments might be as low as thirty dollars at one clinic or as high as a hundred at another. Who could afford such treatments? The Tuberculosis and Health Association concluded, as did others, that clinics treated "large numbers of wage earners."[67]

Still, in 1936, Dr. C. Walter Clarke, whose professional obligations in New York City intersected with those of Jacob Goldberg, characterized vulvovaginitis as a disease of the poor, a view that both the medical community and ASHA endorsed when they republished his comments.[68] As late as 1943, Dr. Lawrence Wharton's gynecology textbook stated that "gonococcal vaginitis . . . is especially prevalent among dispensary patients. An infected servant has often spread the disease to the daughters of her employer."[69] Whether social hygienists were referring to dirt or deviance as the source of infection was unimportant as long as the professional literature continued to insist that incest and infection were both contained within socially marginalized populations.

Institutionalization

In these discourses the voices of the infected girls, whose views are rarely reported, were silenced. Even their bodies, which displayed the physical evidence of their father's assault and held the knowledge of what had occurred, were hidden from view. Doctors and the criminal justice system confined infected girls in various facilities for wayward girls, including reformatories, prisons, hospitals, and convalescent homes. Some doctors and social workers institutionalized girls to ensure they received adequate care. More often, however, doctors, social workers, and parents shared the view that infected girls were a menace to their families and schoolmates. They argued that "isolating" them protected the public health. By 1931, such arguments had the imprimatur of the White House.

In 1931, the published report of the Committee on Communicable Disease Control to the prestigious White House Conference on Child Health and Protection recommended that infected girls be isolated and excluded from schools "or contact with other children" while symptomatic.[70] Local officials across the country modeled their surveillance of infected girls after methods they had used to control unmarried women who were sexually active. Some cities routinely tested for venereal disease women they suspected were prostitutes and, to prevent them from engaging in sex while infected, committed those who tested positive to "quarantine," usually a jail, where they remained often without any medical treatment until they tested negative.[71] After 1919, when rescue workers impatient with prostitutes who sneered at the idea of reform began to recommend incarceration instead as a means to protect society, doctors, parents, social workers, and public officials began to treat infected girls in much the same way.[72]

Girls who were shut away in quarantine for months or even years, like unmarried daughters shipped quietly out of town to a maternity home during a pregnancy, were neither an embarrassment to their families nor a physical reminder that something disturbing, and perhaps neither spoken of nor acknowledged, had occurred. From 1909 to 1914, for instance, the juvenile court in Cincinnati placed infected girls, including girls whose fathers had raped them, in the Catherine Booth Home and Hospital, which the Salvation Army operated primarily for the benefit of unwed mothers.[73] Although some girls may have enjoyed greater security in an institution than in an abusive home situation, protecting them from their fathers was not the reason they were institutionalized. In 1913, pediatrician Richard M. Smith, attending physician at the Children's Medical Department of Boston's Massachusetts General Hospital, singled out infected girls who were "feebleminded" for institutionalization, on the grounds that the girls did not appreciate the danger they posed to others and so were likely to carelessly spread infection.[74] By 1932 health care professionals had extended Smith's recommendation to all infected girls. When experts in pediatric medical care and social work, including Edith Baker, met at the White House Conference on Child Health and Protection, they warned that infected girls "are not sick but they are dangerous to the community and should be treated as such."[75] How had prepubescent girls become such a threat? And how, exactly, did they pose a danger to the community?

The first governmental attempt to isolate infected girls began as the United States prepared for World War I. The military deemed soldiers infected with venereal disease unfit for active duty. Despite a report from the surgeon general of the army that five-sixths of soldiers' infections preceded induction, the military blamed the incidence of infection on the women and girls who flooded the towns around training camps.[76] In 1917 the federal government set up the Commission on Training Camp Activities (CTCA), which it charged with educating servicemen and civilians about venereal disease infection and arresting prostitutes or suspected prostitutes in urban areas.[77] A year later the government created the Interdepartmental Social Hygiene Board to oversee venereal disease control in both the military and civilian populations.[78] The military arrested thousands of women and girls, whether "hardened" prostitutes or "camp girls," adolescent and young girls who were out to have a good time by combining a spirit of adventure with their notion of patriotic duty.[79] Those who tested positive for venereal disease were summarily incarcerated, often without due process.

The American Social Hygiene Association played a pivotal role both in shaping the board's programs and in staffing key positions. Dr. William F. Snow, ASHA's executive director, joined the army's Medical Corps and was among the experts who saw an opportunity to expand the federal government's role in eliminating prostitution and, with it, venereal disease. He served on the Interdepartmental Board and as director of the social hygiene program of the Council of National Defense. Henrietta Additon and Martha P. Falconer were also named to the board, and they helped to develop its program for protective work with girls.[80] Additon was the co-director with William Healy of the Judge Baker Clinic and a renowned protective worker who had worked with the CTCA. Falconer was an ASHA board member, Chicago club woman, and a probation office who later served as superintendent of Sleighton Farm, Pennsylvania's highly regarded reformatory for girls.

Falconer directed the committee on protective work with girls, which the Interdepartmental Board charged with overseeing facilities for civilians "whose detention, isolation, quarantine, or commitment to institutions may be found necessary for the protection of the military and naval forces of the United States against venereal diseases."[81] Falconer's committee convinced "camp towns" to use detention houses as a "clearing house where all women and girls (except hardened cases) who are arrested may be held while awaiting trial, to be studied and treated medically."[82] Over the course of the war and redeployment, the government "admitted" 18,000 girls and women, including more than 15,500 it claimed were infected with venereal disease, to indefinite terms in one of forty-three reformatories, prisons, or "detention houses" that were spread across the country but located primarily in the South.[83]

A lack of funding and objections from communities that did not want facilities for prostitutes in their towns curtailed the detention housing program.[84] But many cities appreciated the offer of federal money to bolster existing institutions by adding quarantine facilities for venereal disease patients, including girls. St. Louis and Memphis worked with their local House of the Good Shepherd; Chattanooga invested in the Florence Crittenden League; San Francisco General Hospital added an isolation ward; and Kalamazoo County, Michigan, paid for improvements to the Catholic Borgess Hospital, the Adrian Industrial School for Girls, and the Fairmont Hospital for contagious diseases. In the South, Columbia, South Carolina, supported the Anna Finstrom Home, a detention house and hospital opened in 1917, and Houston tried to avoid placing younger girls in prisons or temporary detention facilities filled with "wild women of the worst sort" by making improvements to the Dorcas Home for African American girls and the Lodge for white girls.[85]

Investigators from Falconer's committee visited detention facilities across the country to talk with the girls and, when able, their parents.[86] Their reports included data on thirty-three girls they believed had been "innocently" infected, of whom four were incarcerated in reformatories and twenty-nine in detention centers.[87] Seven girls had been sent to the Convalescent Home for Children in Walnut Creek, California, located on fifteen acres near San Francisco. The local juvenile health board opened the home in 1919 "to build up and give expert care to girls under 12 years of age who have chronic venereal disease—gonorrhea only."[88] Social workers who investigated the source of the girls' infections concluded that four girls, ages 5 to 11, had been "infected by their parents," presumably through lack of sanitation. They attributed one girl's infection to her mother, another to having been sexually assaulted by a "friend of the family," and wrote off the source of the last girl's infection as "not known."[89]

Similarly, the source of infection for the eleven girls and boys sent to a detention home in Houston, none of whom investigators considered sexually active, were "not learned."[90] Nor did investigators identify the source of infection for three sisters who, with their mother, were treated for gonorrhea in the quarantine hospital in Kalamazoo. Investigators claimed that two sisters, ages 5 and 13, placed in the detention house in Augusta, Georgia, had been infected by their father's towels. The only girl whose infection investigators attributed to incest was a 3-year-old girl whose father had already been convicted of sexually assaulting her.[91] Still, the Interdepartmental Board concluded that gonorrhea vulvovaginitis was the result of careless parenting and boasted that many communities wanted to make a permanent plan for institutionalized care of infected girls.[92] These cities had discovered that having a quarantine facility for infected girls both reduced the demand for hospital beds and solved a potential public health problem. They expanded the practice of institutionalizing infected girls, including those who had been the victims of sexual assault.[93]

Although prominent doctors like Prince A. Morrow, Alice Hamilton, and L. Emmett Holt had argued that innocently infected girls should not suffer the stigma of an immoral disease, from about 1910 through the 1940s social workers, juvenile authorities, and even some parents placed infected girls in reformatories such as Juvenile Hall in Los Angeles and the San Francisco Juvenile Detention Home.[94] But existing facilities could not absorb them, both because their capacities were already stretched and because of the added expense of having to provide an isolation area where they could keep the infected girls from spreading their infections to the general population. By the late 1930s economic straits related to the depression and an increase in juvenile arrests had seriously degraded the conditions at some of these institutions.[95] Chicago, for instance, tried to quarantine infected girls by incarcerating them in the Juvenile Detention Home. By 1937, however, the home had become dangerously overcrowded, and infected girls were at the center of a public scandal. Critics complained that, because the dormitory for "diseased girls" was too small, infected girls mingled freely with "uncontaminated" girls, making it impossible for the staff to enforce "toilet rules affecting the diseases."[96]

Private and public agencies opened residential homes for infected girls in other cities that also served as maternity homes or were modeled after them. Such facilities provided a less punitive setting than a reformatory and, though generally small and few in number, could be found throughout the country. Maternity homes, including the Florence Crittenden and Salvation Army Booth homes, had policies aimed at avoiding ward epidemics by refusing infected girls, but in practice some of these facilities regularly took them in.[97] Some municipalities, like New York and Houston, opened convalescent homes to which they could transfer girls who had completed their medical treatment but still tested positive for the disease.[98]

In Los Angeles, the county remanded sexually assaulted and infected girls to Juvenile Hall, the county penal institution for minors, until they appeared in court to testify against their assailants. Doing so may have secured their cooperation, but incarcerated girls must have felt both defeated and punished. At her father's trial for incest and statutory rape in 1944, for instance, 16-year-old Mamie Rankins testified that he had coerced her into sexual intercourse and ensured her secrecy for eleven years by threatening to commit her to Juvenile Hall. Mamie took his threat seriously—her father worked as a special patrol officer and carried a gun and handcuffs. When Mamie finally got up the courage to tell school authorities about her father's assaults, she found herself sent to Juvenile Hall anyway, where she was examined, diagnosed with gonorrhea, and held until her father's case came to trial.[99]

The number of infected girls at Juvenile Hall, whether infected "innocently" or otherwise, had become a burden on the facility's resources long before Mamie Rankins arrived. In the 1930s the hall began to transfer infected inmates to the Ruth Home and to the House of the Good Shepherd.[100] Roman Catholic nuns operated the Houses of the Good Shepherd (or Convents of the Good Shepherd), residential facilities strung across the country that accepted unwed mothers, adolescent "delinquents" in need of "reform," and infected girls.[101] The Ruth Home, which accepted infected girls for more than forty years, was unique among such facilities.

Ruth Home

As with Chicago's Frances Home, the mission of the Ruth Home was to provide housing, treatment, and schooling to girls and babies infected with gonorrhea.[102] The Pacific Protective Society, which opened the home as its only facility exclusively for infected girls, was a nonsectarian charitable organization for protective and rescue work with adolescent girls and prostitutes in the Pacific Northwest and California. The first Ruth Home was a small wood-frame cottage in Los Angeles that provided temporary housing for infected girls until they could be placed either with adoptive parents or in domestic service. Demand for services was so great that in 1931 the society closed the cottage and moved to a new home with space for 135 girls and 15 babies.

The society built a fifteen-acre, multibuilding campus in El Monte, a sparsely populated suburb east of Los Angeles. No expense seems to have been too great. Visitors commented that the new campus, which consisted of Spanish-mission-style architecture, picnic facilities, and "artistic landscaping," seemed more like a lovely residential neighborhood than a medical facility.[103] The Pacific Protective Society spent $125,000 on a state-of-the-art hospital that the Pasadena Community Chest praised as without peer and $45,000, including WPA funds, to construct a school and Arts and Crafts building. Four years later it added a thirty-five-bed quarantine facility.[104] By 1935 the Ruth Home was treating an average of 320 girls and babies annually, most of whom were younger than 12 years of age and whose stay ranged from two months to two years.[105] Like Chicago's Frances Home, the Ruth Home offered more than medical services; it provided numerous scouting-type activities, public school teachers for grades one through twelve, and courses in business and cosmetology for the older girls.[106]

Some of the recipients of the home's largesse were girls who would otherwise have been incarcerated in Juvenile Hall, which tested every new inmate for gonorrhea before releasing her into the reformatory's population. Nurse Rhea C. Ackerman began working at Juvenile Hall in 1929 and served as superintendent from 1936 until she retired in 1943.[107] Ackerman's tenure was marked by the number of infected girls who nearly overwhelmed the facility. Over the 1930–31 fiscal year alone, the hall provided costly housing and treatment to nearly five hundred girls infected with gonorrhea.[108] Some of the girls were sexually active adolescents. But many were younger girls who had been raped and infected with gonorrhea, and Ackerman was worried about the deleterious effect of mixing these "innocent" girls with "sex delinquent" teens.[109] After the Ruth Home expanded, Ackerman annually transferred between fifty and one hundred "innocently" infected girls.[110]

With a steady stream of referrals, both from the county and from families who paid for their daughters' stay, by 1935 the Ruth Home was again filled to capacity. This time the society turned it over to Los Angeles County, which had only one other facility providing treatment to infected girls, the county hospital.[111] The home might have expanded indefinitely if not for the discovery of penicillin, which cured gonorrhea infection within seven to ten days.[112] After "rapid treatment" became available in Los Angeles in 1944, any public health justifications for isolating infected girls disappeared, and in 1949 the county closed the home.[113]

Social Hygiene Becomes National Policy

With the onset of the depression, financial problems beset ASHA, and many local chapters closed. But just when the social hygiene movement might have faded away, it achieved sudden and unexpected success. In 1936 President Franklin Roosevelt appointed Dr. Thomas Parran to the office of U.S. surgeon general. Parran had served in the U.S. Public Health Service from 1917 to 1930, rising to chief of the Venereal Disease Division. As governor of New York, Roosevelt had named Parran state health commissioner, a position Parran used to promote social hygiene. By 1934 he had achieved national prominence when he chastised the media for its policy against permitting the words *syphilis* or *gonorrhea* to appear in print or on the air.[114] Parran argued that such "prudery" blocked public health officials from disseminating medical information about diseases that endangered the health of its citizenry.[115] In 1935 President Roosevelt appointed Parran to the committee that drafted the Social Security Act of 1935. The next year he became surgeon general, and the American Public Health Association, of which Parran had been an officer since 1931, elected him president.[116]

Under Parran's leadership, the issue of venereal disease entered both respectable public discussion and federal policy as a pressing medical issue.[117] Instead of trying to convince Americans not to engage in extramarital sex, Parran took a pragmatic approach.[118] A committed New Dealer, Parran wanted to conserve social and economic resources by reducing the direct medical costs incurred in treating infected Americans and the financial losses sustained by business and the national economy when employees were too sick to go to work. Parran successfully lobbied Congress to fund a national anti–venereal disease program for which he gained popular support by publishing articles in magazines like *Reader's Digest* and *Ladies' Home Journal*. Americans responded immediately and positively, agreeing "on the necessity of tearing the shroud of social taboos" about venereal disease.[119] In 1937 Parran published a primer for laypeople, *Shadow on the Land*, in which he laid out the social and economic costs of venereal disease, particularly syphilis. The book became a bestseller, and *Time* put Parran's photograph on its cover.[120]

One year later Congress passed the National Venereal Disease Control Act, which provided $15 million in federal funding to state boards of health for education, research, and diagnostic and treatment facilities.[121] By 1940, the number of people who sought medical advice pertaining to venereal disease had more than tripled.[122] The program's success, however, was limited. It focused almost exclusively on syphilis, even though the Public Health Service estimated in 1939 that four times as many Americans—more than one million—became infected annually with gonorrhea.[123] ASHA president Dr. Edward Keyes Jr., a specialist in gonorrhea, bemoaned the omission but attributed it to the fact that public health officials could not build a program around a disease for which no effective treatment existed.[124]

But the omission had severe ramifications for infected girls. Although doctors' familiarity with diagnosing and treating gonorrhea vulvovaginitis had increased exponentially since they had first discovered the "epidemic," their reluctance to identity father-daughter incest had remained unchanged. No other area of venereal disease research and education had clung so tenaciously to fiction rather than fact. On the eve of World War II, even the country's preeminent public health authorities could not, after finally ruling out the toilet seat as the primary locus of infection, acknowledge incest as a source of infection. In 1938 the New York City Department of Health, with funding from the Public Health Service, the Milbank Memorial Fund, and the New York Foundation, formed the New York Vaginitis Research Project of the Gonococcus Research Committee. The project's advisory committee included New York City health commissioner and former Yale researcher Dr. John L. Rice, ASHA leaders Dr. Keyes and Dr. Walter Clarke, and gonorrhea expert, social hygienist, and University of Pennsylvania urologist Dr. Percy Pelouze.[125]

The Vaginitis Project conducted the first medical research to determine whether girls could acquire gonorrhea vulvovaginitis by sitting on a toilet seat. When investigators placed *Neisseria gonorrhoeae* on a toilet seat and measured how long it remained virulent, they concluded that the bacterium dried before it could infect a child. And when project researchers examined patient records from four hospitals and four child-caring institutions, they could not trace a single infection to contact with a soiled toilet seat. The evidence was compelling. In one hospital, all the children, including girls infected with gonorrhea, used the same toilets, which were still outfitted with old, O-shaped wooden seats. Yet even in this seemingly ideal setting for infection, not one girl had become infected.[126]

The project's report, published in 1940, unequivocally rejected the idea that girls could be infected by casual contacts with toilet seats, mothers, or other children: "The relationship between the child and the infected adult has to be quite intimate to infect the child. The same is true for the transmission of the disease from one child to another; a history of actual sexual contact or manipulation of the genitals occurred not infrequently. However, infection of one child by another through the medium of the toilet seat would appear rare."[127] Vaginitis Project researchers were less confident, however, about rejecting the possibility of infection from damp towels or linens. And even though the committee's report acknowledged that girls became infected from sexual contacts, it mentioned only "sex play" between children, not adult-child contacts or incest.

Thus, when the eleventh edition of *Holt's Diseases of Infancy and Childhood*, the most prestigious pediatrics textbook of the twentieth century, was published in 1940, it repeated the speculation that "gonococcus vaginitis in children is not to be regarded as a venereal disease" because only "an insignificantly small proportion" of infected girls have been assaulted.[128] Not a shred of evidence had been found to prove that Alice Hamilton's and Frederick Taussig's early speculations about dangerous things had been correct. But *Holt's* instructed the next generation of physicians that "in institutions the disease may be spread through faulty sterilization of diapers, clothing, bed linen, thermometers . . . or . . . the hands of nurses. In schools and . . . public places it may be spread by toilet seats . . . The young child may have slept in the same bed with an infected mother or sister; the infection may have occurred through baths, towels, clothing, toilets, etc."[129] With opportunities for infection apparently everywhere, the only preventative measure *Holt's* could suggest was to quarantine infected girls where they would not be a "menace" to others.[130]

Speculation that girls acquired gonorrhea infection from nonsexual contacts had finally become scientific fact; a belief so widely held that even medical examiners considered the forensic value of a diagnosis of gonorrhea vulvovaginitis so compromised as to be useless. In 1937, when Thomas A. Gonzalez, the chief medical examiner of New York City, and his staff published a book on forensic medicine, they acknowledged that many men raped children "because of their comparative helplessness" and that a vaginal discharge was often the first physical sign of what had occurred.[131] But in language disturbingly like that of the medical jurists in the prebacterial era, the medical examiners urged their colleagues not to conclude that an infected girl had been assaulted because "the main difficulty is that a gonococcal vaginitis occurs in female children occasionally as a nonvenereal [not sexually transmitted] infectious disease," a warning repeated in subsequent printings and in the second edition published in 1954.[132]

Despite the remarkable progress of medical science in the late nineteenth and early twentieth centuries, for girls the march of time had ticked backward. Whereas Dr. Burke Ryan had not dared to suggest that casual contact was anything more than an occasional possibility, a 1943 gynecology textbook instructed students that "sexual intercourse is rarely the method of infection in young girls."[133] With the medico-legal connections between assault and infection severed, the need for parents to hide infected daughters in quarantine had passed. Since infected girls were no longer considered at risk for spreading a lingering infection to others, doctors stopped identifying them as a threat to the public health. Interest in the etiology of the disease evaporated as professionals moved on to more pressing—and less discomfiting—subjects.

Epilogue

MARGARET SANGER had taught Dr. Harriet Hardy how to fit diaphragms, and when Hardy became a Radcliffe College physician and the head of its Department of Health Education in 1939, she expected that contraception would be the issue of most importance to her students.[1] She was surprised to discover that students were concerned instead about acquiring venereal disease, even though few were sexually active. Unable to understand why virgins would worry they might be infected with a sexually transmitted disease, Hardy was disheartened to learn that the students still believed the "old wives' tales" that venereal disease could be acquired from doorknobs and public toilets.[2]

Perhaps because physicians and educators like Hardy had begun to disparage such beliefs, within a few years even high school girls were better informed than their Ivy League sisters had been. In 1943 the *Journal of Social Hygiene* reported on the results of a survey given to Wisconsin high school girls. Sex education instructors boasted that, in response to the question whether objects or sexual contacts posed a greater danger of venereal disease infection, 95 percent of the girls "correctly" chose sexual contacts rather than the "rare but formerly much publicized 'unsanitary things.' "[3] That same year ASHA published a pamphlet on communicable diseases for high schools and junior colleges. The section titled "Accidental Gonococcic Infections" reassured students that they should not worry about infection through nonsexual contacts: "There is a widespread but erroneous notion that it is easy to get gonococcic infections from towels, bathtubs, door handles, or other common articles . . . Bacteriologists have found no positive evidence of such danger. If such infections occur they must be very rare."[4]

In the postwar era, educators continued to modify their views about the etiology of girls' infections. Evangeline Morris, associate professor of nursing at Simmons College and former social hygiene supervisor of

Boston's Visiting Nurse Association, told her students in 1946 that the notion that gonorrhea could be acquired from a toilet seat had been discredited but not forgotten. She mused that "a surprising number of intelligent persons whose moral reputations are not at stake continue to endow the gonococcus with amazing powers of locomotion and of survival."[5] The *VD Manual for Teachers*, also published in 1946, allowed for the possibility that infections could spread from parent to child by "common articles such as towels freshly soiled by the discharge" but added that such cases were "very uncommon."[6] And in 1948 Dr. R. A. Vonderlehr, medical director of the Centers for Disease Control, wrote in a highly regarded marriage manual that "most now believe gonorrhea in children is not acquired from inanimate objects but from other children."[7]

After so many years of equivocation, doctors and social hygienists seemed finally to have given up speculation about nonsexual contacts. But what had changed? The availability of "rapid treatment" had eliminated public health concerns about a girl spreading a lingering infection or missing months or years of school. Once girls' infections no longer provoked public health concerns, their bodies no longer represented a menace to their fathers, to their communities, or to a cold-war American society in which the public and private idealization of the male-headed nuclear family had turned the creation and maintenance of the "white home" into a fetish. In an era when nonconformists were demonized as unnatural enemies of the state, there was little room to criticize, let alone explore, any troubling elements of the behavior of the ordinary white fathers on whom Americans' protection from nuclear extinction seemed to depend.

For their part, doctors were excited about having finally found a cure for gonorrhea and concluded that venereal disease had become an insignificant medical and public health problem. Optimistic about the use of antibiotics to control communicable disease, federal and state governments reduced funding for venereal disease prevention and treatment to nominal levels through the 1950s, and ambitious doctors avoided a field that seemed to have become a historical artifact.[8] By the 1970s, however, such optimism had proved wrong. Venereal disease infection began to rise after 1958, and by the late 1970s physicians declared gonorrhea epidemic among sexually active teenagers and adults.[9] And then doctors noticed a high prevalence of gonorrhea infection among girls.[10]

Once again, the infection's appearance in girls with no history of sexual activity mystified doctors, who became newly interested in the etiology of the disease. This time, however, one of the first studies revealed a high correlation between incest and infection. In 1965, investigators from the Los Angeles County Health Department began to study infected girls being treated at the county hospital. Public health physician Geraldine Branch and public health nurse Ruth Paxton were not content to attribute girls' infections to "source unknown." They were determined to identify the source of each girl's infection, and what they learned differed radically from their colleagues' prewar studies: every girl between the ages of 2 and 9 had been sexually assaulted.[11] Not only had Paxton and Branch proved that neither toilet seats nor other schoolgirls nor mothers were to blame. They proved that every girl had been victimized by her father or another male relative, most in their own homes "when their mothers were away."[12] Branch and Paxton uncovered a history of father-daughter incest for every child between the ages of 1 and 9, with one exception.[13]

Branch and Paxton were not alone in their willingness to investigate whether paternal misconduct lay at the root of some children's medical problems. In the 1960s pediatrician C. Henry Kempe conceptualized and popularized the idea of "child abuse" as a medical issue.[14] By the 1970s doctors were beginning to assert their diagnostic authority over the detection of child abuse, and in 1974 Congress passed the Child Abuse Prevention and Treatment Act, which required doctors to report all cases of suspected abuse to the police.[15] But while doctors focused initially on physical abuse, they lacked the training and experience to detect sexual abuse.[16] And the allure of Dr. Burke Ryan's 1851 speculation about nonsexual contacts remained so strong that in 1975 the U.S. Public Health Service issued a strongly worded statement to physicians: "*With gonococcal infection in children, the possibility of child abuse must be considered!*"[17] Two years later, when Dr. Kempe delivered an influential speech to the American Medical Association, he identified child sexual abuse as "another hidden pediatric problem."[18]

Doctors who specialized in treating abused children were already questioning the notion of nonsexual transmission. In 1977 Dr. Suzanne M. Sgroi published a review of the medical literature on the etiology of girls' infections. She found no convincing evidence to support the notion that girls were susceptible to infection from nonsexual contacts, and she argued that the only reason doctors discussed the possibility as viable was because they were reluctant "to entertain the possibility of sexual molestation of a child by an adult," especially "younger children who have few contacts outside the home." In other words, incest.[19] Doctors in the late twentieth century were often no more eager to use the tools made available by advanced medicine to discover incest than their colleagues had been a century earlier.[20] Many avoided ordering tests or asking questions that might lead to the revelation of incest, failing, for instance, to test an uncooperative father. They continued to attribute girls' infections to poor hygiene, toilet seats, or "source unknown" and closed the case.[21]

Still other physicians remained convinced that gonorrhea was not sexually transmitted to girls and continued to try to prove that "fingers, toilet seats, bedsheets, towels and bathwater is somehow magically responsible for gonorrhea infections of the urethra, vagina, rectum and *pharynx* in children."[22] In 1979 the *New England Journal of Medicine* reported on two new studies that tested public toilet seats for the presence of *Neisseria gonorrhoeae*.[23] Other studies documented the "dirty" homes of infected girls and repeated old speculations about bed linens and towels. Still others reiterated the connection between class and disease, emphasizing the poor economic status and the "broken," "disorganized," and dirty homes of the patients they treated, many of whom were impoverished people of color.[24] In cases in which the father admitted the incest, doctors still held mothers responsible for their daughters' infections.[25] Even the "superstitious cure" made a comeback. Discussing the results of a study of infected poor and African American girls, researchers at the University of Maryland School of Medicine in 1980 reminded doctors that a "propensity for infection in very young prepubertal girls has been observed since the turn of the century, and is said to stem from a superstition." They cited Dr. Flora Pollack's 1909 article.[26]

Because physicians are trained to identify the least pathological cause of disease, a physician who believes that father-daughter incest rarely occurs might not consider that possibility unless something in the patient history raises it directly.[27] Similarly, a physician who considers socioeconomic background and class status "predisposing factors" to infection might not suspect incest if the family of an infected girl appears to be stable, healthy, and "ordinary."[28] As Dr. Kempe found, "It is common for children, who are regularly cared for by their pediatrician, to be involved in incest for many years without their physician knowing."[29] Some of doctors' inability to identify child sexual assault may be due to inadequate training or inexperience.[30] However, as late as 1989 one doctor looked to the possibility of fomite or "bathroom spread" in the case of two infected children under age 5 because their "home situations are not suggestive of sexual abuse."[31]

By 1991 doctors' recalcitrance had become so acute that the American Academy of Pediatrics issued a statement reminding physicians that "the diagnosis of sexual abuse and the protection of the child from further harm will depend on the pediatrician's willingness to consider abuse as a possibility."[32] And in 1997 pediatric residencies added a mandatory component on child abuse.[33] Yet improving doctors' ability to detect sexual abuse may not increase their willingness to do so. Because every state requires doctors to report suspected abuse, many medical professionals, citing their "aversion" to the legal system, admit that they avoid investigating a suspicion that, if confirmed, will put them into the same place that physicians have long tried to avoid, becoming a key witness in the legal proceedings that are certain to follow.[34]

The Consequences of the Denial of Incest

Twentieth-century professionals gathered little direct empirical evidence of father-daughter incest among the white middle and upper classes, and much of it is conflicting. In sociologist S. Kirson Weinberg's influential 1955 study, *Incest Behavior*, he calculated that for the years 1910, 1920, and 1930 only one out of every million men committed incest and concluded that incest is "very rare."[35] But other psychiatric and sociological studies published from the 1930s through the 1960s consistently found that 5 to 12 percent of the women surveyed claimed to have been sexually abused by their fathers.[36] Still, however many data were collected to demonstrate the incidence of father-daughter incest, particularly among the white middle and upper classes, researchers viewed its consequences as, if not a positive sign of a girl's psychological development, at least not a negative one.

In 1942, for instance, a psychiatric study in the *American Journal of Orthopsychiatry*, an interdisciplinary mental health journal, argued that because they unconsciously desire it, girls might not suffer any adverse consequences from an incestuous encounter with their fathers.[37] Other prominent experts named "unfavorable heterosexual adjustment" and lesbianism as the chief ill effects of incest.[38] When Alfred Kinsey published *Sexual Behavior in the Human Female* in 1953, he claimed that 6.5 percent of the women interviewed said they had been sexually abused by a father or stepfather.[39] But Kinsey diminished the implication of his finding when he argued that the harm to girls and women, if any, came not from the experience itself but from society's negative reaction to the disclosure. When Dr. Judith Lewis Herman examined Kinsey's data, however, she found that his respondents had used the words "frightening" and "disturbing" to describe the encounter with their fathers, not the aftermath of disclosure.[40]

Sensational publicity about "sex crimes" in the 1930s through the 1950s, including the rape, assault, and murder of girls, might have revealed father-daughter incest throughout American society but achieved instead the opposite effect. Media accounts of child rape and murder put pressure on state and local governments for more aggressive police protection and punishment of "perverts." It was then that the psychiatric profession began to take control over definitions of incest. Politicians looked to psychiatrists for advice, and with doctors and lawyers they jointly constructed the "sexual offender," a "poorly adjusted" or homosexual man, or the odd stranger who preyed on children.

The focus on "stranger danger" displaced the location of the sexual threat to girls by reiterating the protection to be found within the nuclear family. Like the turn-of-the-century doctors who had tried to understand child sexual assault by embracing Richard von Krafft-Ebing's classification of men who molest children as "not ordinary," twentieth-century psychiatrists turned now to psychoanalytic theory. And just as quickly as doctors in the late nineteenth and early twentieth centuries had accepted Krafft-Ebing's conceit to explain, however imperfectly, incest, politicians and the media accepted a psychiatric explanation that could reassure an anxious public that sexual dangers to their daughters were easy to identify and therefore easy to avoid.[41] As historian Jennifer Terry concluded in her study of twentieth-century "sex crime," laws enacted in the late 1930s to protect children from sexual assault "did practically nothing to prosecute coercive incest."[42] And in any case, most investigations into sex crimes looked at the populations that used social welfare services or were confined to mental hospitals or prisons, a method that reinforced associations between poverty, ethnicity, and social depravity.[43]

Psychoanalytic authority over the meaning of father-daughter incest intensified with the outbreak of World War II, at the same time that American psychoanalysts reached the height of their influence in the professional and popular cultures and in the media. As historian Rachel Devlin found in her study of the postwar psychoanalytic literature on father-daughter relationships, between 1940 and the early 1960s, psychoanalysts viewed an adolescent girl's erotic attachment to her father as a sign of healthy development. When, in some cases, actual incest occurred, the consensus among psychoanalysts was that it reflected some girls' emotional and developmental need to transform their incestuous fantasies "into a conscious, deliberate form of expression."[44] American psychoanalysts were not advocating or condoning coercive sexual relationships, and they viewed mother-son incest as destructive.[45] But their theoretical model shifted the onus for the occurrence of father-daughter incest from fathers to daughters. And while they explicated the psychopathology of incestuous mothers, they showed little interest in incestuous fathers. Between 1938 and 1962, not a single article examining the psychopathology of incestuous fathers appeared in the psychiatric literature.[46] As Devlin concluded, postwar psychoanalytic views reaffirmed paternal authority by making a girl's developmental health contingent on a close relationship with her father.[47] Not every father would have fit the idealized model on which this theory depended. But when eminent Kinsey Institute researcher John Gagnon published a study of sexually abused girls in 1965, he reinforced the idea that offending fathers were readily identifiable outsiders. He concluded that most incest offenders were "backward" people from urban slums or isolated rural communities, the "Tobacco Road type milieu wherein incest was regarded as unfortunate but not unexpected."[48]

Girls and Women Speak: Second-Wave Feminism

It was not until the late 1970s that feminists redefined incest from the point of view of the abused girls, as middle- and upper-class women began to publicly reveal not only their histories of childhood incest but also their views of its meaning in their lives.[49] The second-wave feminist movement laid the groundwork for a sea change in social views about gender, including the revelation of father-daughter incest. *People* magazine had not discovered incest, but it had tapped into and expanded a newly popular market for material about father-daughter incest that acknowledged both its occurrence among the white middle and upper classes and its often disturbing impact on a woman's life.[50] Feminists who had conceptualized and built domestic violence shelters and rape hotlines or who had lobbied for new laws defining sexual assault brought to public consciousness overwhelming evidence of the ways in which gender affects society.[51] As women who identified as feminist entered professional and academic fields like psychiatry, sociology, mental health, and history, they offered new analyses of the social conditions that generate incest and the ways in which the experience affects women's lives.[52]

When the Boston Women's Health Collective published *Our Bodies, Ourselves*, the cornerstone of the women's self-help movement, in 1970, it did not mention incest or child sexual abuse.[53] But as mental health professionals began to question and even reject the view that statements of incest are narratives about a woman's inner life, not her lived experience, women began to find the support and direction they needed—in self-help books, in psychotherapy, and in one another's private and public disclosures—for articulating a personal history so inexplicable that many could not even believe themselves that it had happened. Foremost among the new feminist professionals was Harvard psychiatrist Judith Lewis Herman, who began her clinical practice in 1975 and immediately encountered female patients claiming that their fathers had sexually assaulted them. Because Herman had been trained to view actual incest as extremely rare behavior, she sought the advice of experienced clinicians on how to treat these patients. "In every case the veracity of the patient's history was officially questioned. We were reminded by our supervisors, as if this were something everyone knew, that women often fantasize or lie about childhood sexual encounters with adults, especially their fathers."[54]

Herman undertook her own study of father-daughter incest, which she published in 1977 in the feminist academic journal *Signs*, a choice that reflected her argument that a feminist analysis is key to recognizing its occurrence.[55] In 1981 she coauthored her groundbreaking book *Father-Daughter Incest*, in which she made the argument that "to be sexually exploited by a known and trusted [male] adult is a central and formative experience in the lives of countless women."[56] This statement could be true only if incest was not contained within socially marginalized populations but was an integral, though unidentified, part of everyday gender relations. Herman resoundingly declared that to be the case: "Female children are regularly subjected to sexual assaults by adult males who are part of their intimate social world. The aggressors are . . . neighbors, family friends, uncles, cousins, stepfathers, and fathers . . . Any serious investigation of the emotional and sexual lives of women leads eventually to the discovery of the incest secret."[57] Herman's unequivocal critique struck so broadly across society that it might have been easily dismissed. But other mental health professionals were reaching similar conclusions.[58]

A flood of publications in the 1980s, of which psychoanalyst Jeffrey Moussaief Masson's might be the best known, challenged the idea that women's incest narratives were statements of fantasy. Masson, whom Anna Freud had appointed to direct her father's archives, used his access to materials in the archives to pummel Sigmund Freud's theory about incest. In 1984 he published *The Assault on Truth: Freud's Suppression of the Seduction Theory*, a ferocious attack on Sigmund Freud's intellectual honesty and on the foundations of psychoanalytic theory.[59] In 1896 Freud had published the *Aetiology of Hysteria*, in which he claimed that childhood sexual trauma suffered at the hands of nurses, servants, and other children was at the root of "hysteria" in adults. However, Freud later repudiated the *Aetiology*, which his colleagues had scorned, and theorized instead that his patients' statements about childhood traumas had been representations of desire, not fact. But when Masson looked at Freud's records, he claimed that Freud's patients had not been traumatized by female servants or other children but by their fathers, a fact he charged Freud with concealing.[60] Whether or not Masson was correct, his arguments injected a new uncertainty into the meaning of women's incest narratives and legitimated further inquiry into psychoanalytic orthodoxy.[61] As mental health practitioners reconsidered Freud's theory, they also began to listen differently to their patients' accounts of incest.[62] Popular culture reflected women's new interest in the topic. An increasing number of books, articles in popular newspapers and magazines, television programs, and films told about incest from the girl's or woman's point of view.[63]

Novels and films with father-daughter incest plots had always been a part of twentieth-century American culture, since at least 1928, when the silent film *Beggars of Life*, starring Louise Brooks, dramatized the plight of a young woman who is able to stop her father's predatory advances only when she finally shoots him. In 1934 F. Scott Fitzgerald published *Tender Is the Night*, and in 1940 Henry Bellamann published *King's Row*, which was followed two years later by a movie version. Two iconic novels about incest became best sellers in the 1950s, Vladimir Nabakov's *Lolita* (1955) and Grace Metalious's *Peyton Place* (1956). *Peyton Place* was so successful that Hollywood produced a film version the following year. In 1962 alone Hollywood studios released four big-budget productions involving father-daughter incest: *Long Day's Journey into Night*, John Houston's *Freud*, and film versions of *Tender Is the Night* and *Lolita*. Two years later ABC introduced a serialized version of *Peyton Place* for television, which ran for five years. The 1970s opened with Toni Morrison's first novel, *The Bluest Eye*, told from the point of view of a victimized daughter, and in 1974 John Houston directed the blockbuster *Chinatown*, with its explosive father-daughter incest plot running underneath the entire landscape of a modernizing America.

All these films and novels had characterized incest as unpleasant, if not horrific, coercive, and leading to negative and tragic personal and social consequences. But consistent with professional and popular discourses, none had portrayed the incestuous father as an ordinary white middle-class American family man. A turning point occurred after Alice Walker's novel *The Color Purple* became a best seller in 1983.[64] Walker's novel imagined father-daughter incest in a poor, rural, African American family. But a year later, *Something about Amelia*, a made-for-television movie, depicted incest in an otherwise idealized white, urban, middle-class family.[65] The network's head censor had initially balked, ruling the film "unfit for television," but after it aired, media critics praised it lavishly for breaking a "television taboo." They were particularly impressed that the film had avoided "dramatic stereotypes" by casting Ted Danson as a "charming . . . middle-class parent" rather than a "drunkenly brutish monster"—a move that *Newsweek* said gave the film its "most horrific punch." The film was the second-highest-rated program for the 1983–84 network television season and won three Emmy Awards, including best dramatic special.[66] In December 1985 Warner Brothers released a glossy film version of *The Color Purple*, which Steven Spielberg produced and directed and which starred Whoopi Goldberg and Oprah Winfrey, the most popular black women in America, in the lead roles.[67] Then father-daughter incest crossed to the other side of the tracks entirely. In quick succession films about incest among the white middle class, with stars like Barbra Streisand and Richard Dreyfuss (*Nuts*, 1987), Richard Gere and Kim Basinger (*Final Analysis*, 1992), and Kathy Bates and Jennifer Jason Leigh (*Dolores Claiborne*, 1994), hit the multiplex.

But how had a mainstream audience for such "taboo" material been created? In the 1980s feminists worked in tandem with a grassroots "recovery movement" to create the theory and social context in which women could discuss incest and its impact on their lives. Though less cohesive than organized feminism, the "recovery movement" was a culturally powerful phenomenon that provided a framework in which women's public and private incest allegations were shared and supported. The term *recovery* was borrowed from the principles of Alcoholics Anonymous, and adherents identified the roots of many adult problems in unresolved childhood issues created by parental abuse or neglect, particularly from growing up in a "dysfunctional" family, a word that was not necessarily synonymous with "broken" or "disorganized" and which applied across race and class.[68]

With its enthusiasm for self-examination and "clean living," recovery incited either intense devotion or irritation among Americans. But its widespread appeal was undeniable. By 1991 as many Americans had attended a twelve-step meeting as had visited a therapist, and entrepreneurs commodified recovery into a "new age" movement.[69] Writers, artists, psychotherapists, and motivational speakers found an eager mass audience. Sales of books in the recovery genre (including diet, sex, love, self-help, and marriage) peaked between 1986 and 1990, after which books about sexual abuse and incest dominated.[70] After Marilyn Van Derbur Atler made her public disclosure of incest in *People*, other celebrities like Roseanne Barr also described their experiences, with both incest and recovery.[71] As one bookseller put it, "Between a former Miss America and Roseanne Barr, the topic has really come out of the closet."[72]

One self-help book stood out. In 1992 women's bookstore owners picked *The Courage to Heal: A Guide for Women Survivors of Child Sexual Abuse* as the most influential women's book in twenty years.[73] Authors Ellen Bass, an incest survivor, and psychotherapist Laura Davis called it a self-help book for women who had experienced incest.[74] Though controversial, the book was popular. Only five years after its publication *Publishers Weekly* listed it among "Classics to Stock," and in 1995 *USA*

Today placed *The Courage to Heal* first on its list of the "25 best self-help books."[75] But the market for new material on incest went far beyond the recovery or self-help genres. Novels in which incest figured importantly became critical and popular successes, including Jane Smiley's *A Thousand Acres*, which won both the National Book Award and a Pulitzer Prize in 1991, and Dorothy Allison's *Bastard out of Carolina*, which was nominated for the National Book Award in 1992. Both books were best sellers and were made into films with high-profile performers.[76] By 1995 Katie Roiphe captured the frenzy over the new literature on father-daughter incest when she complained in an article in *Harper's* that novels with incest plots had become so commonplace that they had become boring.[77]

Backlash: The Incest Debates of the 1980s and 1990s

As incest became a matter of popular interest and everyday public discussion, it raised disturbing questions about male sexual and paternal behavior. Skeptics raised issues about the reliability of childhood perception, the motivations of women who made the accusations, and the influence of psychotherapy on memory. Most pressing was the issue of whether there was any way to determine whether father or daughter was telling the truth. In the absence of empirical data about incest, and amid a political environment newly interested in promoting "family values" and a backlash against feminism, no answers were forthcoming. From the perspective of girls and women assaulted by their fathers, the acknowledgment and public discussion of father-daughter incest as a gender and political issue was one of the most important accomplishments of feminism in the 1980s. Yet feminists rarely recognized it as such, even as the opposition pointed to the phenomenon as an example of feminist ideology gone awry.[78]

Many critics depicted women's willingness to "break the silence" as a modern-day hysteria incited by feminists who easily manipulated desperate and irrational women.[79] Columbia University psychiatrist Dr. Richard Gardner, who often provided expert testimony on behalf of accused fathers, argued that "feminist groups . . . have their share of fanatics. [They] operate as if they have a lifelong vendetta against men and will never be satisfied until all of them are destroyed. These zealots have found the sex abuse scene to be a perfect opportunity for the expression of their venom."[80] Some feminists were unhappy, too. On the front page of the *New York Times Book Review* in January 1993, psychologist Carol Tavris mocked the outpouring of incest claims as a pathetic "cult of victimhood" and a by-product of the "excesses of feminism."[81] She lambasted the "incest survivor machine," comparing a woman who makes an incest allegation to a woman who alleges that the FBI was "bugging her socks."[82]

The popularity and increasing commercialization of the recovery movement compounded doubts about the motivations and credibility of "incest survivors."[83] Critics charged that the movement defined childhood abuse so broadly as to have made a once serious category meaningless and that it promoted a dishonest attempt to deflect individual responsibility for poor decision making onto one's parents.[84] In a two-part cover article on the recent literature on father-daughter incest in the *New York Review of Books*, Frederick Crews, a scholar of psychoanalytic theory, sneeringly labeled women's incest accusations the "revenge of the repressed" and tied them to both feminism and recovery.[85] He charged the authors of *The Courage to Heal* with exploiting "a public . . . obsession with the themes . . . of . . . the 'dysfunctional family,' and the 'inner child,' " explaining that "while Andrea Dworkin and Susan Brownmiller were hypothesizing that American fathers regularly rape their daughters in order to teach them what it means to be inferior, Bass and Davis set about to succor the tens of millions of victims who must have repressed that ordeal."[86] Using the title "Courage to Hate" to refer to the book, psychologist Elizabeth Loftus and coauthor Katherine Ketcham identified it as the source of innumerable false accusations of incest.[87]

Both Crews and Loftus were on the advisory board of the False Memory Syndrome Foundation (FMSF), which organized a media campaign to discredit accusers that was so effective that by the mid-1990s even the women's magazines had begun to retreat from their initial empathy for what they had only a few years earlier announced was a newly discovered social problem.[88] The FMSF claimed that the urgent social problem was not the apparent widespread occurrence of father-daughter incest but an "epidemic" of "false memories" of incest. Although the term soon became ubiquitous, the medical and psychiatric professions do not recognize "false memory syndrome" as a diagnostic category.[89] The FMSF acknowledges that it invented the term, which the elite academics who joined its board legitimated.[90] Pamela Freyd and her husband, Peter Freyd, a mathematics professor, founded the FMSF in 1992, less than a year after Marilyn Van Derbur Atler appeared on the cover of *People* and within months of a private accusation by one of Peter's daughters that he had touched her inappropriately when she was a child.[91] Jennifer Freyd was an accomplished scientist, wife, and mother. But when it came to weighing her credibility against Peter's, the elder Freyds and their supporters set aside as unimportant the fact that Peter was an admitted alcoholic whose treatment had included multiple hospitalizations and psychiatric care.[92]

Defending the social cachet of middle-class white men accused of incest by attacking the credibility of their daughters was a central organizing concept for the group. The FMSF launched a vicious media campaign that styled accused fathers—or, in their careful wording, accused *families*—as the victims of inexplicably vindictive women, whom Pamela Freyd characterized as "paranoid, delusional, and hysterical people who appear to thrive on hatred, vengeance, and persecution of their parents." In an FMSF newsletter, Freyd wrote that "nothing that the parents ever did can compare to the psychological and emotional torture that these adult children are now delivering to [them] . . . It is as if [they] have literally excoriated their parents and then poured acid all over them and left them to suffer."[93] Those who defended the accused men invoked their social status as reason enough to ridicule the accusations. They argued that it was implausible to believe that men of their sort were capable of imagining, let alone actually assaulting, their daughters.[94] As one board member boasted in 1995, most parents who contacted the organization were college graduates from "functional, intact, successful families."[95] Whether true or not, these demographics reassured the FMSF membership that it was easily distinguished from the class of men capable of incest.

The rhetoric and tactics of the FMSF did little to resolve the issues and a great deal to foment confusion and hostility. The group emphasized two points in its critique: first, that contemporary women's incest accusations are based on feminist theory, which makes them politically motivated and inherently unreliable, and second, that facts discovered or pieced together in psychotherapy are not "objective" but are easily manipulated and lack "scientific" verifiability. Dr. Judith Lewis Herman described childhood incest as a type of traumatic—an extraordinarily painful or frightening—event, like wartime combat, with which some people cope by repressing the memory from consciousness, putting it out of their mind.[96] Herman and Mary Harvey, however, found that the vast majority of women who reported incest never claimed to have "forgotten" or repressed the event from consciousness and that some obtained confirmation of their memories from other sources.[97] But critics ignored such studies and attempted to discredit all incest accusations by distorting the well-documented symptoms of trauma and the psychotherapeutic process.[98]

Just as doctors had repudiated the mothers of infected girls for fabricating stories about sexual assaults, contemporary critics attacked the credibility of women's accusations by arguing that psychotherapists (often women) "implant" false memories of incest in their patients' head. The FMSF argued that, motivated by greed, incompetence, feminism, or lesbianism, therapists who implanted false memories had stirred up a craze of false claims.[99] Richard Ofshe, a Berkeley social psychologist and FMSF board member who testified frequently on behalf of accused fathers, and his coauthor Ethan Watters wrote in 1993 that "many therapists routinely implant false memories of past-life traumas in vulnerable patients and other therapists implant memories of space alien kidnappings and sexual brutalization by ET's evil brothers."[100] Even though it would have been impossible for Ofshe to have had access to enough actual patient-therapist conversations to have supported such a broad conclusion about "routine" implantation, he was a formidable opponent. He shared a Pulitzer Prize in 1979 for his role in exposing unseemly facets of Scientology in California, and in 1994 he co-wrote *Making Monsters: False Memory, Psychotherapy, and Sexual Hysteria*, which the progressive magazine *Mother Jones* endorsed and excerpted.[101]

Critics also attributed a different significance to the fact that so many of the women had undergone psychotherapy: supporters of recovery and other observers saw it as proof of the severity of the consequences of incest and the women's desire to put their troubles behind them, but critics turned this fact into a fundamental reason that the accusations should be viewed as tainted, lacking credibility, and dismissed. The FMSF did so by conflating sensationalized and highly publicized prosecutions of day care owners and operators for sexually abusing the children under their care with women's claims against their own fathers. Beginning in 1983, the media reported scores of stories about ritualistic and satanic child abuse by owners and employees of nursery schools and day care centers in suburban America. In the massive prosecutions that often dragged on for years, children testified about events that seemed as implausible as they did awful. Some observers criticized psychologists for fomenting the accusations, claiming they had used overly suggestive techniques to interview the children.[102] Even though the legal evidence convinced jurors to convict defendants in some of the cases, a media backlash against those people who initially believed the accusations—including a Pulitzer Prize awarded in 1991 to a reporter for the *Los Angeles Times* who severely criticized the paper's coverage of the McMartin case—left the impression that all the allegations had been frivolous, and it curtailed any further investigation into the substance of the issue: the frequency with which children are sexually assaulted.[103]

The FMSF exploited the media backlash against the day care cases by suggesting that women's accusations of father-daughter incest were just as unreliable.[104] It encouraged parents to sue their daughters and their daughters' therapists and provided written materials and the names of lawyers and expert witnesses (from the FMSF board) to help them to do so.[105] They trumpeted lawsuits (ultimately unsuccessful) against the authors and publisher of *The Courage to Heal* and lobbied successfully for legislation to restrict insurance benefits for psychotherapy.[106] And while on the one hand the FMSF disseminated its own fabricated version of psychological theory as "false memory syndrome," on the other it derided the incest allegations as lacking any "scientific" basis. The FMSF insisted that only those allegations of child sexual abuse and incest supported by "scientific proof" should be deemed credible. Its calls for scientific proof seem a reasonable approach to resolving a complex issue. But for the years 1992 through 1996, when I studied the FMSF, it did not concede that a single allegation of incest or child sexual abuse that it had reviewed, whether made by an adult or child, could be substantiated as credible. It argued instead that a woman's inability to verify or recall every detail of a past event (which no memory researcher contended was possible) proved that it had not occurred.[107]

When a multitude of women's voices "broke the conspiracy of silence" about father-daughter incest, they could not shatter ideologies about paternal protection that were inconsistent with women's lived reality. Their claims were met with disbelief, which some white professionals stoked by deploying their supposed allegiance to "science" and "reason" to denigrate the women as "hysterical" and to discredit their statements as ideological overreaching. Critics set up "scientific evidence" as objective and inherently in conflict with women's experiential knowledge or with any theory informed by a gender analysis. As a result, the FMSF decries the idea of incest and child sexual abuse but deems no standard of "scientific proof" credible enough to confirm its reality. Despite a profound cultural shift in social attitudes about talking about sex in general and incest in particular, the first-person acknowledgment by American women of the widespread occurrence of father-daughter incest in the late twentieth century could not overcome Americans' reluctance to see white middle- and upper-class men as so many of their daughters did—as the type of men not only capable of but intent on behaving in ways that were horrifying, revolting, and pitiless.

Science, Gender, and Power: Subjugated Knowledge

As I researched this book, I was struck by the occurrence of father-daughter incest in all parts of American society throughout the nineteenth and twentieth centuries. And I was disturbed by the urgency with which a vast array of professionals rushed to avoid acknowledging its occurrence. All they agreed on was that father-daughter incest was heinous, beastly, and distinctly un-American conduct, and even that consensus was broken when mid-twentieth-century professionals suggested that the effects of father-daughter incest on girls and women were minor. In each era professionals ignored, mislabeled, or disparaged the imminent revelation of father-daughter incest by melding the period's concerns—immigration, sanitation, gender wars—with a science-based response that was often anything but. In the nineteenth century medical jurists and physicians insisted that girls and women lied and that gonorrhea infection, the best evidence corroborating a claim of sexual assault, was unreliable. And they boasted of their unique position in protecting innocent men. In the late nineteenth century and early twentieth, social reformers joined doctors in a determined effort to mislabel that same evidence of incest as the detritus of ignorance or poverty. And they repeated and expanded the notion that girls lied and that they alone were uniquely situated to protect innocent men and, by extension, the nation. In the interwar years professionals found a new tool in psychological theory to support the notion that girls lied. And they designated infected and abused girls menaces to society and locked them up instead of the adult males who left lingering medical evidence of their crimes on their daughters' bodies.

The cultural tensions that gave life to the idea that girls could acquire gonorrhea from toilet seats were part of the context in which Americans read—and embraced—Freud's ideas about sexuality in general and incest in particular. When late twentieth-century critics challenged the validity of psychoanalytic theory, they blamed Freud for creating a culture in which the denial of incest is pervasive. But their anger is misplaced. The culture of denial was in place decades before psychoanalytic theory saturated American culture. Acceptance of Freudian theory was not the first but only the latest in a long series of attempts by white middle- and upper-class Americans to discredit girls and women or to ignore information that exposed as fiction certain ideologies about them. The early twentieth-century epidemic of gonorrhea among girls had threatened to expose the widespread occurrence of incest among the white middle and upper classes. But neither public charges of incest against genteel men in the nineteenth century nor an epidemic of gonorrhea among girls in the early twentieth nor adult women's incest narratives from the midcentury on diminished white middle- and upper-class Americans' commitment, in public policies and institutions, to the idealized white family. It proved a resilient justification both for the social and familial power and privileges accorded to certain Americans based on their race, gender, and class and for the subordination of women to men and the domination of white upper- and middle-class men over all others.

When it became impossible to defend so many men who had apparently abused their paternal authority, professionals simply denied that the behavior had occurred. When professionals discovered an epidemic of white American girls infected with a loathsome disease, they hurried to rebut the repellent implication by revising medical views about the etiology of gonorrhea. They used their professional authority to refute what women and girls were saying about their lived experience. Doing so created a cacophony that drowned out the voices, and knowledge, of the victimized girls and their anxious mothers. Speculation about nonsexual modes of transmission failed to meet any standard of medical science but successfully deflected the revelation of incest. The social hygiene movement supported these efforts by disseminating its views about the source of girls' infections, and Freud's new theories about sexuality and incest further drew attention away from the reality of incest and its consequences for girls and women. Father-daughter incest never disappeared entirely from view; only its occurrence among the white middle and upper classes did. With it went the opportunity to evaluate and critique male sexual behavior and its relation to social power.

We've moved far from the rationale of manly honor and paternal responsibility that undergirds the Founders' decision to retain the male-headed family unit as the foundation of American society. However, shoring up the social power of certain men has, at every turn, been more important than protecting the physical and emotional integrity of girls, who have paid dearly to keep the fabric of American society, its ideologies and social hierarchies, intact. It is a difficult story to believe.

ABBREVIATIONS

Ackerman Collection	Rhea C. Ackerman Collection, Special Collections, Young Research Library, University of California, Los Angeles, California
ASHA Collection	American Social Health Association (formerly American Social Hygiene Association) Collection, Social Welfare History Archives, Walter Library, University of Minnesota, Minneapolis, Minnesota
Bliss Papers	Susan Dwight Bliss Papers, Archives and Special Collections, Augustus C. Long Health Science Library, Columbia University, New York, New York
Booth Records	Records of the Catherine Booth Home and Hospital, Cincinnati, Ohio, Salvation Army Archives and Research Center, Alexandria, Virginia
CDC	Centers for Disease Control and Prevention
COS	Charity Organization Society of New York City
COS Papers	Community Service Society (formerly Charity Organization Society of New York City) Papers, Rare Book and Manuscript Library, Butler Library, Columbia University, New York, New York
Dept. of Health Collection	New York City, Department of Health Collection, New York City Municipal Library, New York, New York
Dummer Papers	Ethel Sturges Dummer Papers, Arthur and Elizabeth Schlesinger Library on the History of Women in America, Radcliffe Institute for Advanced Study, Harvard University, Cambridge, Massachusetts

JSH	*Journal of Social Hygiene*
LAT	*Los Angeles Times*
Mackay-Scott Papers	Ruth Jarvis Mackay-Scott Papers, Arthur and Elizabeth Schlesinger Library on the History of Women in America, Radcliffe Institute for Advanced Study, Harvard University, Cambridge, Massachusetts
MGH	Massachusetts General Hospital, Boston, Massachusetts
MMWR	*MMWR, Morbidity and Mortality Weekly Report* (published by the Centers for Disease Control and Prevention)
MSSH Records	Massachusetts Society for Social Health, Records, 1915–65, MC-203, Arthur and Elizabeth Schlesinger Library on the History of Women in America, Radcliffe Institute for Advanced Study, Harvard University, Cambridge, Massachusetts
NEJM	*New England Journal of Medicine*
NPG	*National Police Gazette* (New York)
NYAM	New York Academy of Medicine
NYT	*New York Times*
OBP	*Old Bailey Proceedings Online*, www.oldbaileyonline.org (accessed July 21, 2007)
Oviatt	Los Angeles Department of Health, Communicable Disease Bulletin Collection, Special Collections and Archives, Oviatt Library, California State University, Northridge, California
Sacramento	California State Archives, Office of the Secretary of State, Sacramento, California
VDI	*Venereal Disease Information*
Wehrbein Draft	Kathleen Wehrbein, Draft of Vaginitis Survey, October 1926, in Susan Dwight Bliss Papers (Committee on Vaginitis), Archives and Special Collections, Augustus C. Long Health Science Library, Columbia University, New York, New York

NOTES

Introduction

1. Marilyn Van Derbur Atler, "The Darkest Secret: Behind the Facade of a 'Perfect' Family, the Unspeakable Crime: Incest," *People Weekly*, June 10, 1991, 88. Atler made the revelation public a month earlier in a talk at the Kempe National Center for the Prevention and Treatment of Child Abuse and Neglect. AP, "A Miss America Says She Was Incest Victim," *NYT*, May 12, 1991. See also Marilyn Van Derbur Atler, *Miss America by Day: Lessons Learned from Ultimate Betrayals and Unconditional Love* (Denver: Oak Hill Press, 2004), and Lenore Terr, *Unchained Memories: True Stories of Traumatic Memories, Lost and Found* (New York: Basic Books, 1994), chap. 5.

2. *60 Minutes*, WCBS-TV, New York, Apr. 17, 1994, Transcript, Radio TV Reports. The segment showed clips from daytime television's *Sally Jesse Raphael*, *Maury Povich Show*, *Oprah Winfrey Show*, *Montel Williams Show*, *Joan Rivers Show*, and *Geraldo*. See Rosaria Champagne, "Oprah Winfrey's *Scared Silent* and the Spectatorship of Incest," *Discourse* 17 (Winter 1994–95): 123. For women's accounts, see, e.g., Linda Katherine Cutting, "Give and Take: A Personal Story of Incest," *NYT Magazine*, Oct. 31, 1993, 52; Barbara Graham and Christiana Ferrari, "Unlock the Secrets of Your Past: Early Childhood Memories," *Redbook*, Jan. 1993, 84; Frances Lear, *The Second Seduction* (New York: Knopf, 1992); Cheryl Tevis, "Silent No More: A Story of Incest in Rural America," *Successful Farming*, Nov. 1992, 20; "Breaking the Silence—Survivors of Incest Speak Out," *CNN Specials*, CNN-TV, Transcript #48-1, Aug. 31, 1992; Linda Saslow, "Growing Numbers Recount Histories of Incest," *NYT*, Aug. 16, 1992; Elizabeth Gleick, "Frances Lear," *People Weekly*, July 13, 1992, 91; Vicki Bane, "A Star Cries Incest," *People Weekly*, Oct. 7, 1991, 84; and Heidi J. LaFleche, "Roseanne's Story: An Expert's View," *People Weekly*, Oct. 7, 1991, 86. *Lear's* magazine devoted its February 1992 issue to incest.

3. Paul McHugh, "The False Memory Craze," *FMSF Newsletter*, Feb. 1, 1995, www.fmsfonline.org/fmsf95.201.html (accessed July 13, 2007).

4. Similar debates arose in Great Britain, Canada, and Australia. See Carol Smart, "Reconsidering the Recent History of Child Sexual Abuse, 1910–1960,"

Journal of Social Policy 29 (2000): 55, and Chris Jenks, *Childhood* (New York: Routledge, 1996), 84–115. Historical work includes Dorothy E. Chunn, "Secrets and Lies: The Criminalization of Incest and the (Re)formation of the 'Private' in British Columbia, 1890–1940," in *Regulating Lives: Historical Essays on the State, Society, the Individual, and the Law*, ed. John McLaren, Robert Menzies, and Dorothy E. Chunn (Vancouver: University of British Columbia Press, 2002), 120; Joan Sangster, *Regulating Girls and Women: Sexuality, Family, and the Law in Ontario, 1920–1960* (Oxford: Oxford University Press, 2001), chap. 2; Louise A. Jackson, *Child Sexual Abuse in Victorian England* (New York: Routledge, 2000); and Karen Dubinsky, *Improper Advances: Rape and Heterosexual Conflict in Ontario, 1880–1929* (Chicago: University of Chicago Press, 1993).

5. See Dorothy Scarborough, "Where the Eighteenth Century Lives On: A Study of Quaint and Primitive Survivals in Vance Randolph's, *The Ozarks*," review of Vance Randolph, *The Ozarks: An American Survival of Primitive Society*, *NYT*, Dec. 27, 1931, 54.

6. E.g., Martha Hodes, *White Women, Black Men: Illicit Sex in the Nineteenth-Century South* (New Haven: Yale University Press, 1997); Leslie J. Reagan, *When Abortion Was a Crime: Women, Medicine, and Law in the United States, 1867–1973* (Berkeley: University of California Press, 1997); Janet Farrell Brodie, *Contraception and Abortion in Nineteenth-Century America* (Ithaca, NY: Cornell University Press, 1994); George Chauncey, *Gay New York: Gender, Urban Culture, and the Making of the Gay Male World, 1890–1940* (New York: Basic Books, 1994).

7. "Boston, September 21," *Northern Post* (Salem, NY), Oct. 10, 1805, [2]; "Court of Sessions," *Farmer's Cabinet* (Amherst, MA), Mar. 24, 1827, [3]; "Court of Sessions," *Raleigh Register, and North-Carolina Gazette*, Mar. 27, 1827, col. C.

8. "The Bowers Held: Both Father and Daughter," *Tacoma Daily News*, Oct. 29, 1892, 7.

9. See Emily K. Abel, *Hearts of Wisdom: American Women Caring for Kin, 1850–1940* (Cambridge: Harvard University Press, 2000); Nancy Tomes, *The Gospel of Germs: Men, Women, and the Microbe in American Life* (Cambridge: Harvard University Press, 1998); Robin Muncy, *Creating a Female Dominion in American Reform, 1890–1935* (New York: Oxford University Press, 1991); and Ruth Schwartz Cowan, *More Work for Mother: The Ironies of Household Technology from the Open Hearth to the Microwave* (New York: Basic Books, 1983).

10. See Timothy J. Gilfoyle, *City of Eros: New York City, Prostitution, and the Commercialization of Sex, 1790–1920* (New York: Norton, 1994).

11. On "dangerous" objects, Alice Hamilton, "Veneral Diseases in Institutions for Women and Girls," *Proceedings of the National Conference of Charities and Corrections* 37 (1910): 53, 53; on "menaces," L. Emmett Holt Jr. and Rustin

McIntosh, eds., *Holt's Diseases of Infancy and Childhood: A Textbook for the Use of Students and Practitioners*, 11th ed. (New York: D. Appleton-Century Co., 1940), 821–22.

12. Abraham L. Wolbarst, "Gonorrhea in Boys," *JAMA* 33 (Sept. 28, 1901): 827, 830.

13. Liz Kelly, *Surviving Sexual Violence* (Minneapolis: University of Minnesota Press, 1988), 138–57.

14. See, e.g., Diana E. H. Russell, introduction to *The Secret Trauma: Incest in the Lives of Girls and Women*, rev. ed. (New York: Basic Books, 1999), and Frederick C. Crews, *The Memory Wars: Freud's Legacy in Dispute* (New York: New York Review of Books, 1995).

15. Elizabeth Pleck, *Domestic Tyranny: The Making of American Social Policy against Family Violence from Colonial Times to the Present* (New York: Oxford University Press, 1987); Linda Gordon, *Heroes of Their Own Lives: The Politics and History of Family Violence, Boston, 1880–1960* (New York: Penguin Press, 1988); Ann Taves, ed., *Religion and Domestic Violence in Early New England: The Memoirs of Abigail Abbott Bailey* (Bloomington: Indiana University Press, 1989). See also Linda Gordon, "Incest and Resistance: Patterns of Father-Daughter Incest, 1880–1930," *Social Problems* 33 (Apr. 1986): 253; Linda Gordon, "The Politics of Child Sexual Abuse: Notes from American History," *Feminist Review* 28 (Sept. 1988): 56; and Linda Gordon and Paul O'Keefe, "Incest as a Form of Family Violence: Evidence from Historical Case Records," *Journal of Marriage and the Family* 46 (Feb. 1984): 27.

16. Irene Quenzler Brown and Richard D. Brown, *The Hanging of Ephraim Wheeler: A Story of Rape, Incest, and Justice in Early America* (Cambridge: Harvard University Press, 2003).

17. Rachel Devlin, " 'Acting Out the Oedipal Wish': Father-Daughter Incest and the Sexuality of Adolescent Girls in the United States, 1941–1965," *Journal of Social History* 38 (2005): 609; Rachel Devlin, *Relative Intimacy: Fathers, Adolescent Daughters, and Postwar American Culture* (Chapel Hill: University of North Carolina Press, 2005). See also Stephen Robertson, *Crimes against Children: Sexual Violence and Legal Culture in New York City, 1880–1960* (Chapel Hill: University of North Carolina Press, 2005); Karen Sánchez-Eppler, *Dependent States: The Child's Part in Nineteenth-Century America* (Chicago: University of Chicago Press, 2005); Louise K. Barnett, *Ungentlemanly Acts: The Army's Notorious Incest Trial* (New York: Hill and Wang, 2000); Philip Jenkins, *Moral Panic: Changing Concepts of the Child Molester in Modern America* (New Haven: Yale University Press, 1998); Peter W. Bardaglio, *Reconstructing the Household: Families, Sex, and the Law in the Nineteenth-Century South* (Chapel Hill: University of North Carolina Press, 1995); Mary Odem, *Delinquent Daugh-*

ters: Protecting and Policing Adolescent Female Sexuality in the United States, 1885–1920 (Chapel Hill: University of North Carolina Press, 1995); Karen Sánchez-Eppler, "Temperance in the Bed of a Child: Incest and Social Order in Nineteenth-Century America," *American Quarterly* 47 (Mar. 1995): 1; and Lisa Duggan, "From Instincts to Politics: Writing the History of Sexuality in the U.S.," *Journal of Sex Research* 27 (Feb. 1990): 95.

18. *Professional and popular work published in the 1990s:* Janice Haaken, *Pillar of Salt: Gender, Memory, and the Perils of Looking Back* (New Brunswick, NJ: Rutgers University Press, 1998); Leigh B. Bienen, "Defining Incest," *Northwestern University Law Review* 92 (Summer 1998): 1501; Janet Walker, "The Traumatic Paradox: Documentary Films, Historical Fictions, and Cataclysmic Past Events," *Signs* 22 (Summer 1997): 803; Elaine Showalter, *Hystories* (New York: Columbia University Press, 1997); Jennifer J. Freyd, *Betrayal Trauma: The Logic of Forgetting Childhood Abuse* (Cambridge: Harvard University Press, 1996); Rosaria Champagne, *The Politics of Survivorship: Incest, Women's Literature, and Feminist Theory* (New York: New York University Press, 1996); Laura S. Brown, "Politics of Memory, Politics of Incest: Doing Therapy and Politics That Really Matter," *Women & Therapy* 19 (Fall 1996): 5; Cynthia Grant Bowman and Elizabeth Mertz, "A Dangerous Direction: Legal Intervention in Sexual Abuse Survivor Therapy," *Harvard Law Review* 109 (Jan. 1996): 549; Janice Haaken, "The Recovery of Memory, Fantasy, and Desire: Feminist Approaches to Sexual Abuse and Psychic Trauma," *Signs* 21 (1996): 1069; Nancy Potter, "The Severed Head and Existential Dread: The Classroom as Epistemic Community and Student Survivors of Incest," *Hypatia* 10 (Spring 1995): 69; Frederick Crews, *Revenge of the Repressed* (New York: New York Review of Books, 1995); Elizabeth Wilson, "Not in This House: Incest, Denial, and Doubt in the White Middle Class Family," *Yale Journal of Criticism* 8 (1995): 35; Terr, *Unchained Memories*; Louise Armstrong, *Rocking the Cradle: What Happened When Women Said Incest* (Reading, MA: Addison-Wesley, 1994); Elizabeth Loftus and Katherine Ketcham, *Myth of Repressed Memory* (New York: St. Martin's Press, 1994); Frederick Crews, "The Revenge of the Repressed," *New York Review of Books*, pt. 1, Nov. 17, 1994, 54, and pt. 2, Dec. 1, 1994, 49; Mark Pendergrast, *Victims of Memory: Incest Accusations and Shattered Lives* (New York: Upper Access, 1994); Richard Ofshe and Ethan Watters, *Making Monsters: False Memory, Psychotherapy, and Sexual Hysteria* (New York: Scribner, 1994); Champagne, "Oprah Winfrey's *Scared Silent*," 123; Janet Liebman Jacobs, *Victimized Daughters: Incest and the Development of the Female Self* (New York: Routledge, 1994); Vicki Bell, *Interrogating Incest: Feminism, Foucault and the Law* (New York: Routledge, 1993); Carol Tavris, "Beware the Incest Survivor Machine," *NYT Book Review*, Jan. 3, 1993, 1; Melba Wilson, *Crossing the Boundary: Black*

Women Survive Incest (Seattle: Seal Press, 1993); Linda Alcoff and Laura Gray, "Survivor Discourse: Transgression or Recuperation?" *Signs* 18 (Winter 1993): 260; Elizabeth Waites, *Trauma and Survival: Post-Traumatic and Dissociative Disorders in Women* (New York: Norton, 1993); Erna Olafson et al., "Modern History of Child Sexual Abuse Awareness: Cycles of Discovery and Suppression," *Child Abuse & Neglect* 17 (1993): 7; Judith Lewis Herman, *Trauma and Recovery* (New York: Basic Books, 1992); Richard A. Gardner, *Sex Abuse Hysteria* (Cresskill, NJ: Creative Therapeutics, 1991); and Linda H. Hollies, "A Daughter Survives Incest: A Retrospective Analysis," in *Black Women's Health Book*, ed. Evelyn C. White (Seattle: Seal Press, 1990), 82.

In the 1980s: Louise DeSalvo, *Virginia Woolf: The Impact of Childhood Sexual Abuse on Her Life and Work* (Boston: Beacon Press, 1989); Ellen Bass and Laura Davis, *The Courage to Heal: A Guide for Women Survivors of Child Sexual Abuse* (New York: Harper and Row, 1988); Diana E. H. Russell, *The Secret Trauma: Incest in the Lives of Girls and Women* (New York: Basic Books, 1986); Jeffrey Moussaief Masson, *The Assault on Truth: Freud's Suppression of the Seduction Theory* (New York: Farrar, Straus and Giroux, 1984); Elizabeth Pleck, "Feminist Responses to 'Crimes against Women,' 1868–1896," *Signs* 8 (Spring 1983): 451; Judith Lewis Herman and Lisa Hirschman, *Father-Daughter Incest* (Cambridge: Harvard University Press, 1981).

First-person and semiautobiographical accounts in the 1980s and 1990s: Katherine Harrison, *The Kiss: A Memoir* (New York: Random House, 1997); Sapphire, *Push* (New York: Knopf, 1996); Sapphire, *American Dreams* (San Francisco: High Risk Books, 1994); Louise DeSalvo, *Vertigo: A Memoir* (New York: Dutton, 1996); Dorothy Allison, *Bastard Out of Carolina* (New York: Dutton, 1992); Toni McNaron, *I Dwell in Possibility* (New York: Feminist Press, 1992); Margaret Randall, *This Is about Incest* (Ithaca, NY: Firebrand Books, 1987); Louise Armstrong, *Kiss Daddy Goodnight: Ten Years Later* (New York: Pocket Books, 1987); and Ellen Bass and Louise Thornton, eds., *I Never Told Anyone: Writings by Women Survivors of Child Sexual Abuse* (New York: Harper and Row, 1983).

Recent academic works: Bettina F. Aptheker, *Intimate Politics: How I Grew Up Red, Fought for Free Speech, and Became a Feminist Rebel* (Emeryville, CA: Seal Press, 2006); Janet Walker, *Trauma Cinema: Documenting Incest and the Holocaust* (Berkeley: University of California Press, 2005); Ann Cvetkovich, *An Archive of Feeling: Trauma, Sexuality, and Lesbian Public Life* (Durham, NC: Duke University Press, 2003); Hughes Evans, "The Discovery of Child Sexual Abuse in America," in *Formative Years: Children's Health in the United States, 1880–2000*, ed. Alexandra Minna Stern and Howard Markel (Ann Arbor: University of Michigan Press, 2002), 233; Janice Doane and Devon Hughes, *Telling Incest: Narratives of Dangerous Remembering from Stein to Sapphire* (Ann

Arbor: University of Michigan, 2001); and Amber L. Hollibaugh, *My Dangerous Desires: A Queer Girl Dreaming Her Way Home* (Durham, NC: Duke University Press, 2000).

See also the heated responses to Bettina Aptheker's accusations against her father, eminent Marxist historian Herbert Aptheker: Christopher Phelps, "Father of History," *The Nation,* Nov. 5, 2007, www.thenation.com/doc/20071105/phelps (accessed Oct. 22, 2007); Christopher Phelps, "Herbert Aptheker: The Contradictions of History," *Chronicle of Higher Education*, Oct. 6, 2006, 12; and Letters to the Editors, "Herbert Aptheker in Public Life and at Home," *Chronicle of Higher Education*, Nov. 10, 2006, 17.

19. Nancy Kellogg and the Committee on Child Abuse and Neglect, "Oral and Dental Aspects of Child Abuse and Neglect," *Pediatrics* 116 (Dec. 2005): 1656, 1565–66; Adaora A. Adimora et al., eds., *Sexually Transmitted Diseases: Companion Handbook*, 2nd ed. (New York: McGraw-Hill, 1994), 383; Allan R. De Jong and Mimi Rose, "Legal Proof of Child Sexual Abuse in the Absence of Physical Evidence," *Pediatrics* 88 (Sept. 1991): 506, 508; Suzanne M. Sgroi, " 'Kids with Clap': Gonorrhea as an Indicator of Child Sexual Assault," *Victimology* 2 (1977): 251, 253–54.

20. Nancy Kellogg and the Committee on Child Abuse and Neglect, "The Evaluation of Sexual Abuse in Children," *Pediatrics* 116 (Aug. 2005): 506, 509; Astrid Heppenstall-Heger et al., "Healing Patterns in Anogenital Injuries: A Longitudinal Study of Injuries Associated with Sexual Abuse, Accidental Injuries, or Genital Surgery in the Preadolescent Child," *Pediatrics* 112 (Oct. 2003): 829, 832–35.

21. Astrid Heger, "Helping the Medical Professional Make the Diagnosis of Sexual Abuse," in *Evaluation of the Sexually Abused Child: A Medical Textbook and Photographic Atlas,* ed. Astrid Heger and S. Jean Emans (New York: Oxford University Press, 1992), 5–7.

22. Kellogg and the Committee on Child Abuse and Neglect, "Evaluation of Sexual Abuse," 509; Heppenstall-Heger et al., "Healing Patterns in Anogenital Injuries," 832–35.

23. John McCann et al., "Healing of Nonhymenal Genital Injuries in Prepubertal and Adolescent Girls: A Descriptive Study," *Pediatrics* 120 (Nov. 2007): 1000; Astrid H. Heger, "Evaluation of Sexual Assault in the Emergency Department," *Topics in Emergency Medicine* 21 (1999): 46, 51.

24. American Academy of Pediatrics, Committee on Child Abuse and Neglect, "Gonorrhea in Prepubertal Children," *Pediatrics* 101 (Jan. 1998): 134–35. See also CDC, "Sexually Transmitted Diseases Treatment Guidelines, 2006," *MMWR,* Aug. 4, 2006, 1, 83, www.cdc.gov/mmwr/preview/mmwrhtml/rr5511a1.htm (accessed Mar. 28, 2008); American Academy of Pediatrics, Committee on Child

Abuse and Neglect, "Guidelines for the Evaluation of Sexual Abuse of Children," *Pediatrics* 87 (Feb. 1991): 254, 256–57.

25. Adimora et al., *Sexually Transmitted Diseases,* 356, 378, 382. Doctors must detect the presence of *N. gonorrhoeae* and test multiple sites (vagina, rectum, throat) to make a differential diagnosis and to rule out the possibility of sexual assault. Daniel R. Mishell Jr. et al., *Comprehensive Gynecology*, 3rd ed. (St. Louis: Mosby, 1997), 266–67; Adimora et al., *Sexually Transmitted Diseases*, 383; Larry J. Copeland, *Textbook of Gynecology* (New York: Saunders, 1993), 611–13. For reasons doctors cannot explain, children almost never acquire syphilis from sexual contact. Margaret Hammerschlag, "Sexually Transmitted Diseases in Sexually Abused Children: Medical and Legal Implications," *Sexually Transmitted Infections* 74 (June 1998): 167; Pamela F. Farrington, "Pediatric Vulvo-Vaginitis," *Clinical Obstetrics and Gynecology* 40 (Mar. 1997): 135; Deborah Stewart, "Sexually Transmitted Diseases," in Heger and Emans, *Evaluation of the Sexually Abused Child*, 151.

26. Sgroi, "Kids with Clap," 252.

27. Adimora et al., *Sexually Transmitted Diseases,* 354. See also David Folland et al., "Gonorrhea in Preadolescent Children: An Inquiry into Source of Infection and Mode of Transmission," *Pediatrics* 60 (Aug. 1977): 153.

28. Adimora et al., *Sexually Transmitted Diseases,* 354; Stewart, "Sexually Transmitted Diseases," 151.

29. CDC, "1998 Guidelines for Treatment of Sexually Transmitted Diseases," *MMWR,* Jan. 23, 1998, 12; Stewart, "Sexually Transmitted Diseases," 146, 150.

30. Naomi F. Sugar and Elinor A. Graham, "Common Gynecologic Problems in Prepubertal Girls," *Pediatrics in Review* 27 (June 2006): 213, 219; Robert A. Shapiro, Charles J. Schubert, and Robert M. Siegel, "Neisseria Gonorrhea Infections in Girls Younger than 12 Years of Age Evaluated for Vaginitis," *Pediatrics* 104 (Dec. 1999): e72, 75, http://pediatrics.aappublications.org/cgi/content/full/104/6/e72 (accessed Feb. 20, 2008). For the elements of a thorough medical examination to evaluate suspected child sexual abuse, including how to determine which genital findings may be symptoms of other diseases or injuries, see Joyce A. Adams, "Medical Evaluation of Suspected Child Sexual Abuse," *Journal of Pediatric and Adolescent Gynecology* 17 (2004): 191.

31. CDC, "1998 Guidelines," 114; American Academy of Pediatrics, Committee on Child Abuse and Neglect, "Gonorrhea," 134–35; Adimora et al., *Sexually Transmitted Diseases,* 356.

32. Infections acquired in the birth canal manifest themselves in the infant's eyes and joints. Adimora et al., *Sexually Transmitted Diseases,* 352–54. Gonorrhea infection in the eyes of a newborn was a leading cause of blindness before silver nitrate eye drops became a standardized part of twentieth-century child-

birth procedures. Laura T. Gutman, "Gonococcal Diseases in Infants and Children," in *Sexually Transmitted Diseases*, ed. King K. Holmes et al., 3rd ed. (New York: McGraw-Hill, 1999), 1145, 1146.

33. Sgroi, "Kids with Clap," 252–53.

34. See, e.g., Francis Huber, "Ward Problems," *Archives of Pediatrics* 28 (Sept. 1911): 752.

35. Stewart, "Sexually Transmitted Diseases," 150; Adimora et al., *Sexually Transmitted Diseases,* 380; Sgroi, "Kids with Clap," 258–63; Folland et al., "Gonorrhea in Preadolescent Children," 156; Helen Britton and Karen Hansen, "Sexual Abuse," *Clinical Obstetrics and Gynecology* 40 (Mar. 1997): 226; James H. Gilbaugh Jr. and Peter C. Fuchs, "The Gonococcus and the Toilet Seat," *NEJM* 301 (July 12, 1979): 91; Michael F. Rein, Letter to the Editor, *NEJM* 301 (Dec. 13, 1979): 1347; cf. Felicity Goodyear-Smith, "What Is the Evidence for Non-Sexual Transmission of Gonorrhoea in Children after the Neonatal Period? A Systematic Review," *Journal of Forensic and Legal Medicine* 14 (2007): 489.

36. CDC, "Sexually Transmitted Diseases Treatment Guidelines, 2006," 1.

37. Stewart, "Sexually Transmitted Diseases," 150. Because some parents delay or refuse a voluntary examination, Sgroi suspects that the actual co-incidence of infection may be higher. Suzanne M. Sgroi, "Pediatric Gonorrhea beyond Infancy," *Pediatrics Annals* 8 (May 1979): 326, 333. See, e.g., *Commonwealth v. Askins*, 18 Mass. App. 927, 928 (1984).

38. Heger, "Helping the Medical Professional," 4; Angelo P. Giardino and Martin A. Finkel, "Evaluating Child Sexual Abuse," *Pediatric Annals* 34 (May 2005): 382; Hughes Evans, "Vaginal Discharge in the Prepubertal Child," *Pediatric Case Reviews* 3 (Oct. 2003): 193; Division of AIDS, STD, and TB Laboratory Research, CDC, *Neisseria gonorrhoeae* (Feb. 10, 2000), www.cdc.gov/ncidod/dastir/gcdir/NeIdent/Ngon.html (accessed Sept. 15, 2000).

39. Division of AIDS, STD, and TB Laboratory Research, CDC, *Neisseria gonorrhoeae* (Feb. 10, 2000). See also Kellogg and the Committee on Child Abuse and Neglect, "Evaluation of Sexual Abuse," 510–11; Giardino and Finkel, "Evaluating Child Sexual Abuse," 382; Evans, "Vaginal Discharge," 193.

40. *Pennsylvania v. Ritchie*, 480 U.S. 39, 60 (1987).

41. U.S. Dept. of Justice, Bureau of Justice Statistics, *Sexual Assault of Young Children As Reported to Law Enforcement: Victim, Incident, and Offender Characteristics* (July 2000), 1, www.ojp.usdoj.gov/bjs/abstract/sayclre.htm (accessed Sept. 26, 2000); CDC, *Rape Fact Sheet: Prevalence and Incidence* (1997), www.cdc.gov/ncipc/factsheets/rape.htm (accessed Sept. 15, 2000). See also David Finkelhor and Jennifer Dziuba-Leatherman, "Children as Victims of Violence: A National Survey," *Pediatrics* 94 (Oct. 1994): 413, 415.

42. David Finkelhor and Richard Ormrod, "Child Abuse Reported to the Po-

lice," *Juvenile Justice Bulletin* (Washington, DC: U.S. Dept. of Justice, Office of Justice Programs, Office of Juvenile Justice and Delinquency Prevention, May 2001), 1–3, 6; Justice Research and Statistics Association, *Domestic and Sexual Violence Data Collection: A Report to Congress under the Violence against Women Act* (Washington, DC: U.S. National Institute of Justice and the Bureau of Justice Statistics, 1996), 11, 30, 35.

43. David Finkelhor and Lisa M. Jones, "Explanations for the Decline in Child Sexual Abuse Cases," *Juvenile Justice Bulletin* (Washington, DC: U.S. Dept. of Justice, Office of Justice Programs, Office of Juvenile Justice and Delinquency Prevention, Jan. 2004), 1.

44. Finkelhor and Ormrod, "Child Abuse Reported," 4. See also David Finkelhor et al., "The Victimization of Children and Youth: A Comprehensive, National Survey," *Child Maltreatment* 10 (Feb. 2005): 5, 10–11, 18.

45. The Department of Justice analyzed approximately 61,000 victim report files made between 1991 and 1996 and another 58,000 files from a victim-identified offender database. Police made arrests in fewer than 20 percent of sex offenses committed against children under 6 years old and in one-third of the cases reported for children between the ages of 6 and 11. See also U.S. Dept. of Health and Human Services, National Center on Child Abuse and Neglect, *Child Maltreatment 1995: Reports from the States to the National Child Abuse and Neglect Data System* (Washington, DC: U.S. Government Printing Office, 1997).

46. See, e.g., John Briere and Diana M. Elliott, "Prevalence and Psychological Sequelae of Self-Reported Childhood Physical and Sexual Abuse in a General Population Sample of Men and Women," *Child Abuse & Neglect* 27 (2003): 1205, 1206, 1216.

47. David Finkelhor, "Current Information on the Scope and Nature of Child Sexual Abuse," *Sexual Abuse of Children* 4 (Summer/Fall 1994): 31, 32, 34, 37, 41, 43, 45–46. See also Jennifer J. Freyd et al., "The Science of Child Sexual Abuse," *Science,* Apr. 22, 2005, 501; David Finkelhor, "How Widespread Is Child Sexual Abuse?" *Children Today* 13 (July–Aug. 1984): 18.

48. Pat Gilmartin, *Rape, Incest, and Child Sexual Abuse: Consequences and Recovery* (New York: Garland, 1994), 47; Jean Goodwin, Doris Sahd, and Richard T. Rada, "False Accusations and False Denials of Incest: Clinical Myths and Clinical Realities," in *Sexual Abuse: Incest Victims and Their Families*, ed. Jean Goodwin, 2nd ed. (Chicago: Year Book Medical Publishers, 1989), 21, 23; David Finkelhor and Larry Barron, "High-Risk Children," in *A Sourcebook on Child Sexual Abuse*, ed. David Finkelhor (Beverly Hills: Sage Publications, 1986), 60, 67–71.

49. See Herman, *Trauma and Recovery,* 96–110, and Jacobs, *Victimized Daughters,* 6–10.

50. CDC, "Perceptions of Child Sexual Abuse as a Public Health Problem—Vermont, September 1995," *MMWR* 46 (Aug. 26, 1997): 801, 802.

51. Finkelhor et al., "Victimization of Children," 14, 16–18; Gail S. Goodman et al., "A Prospective Study of Memory for Child Sexual Abuse: New Findings Relevant to the Repressed-Memory Controversy," *Psychological Science* 14 (Mar. 2003): 113, 115–16; Finkelhor, "Current Information," 47–48; Stefanie Doyle Peters, Gail Elizabeth Wyatt, and David Finkelhor, "Prevalence," in Finkelhor, *Sourcebook*, 15, 27–30; Ruth S. Kempe and C. Henry Kempe, *Common Secret: Sexual Abuse of Children and Adolescents* (New York: W. H. Freeman and Co., 1984), 18–20. See also Lynn Weber Cannon, Elizabeth Higginbotham, and Marianne L. A. Leung, "Race and Class Bias in Qualitative Research on Women," *Gender and Society* 2 (Dec. 1988): 449, 458–60.

52. Ami Lewis and Jacqueline M. Golding, "Sexual Assault History and Eating Disorder Symptoms among White, Hispanic, and African-American Women and Men," *Journal of Public Health* 86 (Apr. 1996): 579, 581.

53. Briere and Elliott, "Prevalence and Psychological Sequelae," 1207.

54. Ibid., 1207, 1217; Kathleen A. Kendall-Tackett, Linda Meyer Williams, and David Finkelhor, "Impact of Sexual Abuse on Children: A Review of Synthesis of Recent Empirical Studies," *Psychological Bulletin* 113 (1993): 164, 167–69.

55. Jennie G. Noll, Penelope K. Trickett, and Frank W. Putnam, "A Prospective Investigation of the Impact of Childhood Sexual Abuse on the Development of Sexuality," *Journal of Consulting and Clinical Psychology* 71 (June 2003): 575; Beth E. Molnar, Stephen L. Buka, and Ronald C. Kessler, "Child Sexual Abuse and Subsequent Psychopathology: Results from the National Comorbidity Survey," *American Journal of Public Health* 91 (May 2001): 753, 757; Kenneth J. Ruggiero, Susan V. McLeer, and J. Faye Dixon, "Sexual Abuse Characteristics Associated with Survivor Psychopathology," *Child Abuse & Neglect* 24 (2000): 951, 961; Kendall-Tackett, Williams, and Finkelhor, "Impact of Sexual Abuse," 170–71.

56. Kellogg and the Committee on Child Abuse and Neglect, "Evaluation of Sexual Abuse," 507; Kendall-Tackett, Williams, and Finkelhor, "Impact of Sexual Abuse," 165–67, 173–74; Heger, "Helping the Medical Professional," 5–7.

57. Kendall-Tackett, Williams, and Finkelhor, "Impact of Sexual Abuse," 165–66.

58. Jacob M. Virgil, David C. Geary, and Jennifer Byrd-Craven, "A Life History Assessment of Early Childhood Sexual Abuse in Women," *Developmental Psychology* 41 (2005): 553; Noll, Trickett, and Putnam, "Prospective Investigation," 575.

59. Mariann R. Weierich and Matthew K. Nock, "Posttraumatic Stress Symptoms Mediate the Relation between Childhood Sexual Abuse and Nonsuicidal

Self-Injury," *Journal of Consulting and Clinical Psychology* 76 (Feb. 2008): 39, 42–44; Kendall-Tackett, Williams, and Finkelhor, "Impact of Sexual Abuse," 167.

60. Vigil, Geary, and Byrd-Craven, "Life History Assessment," 553–55, 59; Natalie Sachs-Ericsson et al., "Childhood Sexual and Physical Abuse and the One-Year Prevalence of Medical Problems in the National Comorbidity Survey," *Health Psychology* 24 (2005): 32, 36–37.

61. Jennie G. Noll et al., "Obesity Risk for Female Victims of Childhood Sexual Abuse: A Prospective Study," *Pediatrics* 120 (July 2007): e61, http://pediatrics.aappublications.org/cgi/content/full/120/1/e61 (accessed Mar. 28, 2008); Lewis and Golding, "Sexual Assault History," 581.

62. Theresa E. Senn et al., "Childhood Sexual Abuse and Sexual Risk Behavior among Men and Women Attending a Sexually Transmitted Disease Clinic," *Journal of Consulting and Clinical Psychology* 74 (2006): 720, 726–28.

Chapter One: Incest in the Nineteenth Century

1. E.g., "San Bernardino County," *LAT,* Apr. 30, 1892; "Local News in Brief, Long Island," *NYT,* Oct. 30, 1870.

2. E.g., "Otero Occurrences," *Santa Fe New Mexican*, Oct. 10, 1899, [1]; "Found Guilty of Incest," *Morning World-Herald* (Omaha), Dec. 2, 1895, 2; "Arrested for Incest," *Aberdeen (SD) Daily News*, Aug. 1, 1895, [2]; "Crime of a Brute," *Morning World-Herald* (Omaha), Feb. 19, 1895, [1]; "An Unnatural Father," *LAT*, Sept. 6, 1893; "Incest Charged," *St. Paul Daily News*, June 29, 1893, col. E; "Bound Over," *Atlanta Constitution*, Aug. 12, 1892, 2; "A Father's Crime," *Atlanta Constitution*, July 22, 1892, 2; "In Pursuit of a Fiend," *Morning Oregonian* (Boise), June 17, 1891, col. C; "They Should Not Have Interfered," *Washington Post*, Apr. 13, 1888, 1; "Hiding from the Justice He Merits," *Dallas Morning News*, Aug. 21, 1887, 14; "Ruined by Her Father," *Atlanta Constitution*, Apr. 15, 1886, 5; "Brief Telegrams," *Wheeling (WV) Register*, Apr. 17, 1885; "A Change; Another Beastly Father," *LAT*, June 11, 1882, 2; "A Father's Crime," *Wheeling (WV) Register*, July 27, 1880, [1]; "Local and Other Matters: An Incestuous Parent," *Deseret (UT) News*, Mar. 10, 1880; "A Case That Judge Lynch Missed," *Wheeling (WV) Register*, Feb. 26, 1879, [1]; "An Inhuman Father," *NYT*, May 27, 1878.

3. "Henrietta, Texas, 20," *Deseret (UT) News*, Feb. 24, 1886, 84; "A Lynching Interrupted," *Washington Post*, Feb. 22, 1886, 1. See also "Local Affairs at the Fort," *Dallas Morning News*, Mar. 1, 1886, 6.

4. See, e.g., "Trial of Chamberlain," *Morning World-Herald* (Omaha), Apr. 12, 1899, 3; "Moscow Matters," *Idaho Daily Statesman* (Boise), Feb. 26, 1895, [2]; "Rockville," *Hartford (CT) Courant*, Mar. 1, 1894, 8; "Bowers Sensation,"

Tacoma Daily News, Jan. 30, 1893, 7; quotation from "The Bowers Held: Both Father and Daughter," *Tacoma Daily News,* Oct. 29, 1892, 7; "A Horrible Crime," *Columbus (GA) Enquirer-Sun*, July 30, 1890, [4]; "An Infamous Parent," *NPG,* Feb. 14, 1880, 10; "The End of the Avery Incest Case," *Chicago Tribune*, Mar. 27, 1862, 4; "Boston, September 21," *Northern Post* (Salem, NY), Oct. 10, 1805, [2]; and *People v. Gillet*, 2 Edm. Sel. Cas. 406 (N.Y. Supp. 1845).

5. "Hartford Correspondence," *NPG*, Mar. 20, 1847, 221; "Hartford Correspondence," *NPG*, Mar. 6, 1847, 205. See also "Rockville," *Hartford (CT) Courant*, Mar. 1, 1894, 8.

6. "[Albert H. Essex]," *Daily Constitution* (Middletown, CT), Feb. 20, 1873, [2].

7. E.g., "England. London, April 8. Incest," *Worcester (MA) Magazine*, July 8, 1790, 2 (Britain); "Incest," *Western Star* (Stockbridge, MA), June 22, 1790, 3 (Britain); "Bermuda," *Salem (MA) Mercury*, Sept. 15, 1789, [2] (Bermuda); "New-York, October 19," *Massachusetts Spy, or, Thomas's Boston Journal*, Oct. 29, 1772, 142 (Boston); "Extract of Letter from London Aug. 2," *Boston Evening Post*, Oct. 2, 1769, [2]; "New-York, September 25," *New-York Gazette*, Sept. 25, 1769, [3] (New Jersey); "London, April 4," *Newport (RI) Mercury*, June 9–16, 1766, [2] (England); "Paris, Octob. 15," *Boston News-Letter*, Dec. 21–28, 1732, [2] (Paris).

8. "Danbury, Saturday-Evening, July 26," *Hartford (CT) Gazette and Universal Advertiser*, July 28, 1794, 3.

9. "Danbury, July 29," *Connecticut Journal* (New Haven), Aug. 13, 1794, [3]; "Connecticut," *Western Star* (Stockbridge, MA), Aug. 12, 1794, [3]; "Moses Johnson," *Oracle of the Day* (Portsmouth, NH), Aug. 8, 1794, [3]; "State of Connecticut," *Norwich (CT) Packet*, Aug. 7, 1794, 3; "Scott," *Columbian Centinel* (Boston), Aug. 6, 1794, [3]; "Connecticut," *Massachusetts Spy: Or, the Worcester Gazette*, Aug. 6, 1794, 3; "Danbury July 26," *Weekly Museum* (New York), Aug. 2, 1794, [3]. Papers spelled the name Johnson or Johnston.

10. "State of Connecticut," *Norwich (CT) Packet*, Aug. 7, 1794, 3.

11. "Scott," *Columbian Centinel* (Boston), Aug. 6, 1794, [3]; "Moses Johnson," *Oracle of the Day* (Portsmouth, NH), Aug. 8, 1794, [3].

12. "Connecticut," *Western Star* (Stockbridge, MA), Sept. 9, 1794, [1]; "Danbury, August 16," *Newport (RI) Mercury*, Sept. 2, 1794, [3]; "State of Connecticut," *Norwich (CT) Packet*, Aug. 28, 1794, [3]; "Danbury, August 16," *Columbian Centinel* (Boston), Aug. 23, 1794, [2]; "Danbury August 16," *Hartford (CT) Gazette and Universal Advertiser*, Aug. 21, 1794, [3]; "Danbury, Saturday-Evening, August 16," *Connecticut Journal* (New Haven), Aug. 20, 1794, [3].

13. "Moses Johnson," *Columbian Centinel* (Boston), Sept. 13, 1794, Supplement [1].

14. E.g., "A Father's Terrible Crime," *NYT*, Jan. 14, 1883, 1.

15. "Ephraim Wheeler," *Portland Gazette and Maine Advertiser*, Oct. 7, 1805, [2]. See also "Damning Evidence against a Father," *Kansas City (MO) Star*, Sept. 23, 1886, [1]; "Bestial Bullock," *Wheeling (WV) Register*, Feb. 10, 1882, [4]; "Incest and Murder," *Daily Commercial Bulletin and Missouri Literary Register* (St. Louis), Sept. 22, 1836, col. B; and "From the *Western Sun*," *Indiana Journal* (Indianapolis), June 30, 1830.

16. "Boston, September 21," *Northern Post* (Salem, NY), Oct. 10, 1805, [2]. See also Irene Quenzler Brown and Richard D. Brown, *The Hanging of Ephraim Wheeler: A Story of Rape, Incest, and Justice in Early America* (Cambridge: Harvard University Press, 2003), 3.

17. Advertisement, *Hampshire Federalist* (Springfield, MA), Apr. 29, 1806, [4]; advertisement, *Providence (RI) Gazette*, Feb. 22, 1806, [1]; advertisement, *Connecticut Herald* (New Haven), Jan. 21, 1806, [3] (republished Jan. 28, [4], Feb. 4, 1806, [4]); advertisement, *Otsego Herald* (Cooperstown, NY), Jan. 9, 1806, [3]; advertisement, *Morning Chronicle* (New York), Jan. 1, 1806, [4]; advertisement, *Commercial Advertiser* (New York), Dec. 23, 1805, [1]; advertisement, *Morning Chronicle* (New York), Dec. 21, 1805, 2; advertisement, *Newburyport (MA) Herald*, Dec. 17, 1805, [3] (republished Dec. 20, [4], Dec. 24, 1805, [4]; Jan. 3, [4], Jan. 10, [1], Jan. 17, [1], Jan. 21, [4], Jan. 28, [1], Feb. 4, 1806, [1]); advertisement, *Daily Advertiser* (New York), Dec. 9, 1805, [3] (republished Jan. 2, 1806, [4]); advertisement, *Republican Spy* (Northampton, MA), Dec. 3, 1805, [3] (republished Dec. 10, [3], Dec. 17, [4], Dec. 24, [4], Dec. 31, 1805, [3]; Jan. 7, [4], Jan. 14, [4], Jan. 28, [4], Feb. 4, [4], Feb. 11, 1806 [4]); advertisement, *Otsego Herald* (Cooperstown, NY), Oct. 17, 1805, [3] (republished Oct. 24, 1805, [3]); advertisement, *Vermont Gazette* (Bennington), Oct. 14, 1805, [4] (republished Nov. 11, 1805, [4]); "Ephraim Wheeler," *Portland Gazette and Maine Advertiser*, Oct. 7, 1805, [2]; advertisement, *Middlebury (VT) Mercury*, Oct. 2, 1805, [3]; advertisement, *Newburyport (MA) Herald*, Oct. 1, 1805, [3] (republished Oct. 4 [1], Oct. 11 [1], Oct. 25 [3], Nov. 22, 1805, [3]).

18. Advertisement, *Newburyport (MA) Herald*, Oct. 1, 1805, [3].

19. "Ephraim Wheeler," *Connecticut Gazette and the Commercial Intelligencer* (New London), Mar. 12, 1806, [3]; "Pittsfield," *Connecticut Courant* (Hartford), Mar. 5, 1806, [3]; "Pittsfield, February 24, 1806," *(Boston) Democrat*, Mar. 5, 1806, [2]; "Boston, Sept. 21," *Kline's Carlisle (PA) Weekly Gazette*, Oct. 18, 1805, [3]. Similarly, after Ira Gardner was convicted in 1833 for killing a stepdaughter who resisted his attempt to rape her, the local paper reported that a "vast multitude of from twelve to fifteen thousand people," including five thousand women, watched the hangman usher him into eternity. "Execution of Gardner," *Ohio Observer* (Hudson), Nov. 9, 1833, col. D.

20. "Springfield Bookstore," *Hampshire Federalist* (Springfield, MA), Apr. 29, 1806, [4].

21. Cornelia Hughes Dayton, *Women before the Bar: Gender, Law, and Society in Connecticut, 1639–1789* (Chapel Hill: University of North Carolina Press, 1995), 275. See also Brown and Brown, *Hanging of Ephraim Wheeler*, 112–13.

22. Diane Miller Sommerville, *Rape and Race in the Nineteenth-Century South* (Chapel Hill: University of North Carolina Press, 2004), 281. Prosecutors in Essex County, New York, filed charges against two men for father-daughter incest, with one conviction. Sean T. Moore, "'Justifiable Provocation': Violence against Women in Essex County, New York, 1799–1860," *Journal of Social History* 35 (2002): 889, 897.

23. "City Crime," *New York Daily Times*, Dec. 31, 1852, 6.

24. "Long Island Court Statistics," *New York Daily Times*, Jan. 2, 1856, 3.

25. "Second District Police Court," *Brooklyn Eagle*, Jan. 5, 1887, 3. See also Stephen Robertson, *Crimes against Children: Sexual Violence and Legal Culture in New York City, 1880–1960* (Chapel Hill: University of North Carolina Press, 2005), 239, 243, 279–80 n. 42.

26. "Sheriff's Report," *LAT*, Jan. 2, 1888, 5.

27. "Police Matters," *LAT*, June 2, 1888, 2.

28. "Riverside," *LAT*, July 11, 1897, 31; "Riverside County," *LAT*, July 9, 1897, 15; "A Father's Baneful Crime," *Morning World-Herald* (Omaha), Nov. 15, 1893, [1]; "An Unnatural Father," *Morning World-Herald* (Omaha), Sept. 19, 1891, [1]; "In Pursuit of a Fiend," *Morning Oregonian* (Boise), June 17, 1891, col. C; "Dye Waives Examination," *Wheeling (WV) Register*, June 12, 1890, [1]; "A Preacher's Flight," *Evening News* (San Jose, CA), May 16, 1889, [2]; "Belton Tragedy," *Dallas Morning News*, July 23, 1891, 5; "Under Three Serious Charges," *Atlanta Constitution*, July 10, 1888, 1; "A Terrible Crime," *Wheeling (WV) Register*, Mar. 8, 1890, [4]; "Arrested Charged with Incest," *Dallas Morning News*, Aug. 20, 1888, 1; "A Horrible Crime," *Aberdeen (SD) Daily News*, Apr. 19, 1888, [1]; "Arrested for Incest," *Dallas Morning News*, Mar. 13, 1888, 2; "Hiding from the Justice He Merits," *Dallas Morning News*, Aug. 21, 1887, 14; "Items from the South," *Aberdeen (SD) Daily News*, Apr. 9, 1886, [2]; "Accused of Incest," *Dallas Morning News*, Oct. 31, 1885, 3; "Atkins's Crime," *Atlanta Constitution*, Nov. 20, 1884, 2; "Disgusting Case of Incest," *Daily Herald* (Grand Forks, ND), Sept. 9, 1883, [4]; "Georgia by Wire," *Atlanta Constitution*, Aug. 18, 1883, 2; "Sioux Commission," *Daily Herald* (Grand Forks, ND), Dec. 23, 1882, [1]; "Most Horrible," *Wheeling (WV) Register*, Aug. 12, 1882, [1]; "Hermosilio Items," *Arizona Weekly Star* (Tucson), Sept. 14, 1882, [4]; "Beallsville," *Wheeling (WV) Register*, May 2, 1880, [4]; "Barnesville," *Wheeling (WV) Register*, Apr. 24, 1880, [1]; "Telegraphic Dots," *Washington Post*, May 13,

1878, 1; "Record of Crime," *Milwaukee Daily Sentinel*, June 28, 1875, col. F; "Multiple News Items, Suicide in Oregon," *Daily Evening Bulletin* (San Francisco), June 14, 1869, col. G; "A Minnesota Horror," *Chicago Tribune*, Oct. 2, 1867, 2; "Another Beastly Father," *LAT*, June 11, 1882, 2; "A Foundling," *Hartford (CT) Daily Courant*, June 13, 1853; "News by the Mails," *New York Daily Times*, Mar. 4, 1853, 3.

29. "The Northwestern States," *Chicago Daily Tribune*, June 6, 1874, 12. See also "Crime," *Chicago Daily Tribune*, May 20, 1875, 5.

30. E.g., "Suicided in Jail," *Grand Forks (ND) Daily Herald*, Apr. 1, 1892, [2]; "A Prisoner Suicides," *Dallas Morning News*, Oct. 21, 1891, 2; "Obituary 1; Starved Himself to Death," *LAT*, May 17, 1889, 4; "An Aged Brute Robbed," *Kansas City (MO) Star*, Oct. 31, 1885, [1]; "A Father's Suicide," *Atlanta Constitution*, Dec. 5, 1884, 1; "World Rid of a Monster," *Wheeling (WV) Register*, Dec. 11, 1882, [1]; "Suicide," *Chicago Daily Tribune*, Aug. 9, 1879, 7; *Hartford (CT) Daily Courant*, Nov. 29, 1873 [1]; "Nevada: Incest, Infanticide, and Attempted Suicide," *Daily Evening Bulletin* (San Francisco), Aug. 28, 1873, col. C; "Dubuque," *Chicago Daily Tribune*, Nov. 4, 1872, 2; "Incest and Suicide," *NPG*, July 18, 1846, 380; "A Man by the Name of Joe!" *Vermont Patriot and State Gazette* (Montpelier), Dec. 18, 1837, col. B; "[Hartwell]," *Eastern Argus* (Portland, ME), June 27, 1837, [2].

31. "Hacked His Own Throat," *Atlanta Constitution*, Jan. 31, 1890, 1.

32. A collection of 65 Los Angeles County trial court transcripts, for instance, for some of the years between 1885 and 1904 included 34 prosecutions for sex crimes, of which only 2 involved father-daughter incest. One additional case involved a man who assaulted his 14-year-old sister-in-law, who lived in his home. *People v. W. P. Hancock*, file #03-1, Jan. 14, 1903; *People v. J. T. E. Johns*, file # 95-8, Dec. 14, 1895; *People v. W. S. Stratton*, file #03-2, Feb. 2, 1903, Early Los Angeles Court Transcripts, 1885–1904, Special Collections and Archives, Oviatt Library, California State University, Northridge, California. The collection does not contain transcripts of every trial during this period, 1885, 1887–1900, and 1903–4.

33. "From the *National Aegis*," *Commercial Register* (Norfolk, VA), Oct. 6, 1802, [2]; "Bulow's Travels," *New-York Herald*, Sept. 25, 1802, [3]; "From the *National Aegis*," *New-York Evening Post*, Sept. 24, 1802, [2].

34. William C. Dowling, *Literary Federalism in the Age of Jefferson: Joseph Dennie and the Port Folio, 1801–1812* (Columbia: University of South Carolina Press, 1999), xiv–xv, 1–2, 19–21.

35. "From the (Phila.) *True American*," *New-Hampshire Sentinel* (Keene), Dec. 4, 1802, [4]; "From the (Phila.) *True American*," *Mercury and New-England Palladium* (Boston), Nov. 23, 1802, [1]; "From the *True American*," *New-York*

Evening Post, Nov. 10, 1802, [2]. See also "Interesting Travels in America," *New-York Herald*, Sept. 29, 1802, [2], and "Interesting Travels in America," *New-York Evening Post*, Sept. 27, 1802, [3].

36. "From the *National Aegis*," *New-York Evening Post*, Sept. 24, 1802, [2]. "Viscous Prussian" from the *True American*'s account; see citations in preceding note.

37. E.g., "A Humane Engineer," *Daily Charlotte (NC) Observer*, Feb. 22, 1896, [1]; "A Child Burned to Death," *Daily Charlotte (NC) Observer*, Nov. 17, 1895, [5]; "Frank L. Palmer Free," *Morning Oregonian* (Boise), Jan. 15, 1895, 8; "Siglar Was Innocent," *Morning World-Herald* (Omaha), June 28, 1892, 2; "Town Talk," *Morning World-Herald* (Omaha), June 17, 1891, 8; "His Daughters against Him," *Wheeling (WV) Register*, Apr. 19, 1889, [1]; "A Strange Incest Case," *Wheeling (WV) Register*, Feb. 19, 1886, [1]; "Falsely Accusing Her Father," *NYT*, May 2, 1885, 5; "A Nice Sort of Daughter," *NPG*, Dec. 23, 1882, 3; "Candidates; Bartered Away Her Father's Reputation," *NPG*, June 26, 1880, 11; *Hartford (CT) Daily Courant*, Nov. 29, 1873, 1; "Crimes and Casualties," *New-Hampshire Patriot* (Concord), Apr. 2, 1873, [3]; "Police Court," *Chicago Tribune*, Feb. 4, 1864, 4; "A Father Charged with Rape upon His Own Daughter," *Chicago Press and Tribune*, Oct. 1, 1858, 1; "Yates," *Barre (MA) Gazette*, Oct. 9, 1846, [2]; "Infernal Depravity—Incest and Rape by a Father on His Own Daughter," *NPG*, Sept. 26, 1846, 21.

38. On adult-child rape, see, e.g., "Brutal Assault on a Little Girl," *NYT*, Dec. 29, 1883, 1 (two German Jews assault 8-year-old girl); "Arrested for Rape," *St. Paul Daily News*, Dec. 22, 1890 (rape of 5-year-old girl and 7-year-old boy); "Criminals and Their Offenses," *NYT*, Nov. 23, 1879, 1 (rape of 7-year-old girl); "Perils of the Park," *NPG*, Sept. 14, 1878, 6 (multiple assaults); "Telegraphic," *Idaho Tri-Weekly Statesman* (Boise), May 20, 1876, [2] (defendant arrested for assaulting four girls in ten days); "Local News," *New Orleans Times*, Oct. 1, 1872, 6 (rape of 7-year-old); "Miscellaneous Telegrams," *NYT*, Sept. 19, 1870, 5 (rape of 6-year-old girl); "Horrible Rape in Newburgh," *Daily Cleveland Herald*, July 9, 1868 (boarder raped 9-year-old girl); "Colorado: A Reign of Terror," *Chicago Tribune*, Feb. 27, 1868, 2 (attempted rape of two girls, ages 7 and 13); "Boston: A Wolf in Sheep's Clothing," *NPG*, May 26, 1867, 3 ("clerical child-debaucher"); "An Atrocious Case of Rape," *Chicago Tribune*, May 11, 1867, 2 (rape of 9-year-old girl); "State Matters," *Hartford (CT) Daily Courant*, Dec. 5, 1865, 2 (attempted rape of 9-year-old girl); "New-York City," *NYT*, Aug. 2, 1858, 8 (German attempted rape of 8-year-old girl); "Brutal Outrage on a Very Young Girl," *New York Daily Times*, July 19, 1855, 2 (13-year-old girl); "Attempt of a Negro to Commit Two Rapes in Massachusetts," *New York Daily Times*, Aug. 2, 1854, 3 (12-year-old white girl); "By the Mails," *Chicago Daily*

Tribune, June 27, 1854, 3 (Catholic priest attempted rape of 14-year-old girl); "Attempt at Rape," *Hartford (CT) Daily Courant*, Aug. 15, 1851, 2 (attempted rape of 8-year-old girl); "Another Warning," *NPG*, Mar. 20, 1847, 22 (rape of 14-year-old in Memphis); "A Black Man Harry," *Virginia Free Press* (Charles Town, WV), Oct. 26, 1837, col. A (rape of 10-year-old); "A Man Was Examined," *Boston Courier*, Aug. 8, 1836, col. D (attempted rape of 5-year-old girl); *American Herald* (Boston), Aug. 7, 1786, [3] (13-year-old girl); "New-London, July 28," *Connecticut Courant and Weekly Intelligencer* (Hartford), July 31, 1786, 3 (same case).

39. E.g., "The Cedarquest Case," *Morning World-Herald* (Omaha), Dec. 23, 1891, 3; "A Lecherous Brute," *Knoxville (TN) Journal*, Aug. 20, 1890, [1]; "Her Wrongs," *NPG*, Oct. 13, 1888, 6; "Under Three Serious Charges," *Atlanta Constitution*, July 10, 1888, 1; "A Beast," *Aberdeen (SD) News*, Apr. 24, 1887, [1]; "Strong Divorce Case," *NYT*, Dec. 9, 1856, 9; "The Law Courts," *Chicago Tribune*, Apr. 28, 1856, 3; "News by the Mails," *NYT*, Feb. 22, 1853, 6.

40. "A Minnesota Horror," *Chicago Tribune*, Oct. 2, 1867, 2. See also "Paying the Penalty," *Atlanta Constitution*, Feb. 7, 1885, 5, and "Alleged Incest and Violent Death of a Farmer—Post-Mortem Examination," *St. Louis Globe-Democrat*, July 25, 1875, 12.

41. E.g., "The Headworks Case," *LAT*, Mar. 28, 1896, 9; "Florida's Horrible Sensation," *Columbus (GA) Daily Enquirer-Sun*, June 22, 1895, [1]; "Belton Tragedy," *Dallas Morning News*, July 23, 1891, 5; "Who Murdered Her?" *NPG*, Dec. 6, 1890, 6; "Her Wrongs," *NPG*, Oct. 13, 1888, 6; "Judge Lynch Delinquent," *Aberdeen (SD) Daily News*, Sept. 26, 1888, [1]; "A Lynching Interrupted," *Washington Post*, Feb. 22, 1886, 1; "Incest and Murder; A Father Ravishes His Six Daughters," *Weekly Eastern Argus* (Portland, ME), May 4, 1871, [1]; "Domestic Intelligence, the Legal Tender Acts," *Galveston Tri-Weekly News*, May 1, 1871, [1]; "The East; the Perkins and Hall Trials," *NYT*, Apr. 29, 1871, 1; "A Fatal Burning Accident in Connecticut; the Joel Perkins Trial," *NYT*, Apr. 28, 1871, 1; "Hartford: A Revolting Case—Trial of a Man for Incest with His Six Daughters and the Murder of Their Babies," *Milwaukee (WI) Sentinel*, Apr. 27, 1871, col. D; "Dreadful Murder Trials in Hartford, Conn.," *NYT*, Apr. 27, 1871, 1; "In General," *Savannah Daily Herald*, Mar. 20, 1866, col. B; "New-Jersey," *NYT*, Oct. 15, 1860, 5; "New York Items," *Hartford (CT) Daily Courant*, Sept. 20, 1860, 3; "Brooklyn Intelligence," *NYT*, Dec. 28, 1858, 8; "Police," *Brooklyn Eagle*, Dec. 21, 1858, 3; "Local Matters, Held for Examination," *Boston Daily Advertiser*, Dec. 25, 1855, col. H; "A Monster in Human Shape," *New York Daily Times*, Aug. 24, 1854, 1; "Varieties," *New York Herald*, Nov. 12, 1845, col. D.

42. E.g., "An Important Case Tried," *Atlanta Constitution*, Aug. 12, 1899, 2; "An Awful Crime," *Wheeling (WV) Register*, June 27, 1892, [1]; "A Horrible

Crime," *Columbus (GA) Enquirer-Sun*, July 30, 1890, [4]; "Paternal Brutality," *Wheeling (WV) Sunday Register*, June 23, 1889, [1]; "Domestic Depravity," *Daily Inter Ocean* (Chicago), Apr. 21, 1887, col. E; "Items from the South," *Aberdeen (SD) News*, Apr. 9, 1886, [2]; "Train Robber Taken," *Kansas City (MO) Evening Star*, Sept. 22, 1885, [1]; "A Father's Crime," *Milwaukee Sentinel*, Jan. 11, 1885, 4; "World Rid of a Monster," *Wheeling (WV) Register*, Dec. 11, 1882, [1]; "Frightful Depravity of a Human Fiend," *NPG*, Mar. 26, 1881, 11; "That Kansas Incest Case," *Kansas City (MO) Evening Star*, Dec. 2, 1880, [1]; "Parkersburg," *Wheeling (WV) Register*, July 18, 1879, [1]; "Incest," *Idaho Tri-Weekly Statesman* (Boise), July 17, 1879, [3]; "A Father's Double Crime," *Wheeling (WV) Daily Register*, Aug. 27, 1878, [1]; "Sickening Case of Incest in Portland—An Unnatural Father Debauches His Own Daughters," *Boston Daily Globe*, Aug. 24, 1876, 4; "Incest, Infanticide, and Attempted Suicide," *Daily Evening Bulletin* (San Francisco), Aug. 28, 1873, col. C; "An Incestuous Brute," *Little Rock (AR) Daily Republican*, July 24, 1873, [1]; "State News," *Milwaukee Daily Sentinel*, Aug. 17, 1863, col. D; "General Intelligence, Unparalleled Depravity," *Boston Investigator*, Mar. 2, 1859, col. C.

43. "News and Other Items," *Pittsfield (MA) Sun*, May 4, 1871, [2]; "Incest and Murder; A Father Ravishes His Six Daughters," *Weekly Eastern Argus* (Portland, ME), May 4, 1871, [1]; "Domestic Intelligence," *Galveston Tri-Weekly News*, May 1, 1871, [1]; "Telegraphic," *Daily Columbus (GA) Enquirer*, Apr. 29, 1871, [2]; "The East; the Perkins and Hall Trials," *NYT*, Apr. 29, 1871, 1; "A Fatal Burning Accident in Connecticut; the Joel Perkins Trial," *NYT*, Apr. 28, 1871, 1; "A Revolting Case," *Chicago Tribune*, Apr. 27, 1871, 1; "A Revolting Case," *Milwaukee Sentinel*, Apr. 27, 1871, [1]; "Dreadful Murder Trials in Hartford, Conn.," *NYT*, Apr. 27, 1871, 1; "Conviction for Murder," *Daily State Gazette* (Trenton, NJ), Apr. 29, 1871, [2]; "New England Items," *Vermont Chronicle* (Bellow Falls), May 6, 1871, 2; "Connecticut. The Child Murder Trial," *Philadelphia Inquirer*, Apr. 28, 1871, [1]; "Telegraphic Summary," *Baltimore Sun*, Apr. 27, 1871, [1].

44. E.g., "Eloped with His Stepdaughter," *NPG*, Aug. 24, 1889, 14; "Bowers Goes Free," *Tacoma Daily News*, Jan. 31, 1893, [1]; "The Bowers Sensation," *Tacoma Daily News*, Jan. 30, 1893, 7; "The Bowers Held," *Tacoma Daily News*, Oct. 29, 1892, 7; "Incest Charged," *Tacoma Daily News*, Oct. 25, 1892, 7; "Beastial [*sic*] Depravity," *Morning World-Herald* (Omaha), Apr. 25, 1891, [1]; "Eloped with His Stepdaughter," *NPG*, Aug. 24, 1889, 14; "The City in Brief," *LAT*, Mar. 10, 1888, 8; "The Moore Case," *LAT*, Feb. 1, 1888, 1; "The Moore Case," *LAT*, Jan. 29, 1888, 1; "The Rape Fiend," *LAT*, Nov. 1, 1887, 1; "A Horror," *LAT*, Oct. 28, 1887, 8; "Divorced from Stepfather," *NPG*, July 12, 1884, 3; "Coast Gleanings; An Oregon Girl Shoots Her Father," *LAT*, June 29, 1880,

4; "A Shameful Alliance," *NPG*, Feb. 21, 1880, 10; "A Sickening Story of Crime," *NPG*, Nov. 30, 1878, 10; "Martial Laws," *New Orleans Times*, Mar. 1, 1876, 2; "Alleged Incest and Violent Death of a Farmer—Post-Mortem Examination," *St. Louis Globe-Democrat*, July 25, 1875, 12; "Another Case of Bigamy," *Farmer's Cabinet* (Amherst, NH), Apr. 28, 1858, [2].

45. "A Revolting Crime," *LAT*, Apr. 23, 1890, 5.

46. E.g., "Trial of Chamberlain," *Morning World-Herald* (Omaha), Apr. 12, 1899, 3; "More about Lynching," *Congregationalist* (Boston), Dec. 30, 1897, 1044; "Kovalev's Confession; A Pool of Iniquity," *LAT*, Nov. 17, 1895, 3; "She Packed the Room with Woman," *Daily Olympian* (Olympia, WA), Jan. 26, 1895, [4]; "Adultery," *Dakota Republican* (Vermillion, SD), Dec. 22, 1870, [2]; "Tenure by Fire," *NYT*, Mar. 20, 1860, 4; "Tenement Traps," *NYT*, Feb. 4, 1860, 4. Criticism of Mormonism emphasized that "incest is occurring constantly." "That Twin Relic of Barbarism," *LAT*, June 19, 1883, 3.

47. E.g., "An Alleged Human Brute," *LAT*, Sept. 15, 1898, 4; "Alleged Unnatural Father," *Morning World-Herald* (Omaha), Mar. 28, 1891, 8; "A Lecherous Brute," *Knoxville (TN) Journal*, Aug. 20, 1890, [1]; "A Most Unnatural Father," *Dallas Morning News*, Mar. 20, 1888, 2; "An Unnatural Father," *NYT*, Aug. 23, 1886, 1; "An Inhuman Brute," *NYT*, Sept. 18, 1885, 3; "A Foundling," *Hartford (CT) Daily Courant*, June 3, 1853. See also *People v. Evans*, 72 Mich. 367, 372 (1888).

48. *NPG*, Aug. 25, 1883, 3.

49. "He Did Right," *NPG*, July 31, 1880, 7.

50. E.g., "Coast Gleanings; An Oregon Girl Shoots Her Father," *LAT*, June 29, 1880, 4.

51. E.g., "Lynching of Negroes," *Washington Post*, Sept. 6, 1897, 3; "Chief Excuse of the Lynchers," *Dallas Morning News*, Nov. 12, 1895, 4; "Southern Ideas of Morality," *Atlanta Constitution*, Sept. 27, 1883, 4 (from the *Burlington [IA] Gazette*); *Daily Constitution* (Atlanta), Oct. 18, 1879, 2; "Western Civilization," *Chicago Daily Tribune*, June 17, 1878, 4; "The Poor Whites of the South," *Daily Cleveland Plain Dealer*, Oct. 17, 1867, col. E.

52. E.g., "Lone Star State," *LAT*, Sept. 5, 1899, 7; "Threats of Lynching," *Atlanta Constitution*, Feb. 21, 1896, 3; "Bowman Not Guilty," *Knoxville (TN) Journal*, Jan. 31, 1896, 3; "Committed for Incest," *Knoxville (TN) Journal*, Nov. 19, 1895, 5; "Strung Him Up," *Wheeling (WV) Register*, June 22, 1895, [1]; "Hanged by a Mob," *State* (Columbia, SC), June 22, 1895, 5; "Incestuous Father Lynched," *Knoxville (TN) Journal*, June 22, 1895, [1]; "Florida's Horrible Sensation," *Columbus (GA) Daily Enquirer-Sun*, June 22, 1895, [1]; "Bestiality of a Negro Father," *Washington Post*, Aug. 3, 1894, 8; "David Brown," *Albuquerque Morning Democrat*, May 27, 1894, [4]; "Taylor Bound Over," *Morning World-*

Herald (Omaha), June 30, 1893, [1]; "In the Southern States," *Dallas Morning News*, Aug. 8, 1892, 5; "Bill Keys Shoots," *Atlanta Constitution*, Aug. 26, 1891, 1; "Belton Tragedy," *Dallas Morning News*, July 23, 1891, 5; "A Horrible Crime," *Columbus (GA) Enquirer-Sun*, July 30, 1890, [4]; "Nine Men Sentenced to Be Hanged," *Chicago Daily Tribune*, July 19, 1890, 1; "Hacked His Own Throat," *Atlanta Constitution*, Jan. 31, 1890, 1; "Arrested for Incest," *Dallas Morning News*, Aug. 15, 1888, 1; "Telegraphic Brevities," *Kansas City (MO) Star*, July 28, 1886, [2]; "Incest Alleged," *Dallas Morning News*, May 16, 1886, 11; "Waco," *Dallas Weekly Herald*, Aug. 13, 1885, [5]; "Hung from a Tree," *Atlanta Constitution*, Mar. 30, 1886, 5; "At the Rope's End," *Atlanta Constitution*, Sept. 16, 1882, 1; "An Ex-Policeman's Depravity," *Washington Post*, Aug. 28, 1881, 4; "A Fiendish Father's Speedy Trial," *Washington Post*, June 2, 1881, 4; "Arraigned for His Crime," *Daily Constitution* (Atlanta), Aug. 23, 1879, 1; "Lynch Law: Terrible Crime and Horrible Punishment," *Philadelphia Inquirer*, Mar. 28, 1879, 4; "Shot with Bullets," *Daily Constitution* (Atlanta), Mar. 26, 1879, 1; "Lynch Law: A Negro Killed by Masked Men in Kentucky," *Philadelphia Inquirer*, Mar. 25, 1879, 4; "Death of the Lynched Raper," *Daily Constitution* (Atlanta), Aug. 14, 1878, 1; "Lynching Was Light," *Constitution* (Atlanta), July 28, 1876, 1.

53. "Burned Himself to Death," *Philadelphia Inquirer*, Nov. 13, 1886, [1]; "Desperate Attempt at Suicide," *Wheeling (WV) Register*, Nov. 13, 1886, [1]; "Swapping the Devil for the Witch," *Columbus (GA) Enquirer-Sun*, Nov. 13, 1886, [1]; "A Negro Pyre," *Grand Forks (ND) Daily Herald*, Nov. 12, 1886, [1]; "Attempted Cremation," *Kansas City (MO) Star*, Nov. 12, 1886, [1].

54. *People v. Lake*, 110 N.Y. 61 (1888).

55. Howard P. Chudacoff, *Age of the Bachelor: Creating an American Subculture* (Princeton, NJ: Princeton University Press, 1999), 187, 201.

56. Ibid., 192, 187–97, 204–7.

57. "Detroit," *NPG*, Apr. 27, 1867, 3.

58. "Bowers Goes Free," *Tacoma Daily News*, Jan. 31, 1893, [1]; "The Bowers Sensation," *Tacoma Daily News*, Jan. 30, 1893. See also "Rockville," *Hartford (CT) Courant*, Mar. 1, 1894, 8; "To Answer for His Crimes," *Aberdeen (SD) Weekly News*, June 1, 1887, [1]; "News by the Mails," *NYT*, Feb. 22, 1853, 6.

59. "The End of the Avery Incest Case," *Chicago Tribune*, Mar. 27, 1862, 4. See also "Latest from Indiana," *NPG*, Oct. 10, 1867, 4; "A Revolting Crime," *Chicago Tribune*, Jan. 7, 1864, 4.

60. E.g., "Peep behind the Scenes," *NPG*, Oct. 14, 1893, 3; "Assaults His Step-Daughter," *NPG*, Sept. 16, 1893, 6; "Accuse Their Father," *NPG*, Sept. 9, 1893, 7; "Eloped with His Stepdaughter," *NPG*, Aug. 24, 1889, 14; "Her Wrongs," *NPG*, Oct. 13, 1888, 6; "Wed His Child," *NPG*, July 28, 1888, 6;

NPG, Aug. 6, 1881, 11; "Tar and Feather Too Good," NPG, June 26, 1880, 11; "Nature Scandalized," NPG, June 11, 1881, 7; "An Inhuman Father," NPG, June 5, 1880, 7; "Lost Sheep!" NPG, May 29, 1880, 11; "A Shameful Alliance," NPG, Feb. 21, 1880, 10; "The Devil's Own," NPG, Apr. 12, 1879, 11; "Henry Bartels, Missing Miscreant," NPG, Jan. 11, 1879, 11; "Hideous Criminality," NPG, Aug. 10, 1878, 11; "A Father's Horrible Crime," NPG, June 8, 1878, 5; "Philadelphia Correspondence," NPG, June 8, 1867, 2; "Detroit: A Man Seduces an Adopted Daughter, Fourteen Years of Age," NPG, Apr. 27, 1867, 3; "The Advantages of Wealth," NPG, June 19, 1847, 324.

61. "Latest From Indiana," NPG, Oct. 10, 1867, 4.

62. "Incest Case," NPG, Apr. 25, 1846, 285.

63. "Legal Opinion of Justice B. W. Osborn, in the Wonderful Case of Incest," NPG, June 20, 1846, 352.

64. "The Incest Case," NPG, June 13, 1846, 339–40; "The Incest Case," NPG, June 6, 1846, 330; "The Incest Case," NPG, May 30, 1846, 323–24; "The Incest Case," NPG, May 28, 1846, 316–17; "The Incest Case," NPG, May 16, 1846, 309; "Further Examination in the Incest Case," NPG, May 16, 1846, 308–9; "The Incest Case," NPG, May 9, 1846, 292–93, 300; "The Incest Case," NPG, May 2, 1846, 293; "The Incest Case," NPG, May 9, 1846, 292–93, 300.

65. "Incest by a Clergyman upon Three Daughters," NPG, Dec. 12, 1846, 106; "Incest Case," NPG, Sept. 12, 1846, 4; "Wonderful Case of Incest," NPG, Aug. 29, 1846, 429; "The Incest Case," NPG, July 4, 1846, 365; "Investigation of the Wonderful Charge of Incest," NPG, June 20, 1846, 352; "The Astonishing Publication," NPG, June 20, 1846, 348; "The Incest Case," NPG, June 20, 1846, 348; "Incest Case," NPG, June 13, 1846, 340; "The Incest Case, Recapitulation," NPG, June 6, 1846, 333.

66. *Poor:* "Total Depravity," *St. Paul Daily News*, May 18, 1893, 8; "Shocking Depravity," NPG, Aug. 18, 1883, 7; "Detroit Wickedness," *Daily Constitution* (Atlanta), Mar. 20, 1881, 1; "New England Specials," *Boston Daily Globe*, Jan. 1, 1881, 1; "Frightful Depravity of a Human Fiend," NPG, Mar. 26, 1881, 11; "The Manca Incest Case," *St. Louis Globe-Democrat*, Jan. 4, 1877, 8; "A Beast Arrested," *New Hampshire Patriot* (Concord), Nov. 10, 1875, [4]; "Incest Case," *Daily Constitution* (Middletown, CT), Mar. 10, 1875, [2]; "The Incest Case in the Police Court," *Chicago Tribune*, Sept. 15, 1863, 4; "Court of General Session," NYT, Feb. 13, 1861, 3.

Wealthy/Respectable: "Story of Incest," *Knoxville (TN) Journal*, Feb. 6, 1898, 8; "A Horrible Story," *Knoxville (TN) Journal*, May 20, 1895, 8; "A Doctor in Trouble," *Morning World-Herald* (Omaha), Nov. 26, 1894, 2; "Incest Charged," *St. Paul Daily News*, June 29, 1893, col. D; "Excitement at Bonham," *Dallas Morning News*, Dec. 3, 1891, 5; "A Prisoner Suicides," *Dallas Morning News*,

Oct. 21, 1891, 2; "Lust Lead to Crime," *Wheeling (WV) Register*, July 27, 1889, [1]; "A Preacher's Flight," *Evening News* (San Jose, CA), May 16, 1889, [2]; "Pacific Coast," *Idaho Daily Statesman* (Boise), Jan. 25, 1889, [1]; "A Grave Charge," *Dallas Morning News*, Oct. 20, 1887, 7; "A Beast," *Aberdeen (SD) News*, Apr. 24, 1887, [1]; "A Horrible Case of Incest," *Aberdeen (SD) News*, Apr. 17, 1887, [1]; "Damning Evidence against a Father," *Kansas City (MO) Star*, Sept. 23, 1886, [1]; "The Stegall Horror," *Dallas Morning News*, Jan. 25, 1886, 8; "An Unnatural Father," *NYT*, Aug. 23, 1886, 1; "Tersified Telegrams," *Summit County Beacon* (Akron, OH), Sept. 23, 1885, [5]; "Tersified Telegrams," *Summit County Beacon* (Akron, OH), June 17, 1885, [5]; "From Savannah," *Macon (GA) Telegraph and Messenger*, June 7, 1885, [1]; "Matters in the West," *Aberdeen (SD) News*, Mar. 27, 1885, [2]; "Arrested for Rape," *Boston Daily Globe*, Apr. 17, 1880, 1; "Incest," *Idaho Tri-Weekly Statesman* (Boise), July 17, 1879, [3]; "A Terrible Crime," *St. Louis Globe-Democrat*, May 26, 1879, 3; "A Drunken Man's Crime," *NYT*, Jan. 1, 1879, 5; "An Inhuman Father," *NYT*, May 27, 1878, 5; "Crime," *Chicago Daily Tribune*, May 20, 1875, 5; "By Mail and Telegraph," *NYT*, Jan. 3, 1873, 1; "A Case of Incest—A Man Marries His Wife's Daughter," *Charleston (SC) Courier, Tri-Weekly*, Apr. 2, 1870, col. E; "New-Jersey," *NYT*, Jan. 4, 1868, 8; "Incest," *Daily Cleveland Herald*, July 9, 1857, col. D; "Startling Disclosure of Crime," *New York Daily Times*, Jan. 8, 1856, 3; "A Monster in Human Shape," *New York Daily Times*, Aug. 24, 1854, 1; "A Case of Bigamy and Incest," *New York Herald*, July 13, 1852; "Incest by a Clergyman upon Three Daughters," *NPG*, Dec. 12, 1846, 106; "Moral Preacher," *Morning News* (New London, CT), Aug. 14, 1846, 2; "Another Reverend Gentleman Uncloaked," *New-Hampshire Gazette*, June 25, 1844, [3]; "Incest and Murder," *Macon (GA) Telegraph*, Sept. 15, 1836, [2].

67. "Man Seduces an Adopted Daughter," *NPG*, Apr. 27, 1867, 3.

68. "Moral Preacher," *Morning News* (New London, CT), Aug. 14, 1846, 2.

69. "Funeral Honors to the Gallant Dead," *Barre (MA) Patriot*, Dec. 18, 1846, [3]; "Incest by a Clergyman upon Three Daughters," *NPG*, Dec. 12, 1846, 106. Three additional indictments alleging he raped three other daughters were continued for trial.

70. "Miscellany," *Essex (MA) Gazette*, Oct. 6, 1832, [4]; "Rev. F. A. Strale," *Boston Investigator*, Oct. 5, 1832, col. C; "Rev. F. A. Strale," *Globe* (Washington, DC), Oct. 3, 1832, col. D; "Rev. F. A. Strale," *New-Hampshire Statesman and State Journal* (Concord), Sept. 29, 1832, col. F; "Rev. F. A. Strale," *Virginia Free Press and Farmers' Repository* (Charles Town, WV), Sept. 27, 1832, col. B; "Rev. F. A. Strale," *Newport (RI) Mercury*, Sept. 22, 1832, [2]; "Rev. F. A. Strale," *Boston Courier*, Sept. 20, 1832, col. C; "Rev. F. A. Strale, "*United States' Telegraph* (Washington, DC), Sept. 18, 1832, col. E.

71. "Found Guilty of Incest," *Morning (Omaha) World-Herald*, Dec. 2, 1895, 2. See also "Wisconsin News," *Milwaukee Sentinel*, July 8, 1884, 8.

72. "Magdalen Reform," *Ohio Observer* (Hudson), June 12, 1834, col. E.

73. "Wethersfield," *Barre (MA) Patriot*, June 10, 1853, [2].

74. E.g., "Excitement at Bonham," *Dallas Morning News*, Dec. 3, 1891, 5; "Incest," *Daily Cleveland Herald*, July 9, 1857, col. D.

75. See Holly Brewer, *By Birth or Consent: Children, Law, and the Anglo-American Revolution in Authority* (Chapel Hill: University of North Carolina Press, 2005), 21–22; 282–83.

76. See Nancy Hathaway Steenburg, *Children and the Criminal Law in Connecticut, 1635–1855* (New York: Routledge, 2005), 168.

77. E.g., "Horrible Depravity," *New Hampshire Patriot and State Gazette* (Concord), Jan. 20, 1858, [2]; "Worse Than Infamous," *Baltimore Patriot*, June 11, 1830, [2].

78. "Winchester; Wednesday; Sheriff; Brooke; Thomas Johnston; Lavinia Johnston," *Farmers' Repository* (Charles Town, WV), June 25, 1817, [3]. See also "Degraded Humanity," *Eagle* (Maysville, KY), July 4, 1817, 4; "Unheard of Crime," *Washington (DC) City Weekly Gazette*, June 28, 1817, 669; and "Unnatural," *Alexandria (VA) Gazette and Daily Advertiser*, June 26, 1817, [2].

79. Brown and Brown, *Hanging of Ephraim Wheeler*, 70, 124. See also Sharon Block, *Rape and Sexual Power in Early America* (Chapel Hill: University of North Carolina Press, 2006), 74–78.

80. E.g., "Vile Crime Charged," *LAT*, Aug. 19, 1899, 4; "Revolting Story," *Morning Herald* (Lexington, KY), Apr. 20, 1896, [1]; "A Horrible Story," *Knoxville (TN) Journal*, May 20, 1895, 8; "Incest," *Penny Press* (Minneapolis), Jan. 18, 1895, 4; "Judge Lynch Delinquent," *Aberdeen (SD) Daily News*, Sept. 26, 1888, [1]; "A Horrible Crime," *Morning Oregonian* (Boise), Oct. 2, 1887, 8; "Horrible Crime Alleged," *Boston Daily Globe*, July 21, 1887, 2; "Shackleford's Frightful Crime," *Atlanta Constitution*, Aug. 21, 1887, 5; "Train Robber Taken," *Kansas City (MO) Evening Star*, Sept. 22, 1885, [1]; "A Father's Unnatural Crime," *NYT*, May 30, 1884, 3; "World Rid of a Monster," *Wheeling (WV) Register*, Dec. 11, 1882, [1]; "Criminals and Their Deeds," *NYT*, Sept. 17, 1882, 9; "A Bunch of Horrors," *NPG*, Feb. 4, 1882, 11; "Fit for the Hangman," *NPG*, July 16, 1881, 2; "Two Brutes in the Shape of Men," *Washington Post*, Mar. 29, 1881, 1; "The Sinful World," *Daily Constitution* (Atlanta), July 14, 1880, 1; "Arrested for Rape," *Boston Daily Globe*, Apr. 17, 1880, 1; "A Terrible Crime: An Ex-Policeman Arrested on a Charge of Incest," *St. Louis Globe-Democrat*, May 26, 1879, 3; "A Father's Fiendish Crime," *NPG*, Dec. 11, 1878, 3; "Crime," *Chicago Daily Tribune*, May 20, 1875, 5; "St. Paul: Contested Land Claim—Arrested for Incest," *Milwaukee Sentinel*, Feb. 1, 1871, col. C; "Charge of Incest,"

NYT, May 15, 1870, 6; "New-Jersey," *NYT*, Jan. 4, 1868, 8; "Local Intelligence," *NYT*, Dec. 9, 1867, 2; "The Incest Case in the Police Court," *Chicago Tribune*, Sept. 15, 1863, 4; "Rape, Incest, and Preparations for Murder," *Omaha Nebraskian*, Sept. 16, 1857, col. G; "City Intelligence," *New York Herald*, May 6, 1845, col. E; "On Thursday Evening Last," *Supporter* (Chillicothe, OH), July 1, 1818, col. A.

81. "Incest Case at Blackfoot," *Salt Lake Semi-Weekly Tribune*, July 30, 1897, 6.

82. "The Sinful World," *Daily Constitution* (Atlanta), July 14, 1880, 1.

83. E.g., the case of a villain from Chillicothe of a very respectable station: "Incest and Murder," *Daily Commercial Bulletin and Missouri Literary Register* (St. Louis), Sept. 22, 1836, col. B; "Incest and Murder," *Macon (GA) Telegraph*, Sept. 15, 1836, [2]; "Wheeling, Aug. 24," *New-York Spectator*, Sept. 5, 1836, col. D; "Multiple News Items," *New York Herald*, Aug. 31, 1836, col. C.

84. "Appalling!" *NPG*, Mar. 30, 1889, 6. See also "Whispers of Scandal," *NPG*, June 3, 1882, 3, and "New-Jersey," *NYT*, Oct. 15, 1860, 5 (after his daughter resisted, defendant tried to kill her and then himself).

85. "A Foul Beast's Crime," *NPG*, June 24, 1882, 10. See also "New York Items," *Hartford (CT) Daily Courant*, Sept. 20, 1860, 3.

86. "Shackleford's Frightful Crime," *Atlanta Constitution*, Aug. 21, 1887, 5; "Ruined by Her Father," *Atlanta Constitution*, Apr. 15, 1886, 5; "A Father's Crime," *Milwaukee Sentinel*, Jan. 11, 1885, 4; "Acworth, Georgia," *Atlanta Constitution*, Aug. 16, 1883, 2; "Frightful Depravity of a Human Fiend," *NPG*, Mar. 26, 1881, 11; "A Fiend as a Father," *NPG*, May 17, 1879, 3; "Nevada: Incest, Infanticide, and Attempted Suicide," *Daily Evening Bulletin* (San Francisco), Aug. 28, 1873, col. C; "Horrible Case of Incest and Child Murder; Shocking Immortality," *NPG*, Oct. 10, 1867, 4; "Compendium," *Boston Investigator*, July 17, 1839, col. E.

87. E.g., "A Foul Beast's Crime," *NPG*, June 24, 1882, 10; "Rohde's Revolting Depravity," *NPG*, Apr. 9, 1881, 7.

88. "Edward Parr," *Chicago Daily*, June 24, 1879, 2.

89. E.g., "Charged with Incest," *Fort Worth Gazette*, June 13, 1891, [1]; "Case Settled," *Wheeling (WV) Sunday Register*, Mar. 16, 1890, 5; "Territorial Items," *Aberdeen (SD) News*, May 11, 1887, [1]; "Georgia by Wire," *Atlanta Constitution*, Aug. 18, 1883, 2; "Parkersburg," *Wheeling (WV) Register*, June 30, 1885, [3]; "A Father's Terrible Crime," *NYT*, Jan. 14, 1883, 1; "Most Horrible," *Wheeling (WV) Register*, Aug. 12, 1882, [1]; "The Devil's Own," *NPG*, Apr. 12, 1879, 11; "Henry Bartels, Missing Miscreant," *NPG*, Jan. 11, 1879, 11;"Another Miscreant Caught," *NPG*, Mar. 20, 1847, 222.

90. "Last Night's Dispatches," *LAT*, June 6, 1882, 2; "Rape and Incest,"

Weekly Arizona Miner (Prescott), June 2, 1882, [2]; "Appropriate Sentence," *Weekly Arizona Miner* (Prescott), June 2, 1882, [2]; "The Apaches," *LAT*, May 30, 1882, 2.

91. "Queer Proposition," *LAT*, Nov. 24, 1895, 34. See also "A Father's Unnatural Crime," *NYT*, May 30, 1884, 3.

92. "The Lybarger Murder," *NPG*, Sept. 21, 1889, 14; "A Father's Unnatural Crime," *NYT*, May 30, 1884, 3; "Detroit Wickedness," *Daily Constitution* (Atlanta), Mar. 20, 1881, 1; "He Did Right: A Brute of a Father Who Realized That He Was Too Vile to Live, and Therefore Committed Suicide," *NPG*, July 31, 1880, 7; "Crime," *Chicago Daily Tribune*, May 20, 1875, 5; "The Northwestern States: Missouri," *Chicago Daily Tribune*, June 6, 1874, 12; "Incest, Infanticide, and Attempted Suicide," *Daily Evening Bulletin* (San Francisco), Aug. 28, 1873.

93. "A Beastly Californian," *Morning Olympian* (Olympia, WA), May 1, 1892, [1]; "Belton Tragedy," *Dallas Morning News*, July 23, 1891, 5; "Who Murdered Her?" *NPG*, Dec. 6, 1890, 6; "A Lecherous Brute," *Knoxville (TN) Journal*, Aug. 20, 1890, [1]; "Father in Jail," *Atlanta Constitution*, May 20, 1890, 1; "The Lybarger Murder," *NPG*, Sept. 21, 1889, 14; "Judge Lynch Delinquent," *Aberdeen (SD) Daily News*, Sept. 26, 1888, [1]; "Items from the South," *Aberdeen (SD) Daily News*, Apr. 9, 1886, [2]; "A Terrible Crime," *St. Louis Globe-Democrat*, May 26, 1879, 3;"A Monster in Human Shape," *New York Daily Times*, Aug. 24, 1854, 1.

94. E.g., "Tough If True," *Grand Forks (ND) Daily Herald*, May 26, 1895, [1]; "A Horrible Charge," *Wheeling (WV) Register*, Feb. 3, 1895, 5; "Incest," *Penny Press* (Minneapolis), Jan. 18, 1895, 4; "Peep behind the Scenes," *NPG*, Oct. 14, 1893, 3; "Incest Charged," *St. Paul Daily News*, June 29, 1893, col. E; "Charged with Incest," *Grand Forks (ND) Daily Herald*, July 22, 1891, [8]; "City Briefs," *LAT*, July 2, 1891, 8; "Another Disnatured Wretch Is under Arrest," *Morning Oregonian* (Boise), Nov. 19, 1889, 6; "Her Wrongs," *NPG*, Oct. 13, 1888, 6; "A Brute," *LAT*, May 2, 1888, 8; "Arrested for Incest," *Dallas Morning News*, Mar. 13, 1888, 2; "Rape and Incest," *Rocky Mountain News* (Denver, CO), Mar. 3, 1888, col. D; "A Horrible Crime," *Morning Oregonian* (Boise), Oct. 2, 1887, 8; "A Beast," *Aberdeen (SD) News*, Apr. 24, 1887, [1]; "A Horrible Case of Incest," *Aberdeen (SD) News*, Apr. 17, 1887, [1]; "A Grave Charge," *Macon (GA) Telegraph*, Aug. 23, 1886, [1]; "Accused of Incest," *Dallas Morning News*, Oct. 31, 1885, 3; "Let Him Be Hanged," *Kansas City (MO) Evening Star*, Jan. 13, 1886, [1]; "Train Robber Taken," *Kansas City (MO) Evening Star*, Sept. 22, 1885, [1]; "Atkins's Crime," *Atlanta Constitution*, Nov. 20, 1884, 2; "Disgusting Case of Incest," *Daily (ND) Herald*, Sept. 9, 1883, [4]; "Incest," *Chicago Daily Tribune*, Dec. 17, 1877, 8; "[Albert H. Essex]," *Daily Constitution* (Middletown, CT), Feb. 20, 1873, [2]; "A Case of Bigamy and Incest," *New York Herald*, July 3, 1852;

"Incest and Rape," *Daily Delta* (New Orleans), Sept. 7, 1848, [3]; *Bergen v. Bergen*, 22 Ill. 187 (1859); *Clark v. Clark*, 29 Ill. App. 257 (3d Dist. 1888).

95. "Otero Occurrences," *Santa Fe New Mexican*, Oct. 19, 1899, [1]; "Suit over Water," *LAT*, Aug. 19, 1899, 4; "Tried to Commit Murder," *Morning World-Herald* (Omaha), Jan. 13, 1899, 2; "Story of Incest," *Knoxville (TN) Journal*, Feb. 6, 1898, 8; "Florida's Horrible Sensation," *Columbus (GA) Daily Enquirer-Sun*, June 22, 1895, [1]; "A Father's Crime," *Morning World-Herald* (Omaha), Dec. 30, 1892, [1]; "An Awful Crime," *Wheeling (WV) Register*, June 27, 1892, [1]; "Raleigh Notes," *Daily News* (Charlotte, NC), Jan. 14, 1889, [4]; "The LeBeau Nastiness," *Aberdeen (SD) News*, Apr. 24, 1887, [1]; "Charged with Incest," *Dallas Morning News*, Jan. 8, 1886, 8; "Parkersburg," *Wheeling (WV) Register*, June 30, 1885, [3]; "Barnesville," *Wheeling (WV) Register*, Apr. 24, 1880, [1]; "Charged with Incest," *Daily Sun* (Columbus, GA), Sept. 3, 1872, [1]; "Adultery," *Dakota Republican* (Vermillion, SD), Dec. 22, 1870, [2]; "A Minnesota Horror," *Chicago Tribune*, Oct. 2, 1867, 2. See also Block, *Rape and Sexual Power*, 103.

96. "Incest," *Chicago Daily Tribune*, Dec. 17, 1877, 8.

97. E.g., "Peep behind the Scenes," *NPG*, Oct. 14, 1893, 3; "An Incestuous Father," *NPG*, July 27, 1878, 7; "A Strange Story," *Boston Daily Globe*, June 13, 1878, 4.

98. "A Father Charged with the Seduction of His Six Daughters, and Attempts to Seduce the Grand-daughter," *Union Dakotaian* (Yankton, SD), Apr. 24, 1869, 1. Thanks to Beverly J. Schwartzberg, who found and hand-copied this report for me.

99. Ibid. Similar cases include "An Incestuous Grandfather," *NPG*, July 27, 1878, 7, and "Alleged Incest and Violent Death of a Farmer," *St. Louis Globe-Democrat*, July 25, 1875, 12.

100. "A Father Charged with the Seduction of His Six Daughters, and Attempts to Seduce the Grand-daughter," *Union Dakotaian* (Yankton, SD), Apr. 24, 1869, 1.

101. Ibid., 1.

102. See Sommerville, *Rape and Race*, 69. For a discussion of differences between cultural proscription and social reality in early America, see Cynthia A. Kierner, *Scandal at Bizarre: Rumor and Reputation in Jefferson's America* (New York: Palgrave Macmillan, 2004), 7.

103. Carole Shammas, *A History of Household Government in America* (Charlottesville: University of Virginia Press, 2002), 20; Laura Edwards, "Law, Domestic Violence, and the Limits of Patriarchal Authority in the Antebellum South," *Journal of Southern History* 65 (Nov. 1999): 733, 759; Steenburg, *Children and the Criminal Law*, 1–4.

104. Brewer, *By Birth or Consent*, 21–22, 282–83; Shammas, *Household Government*, 79; Nancy Cott, *Public Vows: A History of Marriage and the Nation* (Cambridge: Harvard University Press, 2000), 12–22; John Demos, *A Little Commonwealth: Family Life in Plymouth Colony*, 2nd ed. (New York: Oxford University Press, 2000); Edmund S. Morgan, *The Puritan Family: Religion and Domestic Relations in Seventeenth Century New England*, rev. ed. (New York: Harper and Row, 1966), 142–50.

105. Morgan, *Puritan Family*, 146.

106. Shammas, *Household Government*, chap. 5. See also Shawn Johansen, *Family Men: Middle-Class Fatherhood in Early Industrializing America* (New York: Routledge, 2001), chap. 4, and Edwards, "Law, Domestic Violence," 739–42.

107. Community intervention might follow various instances of paternal domestic violence. See Edwards, "Law, Domestic Violence," 754–55, 757–59, 763; Shammas, *Household Government*, 47–50; Block, *Rape and Sexual Power*, 240, 246; Dayton, *Women before the Bar*, 274–76.

108. I reviewed compilations of the earliest laws for Connecticut, Delaware, Georgia, Maine, Maryland, Massachusetts, New Hampshire, New Jersey, New York, North Carolina, Pennsylvania, Rhode Island, South Carolina, Tennessee, Vermont, and Virginia. See also Leigh B. Bienen, "Defining Incest," *Northwestern University Law Review* 4 (Summer 1998): 1501, 1524–31; Bardaglio, *Reconstructing the Household*, 39–44.

109. John D. Cushing, ed., "Incest," in *The Earliest Laws of the New Haven and Connecticut Colonies, 1639–1673* (Wilmington, DE: Michael Glazier, 1977), 35.

110. Bienen, "Defining Incest," 1531; Dayton, *Women before the Bar*, 274; John D. Cushing, ed., "An Act for the Establishment of Religious Worship in This Province According to the Church of England [1701–2]," in *The Laws of the Province of Maryland*, (Wilmington, DE: Michael Glazier, 1978), 17–23; William Hand Brown, ed., "An Act Touching on Marriages [1640]," in *Archives of Maryland: Proceedings and Acts of the General Assembly of Maryland, January 1638–September 1664* (Baltimore: Maryland Historical Society, 1883), 97; Rhode Island, "An Act Regulating Marriage and Divorce [1798]," in *The Public Laws of Rhode-Island and Providence Plantations* (Providence: Carter and Wilkinson, 1798), 477–81.

111. Grossberg, *Governing the Hearth*, 145; Dayton, *Women before the Bar*, 233; Bardaglio, *Reconstructing the Household*, 201–4.

112. See Bardaglio, *Reconstructing the Household*, 45–48; Steenburg, *Children and the Criminal Law*, 178–80; Amasa J. Parker, ed., "People vs. James Haridan," May 1852, *Reports of Decisions in Criminal Cases Made at Term, at*

Chambers and in the Courts of Oyer and Terminer of the State of New York, vol. 1 (New York: Banks and Bros., 1860), 43. Courts across the country relied on *Harridan* to dismiss incest cases in which the daughter had not consented to the sexual contact.

113. *State v. Smith*, 30 La. Ann. 846, 848 (1878).

114. Department of the Interior, Census Office, *Report on Crime, Pauperism, and Benevolence in the United States at the Eleventh Census, 1890*, pt. II, General Tables (Washington, DC: U.S. Government Printing Office, 1895), 372.

115. Ibid.

116. Kathleen Ruth Parker, "Law, Culture, and Sexual Censure: Sex Crime Prosecutions in a Midwest County Court, 1850–1950" (Ph.D. diss., Michigan State University, 1993), 238.

117. I searched from 1800 to 1899 in the Lexis-Nexis Academic Universe and Westlaw electronic databases. These databases include all state supreme court opinions, many appellate court decisions, and a few trial court opinions, but they are not comprehensive. See also Bienen, "Defining Incest," 1545.

118. See also Bardaglio, *Reconstructing the Household*, 203–12, 234; Bienen, "Defining Incest," 1502, 1538–39.

119. *Incest: People v. Kaiser*, 119 Cal. 456 (1895) (father, affirmed); *People v. Scott*, 59 Cal. 341 (1881) (father, reversed); *People v. Patterson*, 102 Cal. 239 (1894) (father, affirmed); *People v. Benoit*, 97 Cal. 249 (1893) (father, reversed).

Rape: People v. Lambert, 120 Cal. 170 (1898) (father, reversed); *People v. Fultz*, 109 Cal. 258 (1895) (father, affirmed); *People v. Lenon*, 79 Cal. 625 (1889) (stepfather, affirmed); *People v. Hamilton*, 46 Cal. 540 (1873) (stepfather, reversed).

Assault: People v. Manchego, 80 Cal. 306 (1889) (stepfather, affirmed).

120. See also. Bienen, "Defining Incest," 1545.

121. E.g., *Dean v. Raplee*, 145 N.Y. 319 (1895) (14-year-old orphan sued guardian who repeatedly raped her).

122. E.g., *Jackson v. State*, 28 Tex. Ct. App. 108 (1889); *Clark v. Clark*, 29 Ill. App. 257 (3d Dist. 1888); "Big Libel Suits," *Chicago Daily Tribune*, Jan. 15, 1881, 6; *People v. Parton*, 49 Cal. 632 (1875); *Dukes v. Clark*, 2 Blackf. 20 (Indiana, 1826).

123. E.g., John D. Cushing, ed., "Capital Laws, Paragraph 9," in *Earliest Laws of the New Haven and Connecticut Colonies*, 9; John D. Cushing, ed., "Capital Laws, Chapter IV.14," *Laws of the Pilgrims: A Facsimile Edition of the Book of the General Laws of the Inhabitants of the Jurisdiction of New-Plimouth, 1672 and 1685* (Wilmington, DE: Michael Glazier, 1977), 10; Bienen, "Defining Incest," 1547–49; Steenburg, *Children and the Criminal Law*, 167–82; *LAT*, Jan. 28, 1893, 4; "The Courts," *LAT*, Nov. 16, 1892, 5; "The Courts," *LAT*, Sept. 15, 1892, 10.

124. E.g., "Concurrence of Rape and Adultery or Incest," *Albany Law Journal: A Weekly Record of the Law and the Lawyers* 25 (June 24, 1882): 484; *Commonwealth v. Goodhue*, 43 Mass. 193 (1840).

125. E.g., "Criminal Law," *Albany Law Journal: A Weekly Record of the Law and the Lawyers* 25 (Feb. 25, 1882): 159; "Trial For Rape," *Boston Cultivator* 2 (Dec. 5, 1840): 2.

126. See Robertson, *Crimes against Children*, 49; Steenburg, *Children and the Criminal Law*, 161–62, 170–71, 180–81. For issues in rape prosecutions peculiar to girls, see Sommerville, *Race and Rape*, chap. 2.

127. This standard fails to consider surprise attacks or circumstances like drugging that make it difficult for a woman to forcefully resist. Robertson, *Crimes against Children*, 33–35.

128. Christine Stansell, *City of Women: Sex and Class in New York, 1789–1860* (New York: Knopf, 1986), 183, 182–83. Fathers also asserted this defense. *People v. Grauer*, 42 N.Y.S. 721 (1896).

129. E.g., "An Ugly Affair," *Santa Fe Daily New Mexican*, Nov. 7, 1894, [4]; "A Vile Crime," *NPG*, July 31, 1886, 7; "Parkersburg," *Wheeling (WV) Register*, June 30, 1885, [3]; "Detroit Wickedness," *Daily Constitution* (Atlanta), Mar. 20, 1881, 1; "Beallsville," *Wheeling (WV) Register*, May 26, 1880, [4]; "News of the State," *Hartford (CT) Daily Courant*, Sept. 4, 1874, 1.

130. "Otero Occurrences. Incest Cases in the New County—Other Matters," *Santa Fe New Mexican*, Oct. 19, 1899, [1].

131. Linda Gordon found similar patterns, but her sample was limited to poor and immigrant families who sought social service intervention in Boston. Linda Gordon, *Heroes of Their Own Lives: The Politics and History of Family Violence, Boston 1880–1960* (New York: Viking, 1988).

132. "Around the Courthouse," *Dallas Morning News*, Sept. 26, 1896, 3 (6 years); "Waived Examination," *Aberdeen (SD) Daily News*, June 10, 1893, [3] (3 years); "A Brute Arrested," *Wheeling (WV) Sunday Register*, Aug. 26, 1890, [4] (8 years); "They Should Not Have Interfered," *Washington Post*, Apr. 13, 1888, 1 (3 years); "A Beastly Father," *Dallas Morning News*, Feb. 12, 1886, 1 (6 years); "Most Horrible," *Wheeling (WV) Register*, Aug. 12, 1882, [1] ("years," two babies born); "A Father's Crime," *Wheeling (WV) Register*, July 27, 1880, [1] (8 years); "A Terrible Crime," *St. Louis Globe-Democrat*, May 26, 1879, 3 (4 years); "A Revolting Crime," *Chicago Tribune*, Jan. 7, 1864, 4 (3 years).

133. "New Mexico News," *Santa Fe Daily New Mexican*, Nov. 14, 1895, [1]; "Heinous Practices," *LAT*, Sept. 12, 1895, 8; "South Dakota," *Aberdeen (SD) Daily News*, June 27, 1895, [2]; "Horrible Case," *Idaho Daily Statesman* (Boise), June 7, 1895, [6] (four daughters, over 3 years); "Tough If True," *Grand Forks (ND) Daily Herald*, May 26, 1895, [1]; "A Horrible Story," *Knoxville (TN)*

Journal, May 20, 1895, 8; "Sentenced to Fifteen Years," *Tacoma Daily News*, Apr. 23, 1895, [1]; "Crowd Watching," *Dallas Morning News*, Apr. 11, 1895, 5; "Charged with Incest," *Morning World-Herald* (Omaha), Dec. 1, 1894, [2]; "A Texan's Crime," *Atlanta Constitution*, Sept. 7, 1883, 5; "Could Anything Be Worse?" *Knoxville (TN) Journal*, Jan. 27, 1893, [1]; "A [*sic*] Inhuman Father," *Morning World-Herald* (Omaha), Aug. 11, 1892, 5; "Two Warrants," *Atlanta Constitution*, June 24, 1892, 3; "A Beastly Californian," *Morning Olympian* (Olympia, WA), May 1, 1892, [1]; "An Incestuous Brute," *LAT*, Feb. 11, 1892, 6; "Sensational Arrests," *Dallas Morning News*, June 15, 1889, 1; "Charged with a Terrible Crime," *Atlanta Constitution*, Aug. 1, 1888, 2; "Pacific Coast," *Idaho Tri-Weekly Statesman* (Boise), June 27, 1888, [1]; "News Notes," *Kansas City (MO) Star*, Aug. 24, 1887, [1]; "Fourteen Years in Prison," *Boston Daily Globe*, July 23, 1887, 2; "A Beast," *Aberdeen (SD) News*, Apr. 24, 1887, [1]; "His Just Desserts," *Duluth (MN) Daily Tribune*, Aug. 27, 1885, [1]; "A Texan's Crime," *Atlanta Constitution*, Sept. 7, 1883, 5; "Sioux Commission Closed," *Grand Forks (ND) Daily Herald*, Dec. 23, 1882, [1]; "Bestial Bullock," *Wheeling (WV) Register*, Feb. 10, 1882, [4]; "An Improving Outlook at Albany," *Washington Post*, Jan. 13, 1882, 1; "Missouri and Kansas," *Kansas City (MO) Evening Star*, Dec. 1, 1880, [2]; "Sickening Case of Incest in Portland—An Unnatural Father Debauches His Own Daughters," *Boston Daily Globe*, Aug. 24, 1876, 4; "Incest," *Daily Cleveland Herald*, July 9, 1857; "Interesting Items," *Hartford (CT) Daily Courant*, Oct. 8, 1863, 2.

134. "Revolting Story," *Morning Herald* (Lexington, KY), Apr. 20, 1896, [1]; "South Dakota," *Aberdeen (SD) Daily News*, June 27, 1895, [2]; "Horrible Case," *Idaho Daily Statesman* (Boise), June 7, 1895, [6]; "Suicided in Jail," *Grand Forks (ND) Daily Herald*, Apr. 1, 1892, [2]; "An Incestuous Beast," *Wheeling (WV) Register*, Feb. 19, 1892, [1]; "Lust Lead to Crime," *Wheeling (WV) Register*, July 27, 1889, [1]; "Appalling!" *NPG*, Mar. 30, 1889, 6; "Judge Lynch Delinquent," *Aberdeen (SD) Daily News*, Sept. 26, 1888, [1]; "Under Three Serious Charges," *Atlanta Constitution*, July 10, 1888, 1; "A Change," *LAT*, June 11, 1882, 2; "Detroit Wickedness," *Daily Constitution* (Atlanta), Mar. 20, 1881, 1; "A Shameful Alliance," *NPG*, Feb. 21, 1880, 10; "A Case of Incest—A Man Marries His Wife's Daughter," *Charleston (SC) Courier, Tri-Weekly*, Apr. 2, 1870, col. E; "The Incest Case in the Police Court," *Chicago Tribune*, Sept. 15, 1863, 4.

135. "A Long Sentence," *Chicago Daily Tribune*, Sept. 6, 1883, 7; "Assaults His Step-Daughter," *NPG*, Sept. 16, 1893, 6.

136. E.g., *State v. Baker*, 136 Mo. 74, 82 (1896). See also Bardaglio, *Reconstructing the Household*, 208–12; Dayton, *Women before the Bar*, 276–80.

137. *Bailey v. Commonwealth*, 82 Va. 107, 112 (1886).

138. *Curby v. Territory of Arizona*, 4 Ariz. 371 (1895).

139. See Sommerville, *Rape and Race*, 44–45; Bienen, "Defining Incest," 1546–56.

140. Theodoric Romeyn Beck, *Elements of Medical Jurisprudence*, ed. John Darwall, 3rd ed. (London: Longman, Rees, Orme, Brown, and Green, 1829), 56, 59; Robertson, *Crimes against Children*, 39.

141. "Hanging of a Man in North Carolina for Rape," *Chicago Tribune*, Sept. 4, 1867, 2.

142. E.g., "Worse Than a Brute," *Wheeling (WV) Register*, Aug. 4, 1895, [1]; "Nine Men Sentenced to Be Hanged," *Chicago Daily Tribune*, July 19, 1890, 1; "Hanged for Incest," *Albuquerque Morning Democrat*, Jan. 11, 1890, [1]; "Execution," *American Phrenological Journal* 16 (Sept. 1852): 65; "Execution," *Southern Patriot* (Charleston, SC), Dec. 24, 1845, [2]; "Arkansas, A.T., Saturday, January 22, 1820," *Arkansas Gazette* (Arkansas Post), Jan. 22, 1820, [3] (sentenced to death).

143. E.g., "Threats of Lynching," *Atlanta Constitution*, Feb. 21, 1896, 3 (20 years); "Executive Clemency," *LAT*, July 12, 1897, 9 (15 years); "Garrett Arraigned," *LAT*, Dec. 15, 1895, 26 (maximum penalty, 10 years Folsom); "Spirited Her Away," *Morning World-Herald* (Omaha), Mar. 21, 1895, 4 (15 years); "Frank L. Palmer Free," *Morning Oregonian* (Boise), Jan. 15, 1895, 8 (old man given 25 years); "Northwestern News," *Tacoma Daily News*, July 31, 1894, [4] (25 years); "Ventura County," *LAT*, Nov. 19, 1893, 20 (maximum penalty, 10 years San Quentin); "Fifteen Years for Incest," *Grand Forks (ND) Daily Herald*, Mar. 16, 1892, [2] (15 years hard labor); "To the Pen for Life," *Morning World-Herald* (Omaha), Oct. 28, 1891, 6; "Case Reversed," *Knoxville (TN) Journal*, Mar. 4, 1891, [1] (21 years); "The Punishment Severe," *Morning World-Herald* (Omaha), Oct. 24, 1890, [1] (30 years hard labor); "Burned in the Mine," *Macon (GA) Telegraph*, Aug. 6, 1889, [1] (life); "Tersified Telegrams," *Summit County Beacon* (Akron, OH), Sept. 23, 1885, [5] (20 years); "Falsely Accusing Her Father," *NYT*, May 2, 1885, 5 (life); "A Long Sentence," *Chicago Daily Tribune*, Sept. 6, 1883, 7 (30 years for the rape of one daughter; trial for rape of second daughter pending); "A Beast Sentenced—Indians," *LAT*, May 30, 1882, 2 (life); "That Iowa Case," *Chicago Daily Tribune*, June 27, 1875, 13 (10 years solitary); [Wm. N. Gove], *Bangor (ME) Daily Whig and Courier*, Mar. 20, 1872, col. G (life); "Another Hideous Crime—A Father Commits a Rape upon His Own Daughter!" *NPG*, Apr. 20, 1867, 3 (life); "Multiple News Items," *Lowell (MA) Daily Citizen and News*, July 3, 1858, col. B (life); "Reported for the Courant," *Hartford (CT) Daily Courant*," Mar. 17, 1847, 2 (life); "Moses Goodhue," *Haverhill (MA) Gazette*, Dec. 5, 1840, [2] (3 days solitary plus 20 years hard labor); "Indiana v. John Allen," *Indiana Journal* (Indianapolis), June 30, 1830 (intent to rape, 10 years); "Court of Sessions," *Farmer's Cabinet* (Amherst, NH),

Mar. 24, 1827, [3] (life); "Newbury," *Village Register, and Norfolk County Advertiser* (Dedham, MA), Dec. 5, 1823, [26] (life);"Unnatural Crime," *Alexandria (VA) Herald*, Aug. 5, 1818, [2] (48-year-old defendant, 14 years hard labor).

144. "Multiple News Items," *Daily Atlas* (Boston), Apr. 14, 1843, col. A.

145. "Superior Court," *Hartford (CT) Daily Courant*, Mar. 17, 1847, 2; cf. "News of the State," *Hartford (CT) Daily Courant*, Apr. 7, 1884, 2 (guilty plea, 4 years).

146. "Las Vegas Happenings," *Santa Fe New Mexican*, Feb. 17, 1898, [4].

147. "Goes Up for Thirty Years," *Duluth (MN) News*, Oct. 12, 1892, [1]. See also "A Proposed Arizona Railroad," *LAT*, Sept. 2, 1897, 3 (54-year-old, 30 years); "Concord and Vicinity," *New Hampshire Patriot* (Concord), Apr. 25, 1877, [2] (54-year-old, 27 years).

148. *Atlanta Constitution*, June 30, 1875, 4; "That Iowa Case," *Chicago Daily Tribune*, June 27, 1875, 13.

149. E.g., "Pavilades Insolvent," *LAT*, Dec. 12, 1895, 9 (jury deliberated five minutes); "Court of Sessions," *Raleigh Register, and North-Carolina Gazette*, Mar. 27, 1827, col. C.

150. "Arkansas, A.T.," *Arkansas Gazette* (Arkansas Post), Jan. 22, 1820, [3].

151. *Tuberville v. State*, 4 Tex. 128, 130 (1849).

152. Ibid., 130.

153. Ibid.

154. Ibid.

155. See Gail Bederman, *Manliness and Civilization: A Cultural History of Gender and Race in the United States, 1880–1917* (Chicago: University of Chicago Press, 1995), 5–12.

156. E.g., *Clements v. State*, 34 Tex. Crim. 616 (1895); *Johnson v. State*, 20 Tex. Ct. App. 609 (1886).

157. The court admitted that evidence of a family's "degraded" condition did not prove incest and reversed the conviction. *Owens v. Nebraska*, 32 Neb. 167, 172 (1891).

158. Department of the Interior, Census Office, *Report on Crime*, 373–77.

159. Parker, "Law, Culture, and Sexual Censure," 256.

160. E.g., "Otero Occurrences," *Santa Fe New Mexican*, Oct. 19, 1899, [1]; "Las Vegas Notes," *Santa Fe New Mexican*, July 27, 1899, [4]; "An Alleged Human Brute," *LAT*, Sept. 15, 1898, 4; "Arrested for Incest," *Morning World-Herald* (Omaha), July 16, 1898, 3; "A Girl's Confession," *LAT*, Sept. 15, 1895, 9; "Heinous Practices," *LAT*, Sept. 12, 1895, 8; "Tough If True," *Grand Forks (ND) Daily Herald*, May 26, 1895, [1]; "Nubs of News," *Grand Forks (ND) Daily Herald*, Apr. 26, 1895, 2; "A Horrible Charge," *Wheeling (WV) Register*,

Feb. 3, 1895, 5; "Bestiality of a Negro Father," *Washington Post*, Aug. 3, 1894, 8; "Father's Baneful Crime," *Morning World-Herald* (Omaha), Nov. 15, 1893, [1]; "Accuse Their Father," *NPG*, Sept. 9, 1893, 7; "Taylor Bound Over," *Morning World-Herald* (Omaha), June 30, 1893, [1]; "Total Depravity," *St. Paul Daily News*, May 18, 1893, 8; "Excitement at Bonham," *Dallas Morning News*, Dec. 3, 1891, 5; "An Unnatural Father," *Morning World-Herald* (Omaha), Sept. 19, 1891, [1]; "Charged with Incest," *Grand Forks (ND) Daily Herald*, July 22, 1891, [8]; "A Brute Arrested," *Wheeling (WV) Sunday Register*, Aug. 26, 1890, [4]; "Steward Incest Case," *Wheeling (WV) Register*, Mar. 13, 1890, [1]; "Ruined by Her Step-Father," *Wheeling (WV) Register*, Apr. 26, 1889, [1]; "Pacific Coast," *Idaho Daily Statesman* (Boise), Jan. 25, 1889, [1]; "Charged with Incest—Injured," *Dallas Morning News*, Sept. 8, 1888, 2; "They Should Not Have Interfered," *Washington Post*, Apr. 13, 1888, 1; "A Most Unnatural Father," *Dallas Morning News*, Mar. 20, 1888, 2; "Shackleford's Frightful Crime," *Atlanta Constitution*, Aug. 21, 1887, 5; "Terrible Story of Incest," *Dallas Morning News*, July 2, 1886, 1; "A Beastly Father," *Dallas Morning News*, Feb. 12, 1886, 1; "Brief Telegrams," *Wheeling (WV) Register*, Apr. 17, 1885, [1]; "Charged with Incest," *Dallas Weekly Herald*, Aug. 28, 1884, [6]; "Georgia Press," *Macon (GA) Telegraph and Messenger*, Sept. 9, 1883, [3]; "Most Horrible," *Wheeling (WV) Register*, Aug. 12, 1882, [1]; "A Fiendish Father's Speedy Trial," *Washington Post*, June 2, 1881, 4; "Candidates," *NPG*, June 26, 1880, 11; "Beallsville," *Wheeling (WV) Register*, May 26, 1880, [4]; "A Terrible Crime: An Ex-Policeman Arrested on a Charge of Incest," *St. Louis Globe-Democrat*, May 26, 1879, 3; "From the Four Courts," *St. Louis Globe-Democrat*, Nov. 14, 1878, col. B; "A Father's Double Crime," *Wheeling (WV) Daily Register*, Aug. 27, 1878, [1]; "A Strange Story," *Boston Daily Globe*, June 13, 1878, 4; "Incest," *Chicago Daily Tribune*, Dec. 17, 1877, 8; "Criminal," *Chicago Daily Tribune*, Oct. 6, 1876, 8; "A Beast Arrested," *New Hampshire Patriot* (Concord), Nov. 10, 1875, [4]; "Michigan Morals, A Father Seduces His Stepdaughter," *Inter Ocean* (Chicago), July 12, 1875, col. F; "Local Items," *Constitution* (Middletown, CT), Apr. 21, 1875, [2]; "The Northwestern States," *Chicago Daily Tribune*, June 6, 1874, 12; "Crime," *Chicago Daily Tribune*, Feb. 26, 1874, 8; "Crimes and Casualties," *New-Hampshire Patriot* (Concord), Oct. 22, 1873, [3]; "Latest from Indiana," *NPG*, Oct. 10, 1867, 4; "Wisconsin Items," *Milwaukee Daily Sentinel*, Jan. 23, 1867, col. D; "Neighborhood News," *Daily Cleveland Herald*, Aug. 18, 1865, col. D; "Multiple News Items, Incest," *North American and United States Gazette* (Philadelphia), Oct. 30, 1854, col. I; "Wethersfield", *Barre (MA) Patriot*, June 10, 1853, [2]; "A Foundling," *Hartford (CT) Daily Courant*, June 3, 1853, 2; "A Case of Bigamy and Incest," *New York Herald*, July 3, 1852, col. B; "Charge of

Incest," *New York Daily Herald*, Mar. 25, 1845. See also *Commonwealth v. Kammerdiner*, 165 Pa. 222 (1895), and *Dodson v. State*, 24 Tex. App. 514 (1887).

161. Nell Irvin Painter, "Soul Murder and Slavery: Toward a Fully Loaded Cost Accounting," in *U.S. History as Women's History: New Feminist Essays*, ed. Linda K. Kerber et al. (Chapel Hill: University of North Carolina Press, 1995), 130, 145.

162. Ann Taves, ed., *Religion and Domestic Violence in Early New England: The Memoirs of Abigail Abbott Bailey* (Bloomington: Indiana University Press, 1989).

163. Ibid., 70–83, 87–88.

164. Ibid., 70–83. See also Block, *Rape and Sexual Power*, 96.

165. Taves, *Religion and Domestic Violence*, 77–78.

166. Ibid., 172–78.

167. Brown and Brown, *Hanging of Ephraim Wheeler*.

168. Irene Quenzler Brown and Richard D. Brown, "Tales from the Vault," *Common-Place* 1 (Sept. 2000), www.common-place.org/vol-01/no-01/tales/.

169. Brown and Brown, *Hanging of Ephraim Wheeler*, 4, 190–91.

170. See Irene Quenzler Brown and Richard D. Brown, "The Republic of Letters," *Common-Place* (Sept. 27, 2000), www.common-place.org/republic/output .html.

171. Parker, "Law, Culture, and Sexual Censure," 252. See, e.g., "Taylor on Trial," *Morning World-Herald* (Omaha), Oct. 22, 1891, 4; "A Scoundrel Put in Jail," *NYT*, Aug. 31, 1881, 8; "Queer Freaks," *NPG*, Mar. 12, 1881, 11; "Arrested for Rape," *Boston Daily Globe*, Apr. 17, 1880, 1.

172. Judith Lewis Herman, *Trauma and Recovery* (New York: Basic Books, 1992), 7–32; Judith Lewis Herman and Lisa Hirschman, *Father-Daughter Incest* (Cambridge: Harvard University Press, 1981), 3. See also Erna Olafson et al., "Modern History of Child Sexual Abuse and Awareness: Cycles of Discovery and Suppression," *Child Abuse and Neglect* 17 (1993): 7, 17.

173. See Joseph E. Illick, *American Childhoods* (Philadelphia: University of Pennsylvania Press, 2002), 84–91; Elizabeth Pleck, *Domestic Tyranny: The Making of American Social Policy against Family Violence from Colonial Times to the Present* (New York: Oxford University Press, 1987), 69–76, 81; and Anthony M. Platt, *The Child Savers: The Invention of Delinquency* (Chicago: University of Chicago Press, 1969).

174. See Gordon, *Heroes*, 12–22, 215–26; Pleck, *Domestic Tyranny*, chap. 4.

175. Gordon, *Heroes*, 291–93.

176. Robertson, *Crimes against Children*, 27–31; Gordon, *Heroes*, chap. 2; Pleck, *Domestic Tyranny*, 74–87, 43–45.

177. See Gordon, *Heroes*, 46; Platt, *Child Savers*, 36–43; and Lela B. Costin,

Howard Jacob Karger, and David Stoesz, *Politics of Child Abuse in America* (New York: Oxford University Press, 1996), 48–75.

178. See Karen Sánchez-Eppler, *Dependent States: The Child's Part in Nineteenth-Century American Culture* (Chicago: University of Chicago Press, 2005), xv–xxii, chap. 2; Robertson, *Crimes against Children*, 17–21, 23–25; Michael Willrich, *City of Courts: Socializing Justice in Progressive Era Chicago* (New York: Cambridge University Press, 2005), 73–74; Illick, *American Childhoods*, 55–75.

179. Dierdre M. Moloney, *American Catholic Lay Groups and Transatlantic Social Reform in the Progressive Era* (Chapel Hill: University of North Carolina Press, 2002), 167–74; Sarah Deutsch, *Women and the City: Gender, Space, and Power in Boston, 1870–1940* (New York: Oxford University Press, 2000), 54–71; Janis Appier, *Policing Women: The Sexual Politics of Law Enforcement and the LAPD* (Philadelphia: Temple University Press, 1998), 11–22; Pleck, *Domestic Tyranny*, chap. 5; Robyn Muncy, *Creating a Female Dominion in American Reform, 1890–1935* (New York: Oxford University Press, 1994), 3–4.

180. Elbridge T. Gerry, "Must We Have the Cat-o'-Nine Tails?" *North American Review* 160 (Mar. 1895): 318. On Gerry, see Pleck, *Domestic Tyranny*, 72–73.

181. See Gordon, *Heroes*, 215.

182. See Carroll Smith-Rosenberg, "Beauty, the Beast, and the Militant Woman: A Case Study of Sex Roles and Social Stress in Jacksonian America," in *Disorderly Conduct: Visions of Gender in Victorian America* (New York: Oxford University Press, 1985), 109, 115–19, 124–25; Suzanne M. Marilley, *Woman Suffrage and the Origins of Liberal Feminism in the United States, 1820–1920* (Cambridge: Harvard University Press, 1996), chap. 4; Anne Firor Scott, *Natural Allies: Women's Associations in American History* (Urbana: University of Illinois Press, 1993), 37–43; Pleck, *Domestic Tyranny*, 89–90; and Paula Baker, "The Domestication of Politics: Women and American Political Society, 1780–1920," *American Historical Review* 89 (June 1984): 620.

183. Pleck, *Domestic Tyranny*, chap. 3; Gordon, *Heroes*, 254.

184. A history of domestic abuse, including incest, was one of five most frequently noted events contributing to a girl's institutionalization. Barbara M. Brenzel, *Daughters of the State: A Social Portrait of the First Reform School for Girls in North America, 1856–1905* (Cambridge: MIT Press, 1983), 129–31. See also Marilynn Wood Hill, *Their Sisters' Keepers: Prostitution in New York City, 1830–1870* (Berkeley: University of California Press, 1993), 77.

185. Elizabeth Pleck, "Feminist Responses to 'Crimes against Women,' 1869–1896," *Signs* 8 (1983): 451, 452–59; Pleck, *Domestic Tyranny*, 102.

186. E.g., Lucy Stone, "Crimes against Women," *Woman's Journal*, July 16, 1877, 188.

187. Barbara Leslie Epstein, *Politics of Domesticity: Women, Evangelism, and Temperance in Nineteenth-Century America* (Middletown, CT: Wesleyan University Press, 1981), 125–26.

188. Deutsch, *Women and the City*, 63, 54–55; Sharon E. Wood, *The Freedom of the Streets: Work, Citizenship, and Sexuality in a Gilded Age City* (Chapel Hill: University of North Carolina Press, 2005), 34–36; Willrich, *City of Courts*, 82–84; Moloney, *American Catholic Lay Groups*, 91–95; Pleck, *Domestic Tyranny*, 94–95.

189. Jane E. Larson, "'Even a Worm Will Turn at Last': Rape Reform in Late Nineteenth-Century America," *Yale Journal of Law and Humanities* 9 (1997): 1, 5; Anastasia Sims, *The Power of Femininity in the New South: Women's Organizations and Politics in North Carolina, 1880–1930* (Columbia: University of South Carolina Press, 1997), 70–71. See Robertson, *Crimes against Children*, 76, 87–92.

190. Stansell, *City of Women*, 69, 183; Timothy J. Gilfoyle, *City of Eros: New York City, Prostitution, and the Commercialization of Sex, 1790–1920* (New York: Norton, 1992), 68–69; Wood, *Freedom of the Streets*, 135.

191. Gilfoyle, *City of Eros*, 68.

192. Ibid., 349–50.

193. Robertson, *Crimes against Children*, 40–42.

194. Sims, *Power of Femininity*, 72–73.

195. Catherine Gilbert Murdock, *Domesticating Drink: Women, Men, and Alcohol in America, 1870–1940* (Baltimore: Johns Hopkins University Press, 2002), 16; Moloney, *American Catholic Lay Groups*, 46–57.

196. Elaine Franz Parsons, *Manhood Lost: Fallen Drunkards and Redeeming Women in the Nineteenth-Century United States* (Baltimore: Johns Hopkins University Press, 2003), 55, 33–36, 54–56; Moloney, *American Catholic Lay Groups*, 57–62.

197. See Gordon, *Heroes*, 174; Carol Mattingly, *Well-Tempered Women: Nineteenth Century Temperance Rhetoric* (Carbondale: Southern Illinois University Press, 1998), 148; "The Devil's Own," *NPG*, Mar. 4, 1882, 3; "Marked for Death," *NPG*, Mar. 6, 1880, 11.

198. Gordon, *Heroes*, 218, 174.

199. E.g., "Arrested for Incest," *Salt Lake Semi-Weekly Tribune*, July 15, 1898, 6; "The Stegall Horror," *Dallas Morning News*, Jan. 25, 1886, 8; "A Father's Crime," *Wheeling (WV) Register*, July 27, 1880, [1]; "Marked for Death," *NPG*, Mar. 6, 1880, 11; "An Inhuman Father," *Inter Ocean* (Chicago), Sept. 14, 1876, col. D; "Mark the Villain," *Lynchburg Virginian*, July 18, 1833, col. B; "A Drunken Man's Crime," *NYT*, Jan. 1, 1879, 5; "Multiple News Items, Most Horrid Fruits of Rum," *Vermont Watchman and State Journal* (Montpelier), July 6,

1855, col. D; "Multiple News Items, a Chapter of Horrors—Rum's Doings," *Vermont Watchman and State Journal* (Montpelier), July 28, 1853, col. D; "Horrible!!!" *New-York Spectator*, Mar. 20, 1827, col. B.

200. Murdock, *Domesticating Drink*, 17; Epstein, *Politics of Domesticity*, 114.

201. Larson, "Even a Worm Will Turn," 26–27; Franz, *Manhood Lost*, 176, 167–75.

202. Larson, "Even a Worm Will Turn," 4, 7.

203. Murdock, *Domesticating Drink*, 16–18, 21.

204. "She Packed the Room with Women," *Daily Olympian* (Olympia, WA), Jan. 26, 1895, [4]; *The Bay of San Francisco: The Metropolis of the Pacific Coast and Its Suburban Cities: A History*, vol. 1 (Chicago: Lewis Publishing Co., 1892), 479–80.

205. "Trial of Chamberlain," *Morning World-Herald* (Omaha), Apr. 12, 1899, 3; "Court at Vermillion," *Morning World-Herald* (Omaha), Apr. 7, 1899, 5; "Accused by His Daughter," *Morning World-Herald* (Omaha), Apr. 1, 1899, [2]. See also "A Pool of Iniquity," *LAT*, Nov. 17, 1895, 3; "The Courts," *LAT*, Nov. 15, 1892, 5; and "Who Murdered Her?" *NPG*, Dec. 6, 1890, 6.

206. See Larson, "Even a Worm Will Turn," 19.

207. Ibid., 14; cf. Wood, *Freedom of the Streets*, chap. 6 (defense successfully characterized 10-year-old rape victim as a prostitute). See also Ellen Carol DuBois and Linda Gordon, "Seeking Ecstasy on the Battlefield: Danger and Pleasure in Nineteenth-Century Feminist Thought," in *Pleasure and Danger: Exploring Female Sexuality*, ed. Carole S. Vance (Boston: Routledge and Kegan Paul, 1984), 31, 38.

208. Larson, "Even a Worm Will Turn," 37.

209. "Since Our Last Issue," *Union Signal* (Evanston, IL), July 5, 1888, 1, cited in Larson, "Even a Worm Will Turn," 17.

210. Parker, "Law, Culture, and Sexual Censure," 257. Parker found additional cases of incest among her sample of more than five hundred cases prosecuted as forcible rape, attempted rape, and indecent liberties with a child.

211. See Larson, "Even a Worm Will Turn," 19.

212. Sánchez-Eppler argues that the incest stories "represent incest, but they do not acknowledge it." Sánchez-Eppler, *Dependent States*, 95; Larson, "Even a Worm Will Turn," 12; Alison M. Parker, " 'Hearts Uplifted and Minds Refreshed': The Woman's Christian Temperance Union and the Production of Pure Culture in the United States, 1880–1930," *Journal of Women's History* 11 (1999): 135, 140.

213. Deutsch, *Women and the City*, 64, 67.

214. Murdock, *Domesticating Drink*, 24–25; Parsons, *Manhood Lost*, 12–13, 54–57, 62.

Chapter Two: Medicine and the Law Weigh In

1. Thomas Percival, *Medical Ethics: Or, A Code of Institutes and Precepts Adapted to the Professional Conduct of Physicians and Surgeons* (Manchester: S. Russell, 1803), 102–3, 231–34.

2. Ibid., 102–3, 231–32.

3. Ibid., 232–33.

4. U.S. National Library of Medicine and National Institutes of Health, "Medical Encyclopedia," *MedlinePlus* (Sept. 25, 2006) s.v. "typhus," www.nlm.nih.gov/medlineplus/ency/article/001363.htm (accessed July 21, 2007).

5. Richard Temple, *Practice of Physic: Wherein Is Attempted a Concise Exposition of the Characters, Symptoms, Causes of Diseases, and Method of Cure . . .* (London: J. Johnston, 1792), 22.

6. Ibid., 20–24.

7. Ibid., 24–27; John Ferriar, *Medical Histories and Reflections*, vol. 1 (Warrington: printed by W. Eyres, for T. Cadell, London, 1792), 120, 135.

8. Thomas Percival, *Observations on the State of Population in Manchester, and Other Adjacent Places* (Manchester: written for the Royal Society and printed for the author, 1773), 3; Stanford T. Shulman, "History of Pediatric Infectious Diseases," *Pediatric Research* 55 (2004): 163, 165; Donal Sheehan, "The Manchester Literary and Philosophical Society," *Isis* 33 (Dec. 1941): 519, 520.

9. Ferriar, *Medical Histories*, 117. See also "Manchester Infirmary," *Times* (London), Jan. 1, 1877, col. A.

10. Ferriar, *Medical Histories*, 135–44; *Oxford Dictionary of National Biography* (Oxford: Oxford University Press, 2004), s.v. "Ferriar, John (1761–1815)" (by K. A. Webb), www.oxforddnb.com/view/article/9368 (accessed July 21, 2007).

11. Ferriar, *Medical Histories*, 133.

12. M. Dupuytren, "Facts in Surgery and Legal Medicine," *North American Review* 1 (Dec. 1834): 195. See also William R. Wilde, "History of the Recent Epidemic of Infantile Leucorrhoea, with an Account of Five Cases of Alleged Felonious Assault Recently Tried in Dublin," *Medical News* 11 (Oct. 1853): 153.

13. Percival, *Medical Ethics*, 233.

14. Thomas Percival, *Medical Jurisprudence; or, A Code of Ethics and Institutes Adapted to the Professions of Physic and Surgery* (Manchester: for private circulation, 1794), www.collphyphil.org/HMDLSubweb/Pages/P/PercivalT/medjurPgAccess.htm (accessed July 21, 2007); [Royal] Society in Edinburgh, *Medical Commentaries for the Year MDCCXCIV, Exhibiting a Concise View of the Latest and Most Important Discoveries in Medicine and Medical Philosophy, Decade Second*, vol. 9, collected and published by Andrew Duncan (Edinburgh: printed

for G. Mudie, and G. G. and J. Robinson, and J. Johnson, London, 1795), 374–75.

15. *Oxford Dictionary of National Biography*, s.v. "Percival, Thomas (1740–1804)" (by Albert Nicholson, rev. John V. Pickstone), www.oxforddnb.com/view/article/21921 (accessed July 21, 2007); Udo Thiel, ed., introduction to *Philosophical Writings of Thomas Cooper* (Bristol, England: Thoemmes Continuum, 2000), v–xix.

16. Percival, preface to *Medical Ethics*, 1–7; Chester Burns, "Reciprocity in the Development of Anglo-American Medical Ethics, 1765–1865," in *Codification of Medical Morality: Historical and Philosophical Studies of the Formalization of Western Medical Morality in the Eighteenth and Nineteenth Centuries*, vol. 2: *Anglo-American Medical Ethics and Medical Jurisprudence in the Nineteenth Century*, ed. Robert Baker (Dordrecht: Kluwer Academic Publishers, 1995), 135–37.

17. John Bell et al., "Note to Convention, Minutes of the Proceedings of the National Medical Convention Held in the City of Philadelphia, in May 1847," in Baker, *Codification*, 2:72; Burns, "Reciprocity, 1765–1865," in Baker, *Codification*, 2:135–43; Robert Baker, "The Historical Context of the American Medical Association's 1847 Code of Ethics," in *Codification*, 2:47–63; American Medical Association, "E-History" (Aug. 1, 2005), www.ama-assn.org/ama/pub/category/8291.html (accessed July 21, 2007).

18. Percival, *Medical Ethics*, 231–34.

19. Ibid., 102.

20. Ibid., 101; Sir Matthew Hale, *History of the Pleas of the Crown*, rev. ed., ed. George Wilson, vol. 1 (London: T. Payne et al., 1778), 635; *Oxford Dictionary of National Biography*, s.v. "Hale, Sir Mathew (1609–1676)" (by Alan Cromartie), www.oxforddnb.com/view/article/11905 (accessed July 21, 2007).

21. Percival, *Medical Ethics*, 102.

22. Ibid., 103.

23. Ibid., 233–34.

24. Thomas Rogers Forbes, *Surgeons at the Bailey: English Forensic Medicine to 1878* (New Haven: Yale University Press, 1985), iv; B. T. Davis, "George Edward Male MD—the Father of English Medical Jurisprudence," *Proceedings of the Royal Society of Medicine* 67 (Feb. 1974): 117, 118, www.pubmedcentral.nih.gov/picrender.fcgi?artid=1645272&blobtype=pdf (accessed July 21, 2007).

25. George Edward Male, "An Epitome of Juridical or Forensic Medicine: For the Use of Medical Men, Coroners, and Barristers," in *Tracts on Medical Jurisprudence*, ed. Thomas Cooper (Philadelphia: James Webster, 1819), 111, 229–30.

26. John Hunter, *A Treatise on the Venereal Disease* (London: sold at No. 13, Castle-Street, Leicester-Square, 1786), 62; *Oxford Dictionary of National Biog-*

raphy (2004; online ed., 2006), s.v. "Hunter, John (1728–1793)" (by Jacob W. Gruber), www.oxforddnb.com/view/article/14220 (accessed July 21, 2007).

27. Kinder Wood, "History of a Very Fatal Affection of the Pudendum of Female Children," *Eclectic Repertory and Analytical Review, Medical and Philosophical* 7 (Jan. 1817): 11, 17.

28. Ibid., 11–12, 14.

29. Ibid., 17–19.

30. Thiel, introduction to *Philosophical Writings of Thomas Cooper*, v–xix; Male, "Epitome," 230.

31. *Oxford Dictionary of National Biography* (2004; online ed., 2006), s.v. "Cooper, Thomas (1759–1839)," (by Stephen L. Newman), www.oxforddnb .com/view/article/6231?docPos=5 (accessed July 22, 2007); John Osborne and James Gerencser, "Their Own Words; about the Author: Thomas Cooper," Dickinson College (July 1, 2003), http://deila.dickinson.edu/theirownwords/author/ CooperT.htm (accessed July 21, 2007); John Osborne, "Thomas Cooper," *Encyclopedia Dickinsonia*, Dickinson College Chronicles (Aug. 2001), http://chroni cles.dickinson.edu/encyclo/c/ed_cooperT.htm (accessed July 21, 2007); Library and Archive Catalogue, Royal Society, s.v. "Cooper, Thomas (1759–1840)," "Certificate of First Unsuccessful Candidature," EC/1789/07, www.royalsoc.ac .uk/DServe/dserve.exe?dsqIni=Dserve.ini&dsqApp=Archive&dsqDb=Catalog& dsqSearch=RefNo=='EC/1789/07'&dsqCmd=Show.tcl (accessed July 21, 2007); "Certificate of Second Unsuccessful Candidature," EC/1791/21 (1791), www .royalsoc.ac.uk/DServe/dserve.exe?dsqIni=Dserve.ini&dsqApp=Archive&dsqCmd =Show.tcl&dsqDb=Catalog&dsqSearch=(PersonCode=='NA2474')&dsqPos=1 (accessed July 21, 2007). Cooper had been an officer in the philosophical society, but he led a contentious political life, which may have been a factor in the Royal Society twice rejecting him for membership even though his nominators included Thomas Percival and Joseph Priestly.

32. James C. Mohr, *Doctors and the Law: Medical Jurisprudence in Nineteenth-Century America* (Baltimore: Johns Hopkins University Press, 1993), 8–11, 30.

33. Edward Mansfield Brockbank, *Sketches of the Lives and Work of the Honorary Medical Staff of the Manchester Infirmary: From Its Foundation in 1752 to 1830, When It Became the Royal Infirmary* (Manchester: University of Manchester Press, 1904), 189.

34. Theodoric Romeyn Beck, *Elements of Medical Jurisprudence*, 3rd ed., brought down to the present time by John Darwall (London: printed for Longman, Rees, Orme, Brown, and Green; S. Highley; Simpkin and Marshall; J. Smith; and W. Blackwood, Edinburgh, 1829).

35. See, e.g., William Acton, *A Complete Practical Treatise on Venereal Dis-*

eases: And Their Immediate and Remote Consequences. Including Observations on Certain Affections of the Uterus, Attended with Discharges, 2nd American ed. (New York: Redfield, 1848), 143–56.

36. Jane F. Thraikill, "Killing Them Softly: Childbed Fever and the Novel," *American Literature* 71 (1999): 679; Richard H. Shyrock, "Medical Sources and the Social Historian," *American Historical Review* 41 (Apr. 1936): 458, 472; John Bell, "Obituary Notice of Charles D. Meigs, M.D.," *Proceedings of the American Philosophical Society* 13 (1873): 170.

37. Charles D. Meigs, *Females and Their Diseases: A Series of Lectures to His Class* (Philadelphia: Lea and Blanchard, 1848), 101. On Meigs's willingness to add fictional or unsupported claims into his lectures, see Susan Wells, *Out of the Dead House: Nineteenth-Century Women Physicians and the Writing of Medicine* (Madison: University of Wisconsin Press, 2001), 22–28.

38. Robert Liston, *Elements of Surgery*, ed. Samuel D. Gross (Philadelphia: Ed. Barrington and Geo. D. Haswell, 1842), 531; *Oxford Dictionary of National Biography*, s.v. "Liston, Robert (1794–1874)" (by D'A. Power, rev. Jean Loudon), www.oxforddnb.com/view/article/16772 (accessed July 21, 2007).

39. Barbara J. Niss, "Archives History: Mount Sinai: The Evolution of a Mission," Archives Division, Gustave L. and Janet W. Levy Library, Mount Sinai School of Medicine, www.mssm.edu/library/services/archives/history.shtml (accessed July 21, 2007).

40. John Morris, "Gonorrhoeae in Women," *Virginia Medical Monthly* 5 (1878): 373–74.

41. George Xu, "History of the Gram Stain and How It Works," *Gram Stain* (University of Pennsylvania Health System, Oct. 31, 1997), www.uphs.upenn.edu/bugdrug/antibiotic_manual/Gram1.htm (accessed July 21, 2007); U.S. National Library of Medicine and the National Institutes of Health, *Medline Plus Medical Encyclopedia* (Feb. 10, 2006), s.v. "Endocervical Gram Stain," www.nlm.nih.gov/medlineplus/ency/article/003753.htm (accessed July 21, 2007).

42. S. Pancoast, *Ladies' Medical Guide: An Instructor, Counsellor and Friend, Indispensable to Mothers and Daughters*, rev. and enl. ed. (Philadelphia: John E. Potter and Co., 1886), 404; see also Henry C. Coe, "Gynecology: Gonorrhoea in Women from a Medico-Legal Standpoint," *American Journal of the Medical Sciences* 112 (July 1896): 117, 118 (summary of Neisser, *Centralblatt für Gynäkologie*, 1896, no. 14).

43. E.g., A. K. Paine, "Gonorrhea in the Female," *Commonhealth* [Bulletin of the Massachusetts Department of Health] 15 (July–Sept., 1928): 51, 54; B. K. Rachford, "Epidemic Vaginitis in Children," *American Journal of the Medical Sciences* 153 (Feb. 1917): 207; "The Gonococcus in the Diagnosis of Vulvitis," *Medical and Surgical Reporter* 64 (Jan. 24, 1891): 109.

44. See Allan M. Brandt and David Shumway Jones, "Historical Perspectives on Sexually Transmitted Diseases: Challenges for Prevention and Control," in *Sexually Transmitted Diseases*, ed. King K. Holmes et al., 3rd ed. (New York: McGraw-Hill, 1999), 15, 18; Edward Shorter, *Women's Bodies: A Social History of Women's Encounter with Health, Ill-Health, and Medicine* (New Brunswick, NJ: Transaction Publishers, 1991), 265; Rudolph H. Kampmeier, "Early Development of Knowledge of Sexually Transmitted Diseases," in *Sexually Transmitted Diseases*, ed. King K. Holmes et al. (New York: McGraw-Hill, 1984), 28; Michael M. Davis, "What the Campaign against Venereal Disease Demands of Hospitals and Dispensaries," *American Journal of Public Health* 6 (1916): 346.

45. Charles E. Rosenberg, *The Care of Strangers: The Rise of America's Hospital System* (New York: Basic Books, 1987), 23, 29, 40, 120–21, 304–5, 322–23; Allan M. Brandt, *No Magic Bullet: A Social History of Venereal Disease in the United States since 1880* (New York: Oxford University Press, 1987), 43–45; John S. Haller Jr. and Robin M. Haller, *The Physician and Sexuality in Victorian America* (Urbana: University of Chicago Press, 1974), 263–64; Ferdinand C. Valentine and Terry M. Townsend, "How the General Practitioner Should Treat Gonorrhea," *JAMA* 43 (Oct. 29, 1904): 1279; Felix Pascalis, "To the Attending Physicians of the New-York Public Dispensary," *Medical Repository of Original Essays and Intelligence* 4 (Jan. 1, 1818): 30.

46. Samuel Akerly, "Observations on the Effects of Hemorrhage and Some Other Remedies in the Cure of Gonorrhoea; Communicated to Dr. Miller by Dr. Samuel Akerly, One of the Physicians of the Dispensary, in the City of New-York," *Medical Repository of Original Essays and Intelligence* 6 (Aug.–Oct. 1808): 116.

47. See Rosenberg, *Care of Strangers*, 15–17, 35–36, 40; William W. Sanger, *History of Prostitution: Its Extent, Causes and Effects throughout the World* (New York: Harper and Brothers, 1858; online ed., New York: American Medical Press, 1895), 633–40, http://pds.lib.harvard.edu/pds/view/2581510?n=8&s=4 (accessed July 22, 2007); "Statistics of Licentiousness," *Brooklyn Eagle*, June 27, 1857, 2; "The Penitentiary Hospital," *New York Daily Times*, June 21, 1853, 1; "Sketches from the Life School: Blackwell's Island—The Hospital," *New York Daily Times*, Sept. 24, 1852, 3.

48. Minutes, Executive Committee of the Boston Dispensary, 1844, box 1, Papers of the Boston Dispensary, Rare Books and Special Collections Department, Countway Library of Medicine Rare Books, Harvard University, Cambridge, Massachusetts; Davis, "What the Campaign against Venereal Disease Demands," 346. See also "Report of the Committee on the City Hospital," *Boston Courier*, Nov. 22, 1849, col. D.

49. Rosenberg, *Care of Strangers*, 276.

50. Mary Spongberg, *Feminizing Venereal Disease* (New York: New York University Press, 1997), 17–20; Brandt, *No Magic Bullet*, 31–32; John C. Burnham, "Medical Inspection of Prostitutes in America in the Nineteenth Century: The St. Louis Experiment and Its Sequel," *Bulletin of the History of Medicine* 45 (May–June 1971): 203.

51. Benjamin H. Coates, "Brief Account of the History and Present State of the Pennsylvania Hospital," *Philadelphia Journal of the Medical and Physical Sciences* 9 (1824): 30, 46–47.

52. Andrew F. Currier, "Vulvo-Vaginitis in Children," *Medical News* 55 (July 6, 1889): 3, 4; Beth, San Francisco, "Needed Medical Reform," *Woman's Journal*, Feb. 5, 1876, 42; William Goodell, "Original Lectures: Two Clinical Lectures on the Causes, the Prevention, and the Cure of Laceration of the Perineum," *Philadelphia Medical Times* 6 (Nov. 13, 1875): 73, 76; Michael Ryan, *A Manual of Medical Jurisprudence, Compiled from the Best Medical and Legal Works: Being an Analysis of a Course of Lectures on Forensic Medicine, Annually Delivered in London*, 1st American ed. (Philadelphia: Cary and Lea, 1832), 161.

53. Mary Louise Marshall, "Some Nineteenth-Century Fee Bills," *Bulletin of the History of Medicine* 6 (1938): 62, 66, 69, 74, 79.

54. Silas Durkee, *Treatise on Gonorrhoeae and Syphilis* (Boston: John P. Jewett and Co., Cleveland: Henry P. B. Jewett, 1859), v.

55. Ibid., 4; see also Felix Pascalis, "Original Communications: Observations and Practical Remarks on the Nature, Progress and Operations of the Venereal Disease," *Medical Repository of Original Essays and Intelligence, Relative to Physic, Surgery, Chemistry, and Natural History* 1 (Jan. 1, 1813): 1.

56. See, e.g., *Venzke v. Venzke*, 94 Cal. 225 (1892), and *White v. Nellis*, 31 N.Y. 405, 407 (1865): "It is plain that a seducer who imparts to his victim a vile, contagious disease, does an act as abhorrent to morality, and more so to nature, than one who gets her with child" (407).

57. Hunter, *Treatise on the Venereal Disease*, 62–63.

58. Liston, *Elements of Surgery*, 463.

59. Durkee, *Treatise*, 4.

60. Morris, "Gonorrhoeae in Women," 376.

61. "Gonorrhoea in Children," *Medical and Surgical Reporter* 57 (Dec. 24, 1887): 853.

62. I have used masculine pronouns exclusively because, although there were a few female physicians at the time, most were men.

63. Rosenberg, *Care of Strangers*, 156–58; Robert L. Pittfield, "The Use of Simple Microscopical Methods by the General Practitioner," *Medical News* 81 (Sept. 13, 1902): 496.

64. Editor, "Opinion and Comment," *Monthly Bulletin of the Dept. of Health*

of the City of New York 2 (Jan. 1912): 3; Papers of the New York Academy of Medicine, Venereal Disease–Section on Public Health, Venereal Diseases 1909–14 folder, Rare Book and History of Medicine Collection, Malloch Rare Book Room, New York Academy of Medicine, New York, New York.

65. Alfred S. Taylor, *Medical Jurisprudence*, ed. R. Egglesfield Griffith (Philadelphia: Lea and Blanchard, 1845), iv. All citations are from this edition (the first American edition).

66. *Oxford Dictionary of National Biography*, s.v. "Taylor, Alfred Swaine (1806–1880)" (by M. P. Earles), www.oxforddnb.com/view/article/27017 (accessed July 21, 2007); Alfred Swaine Taylor, *Elements of Medical Jurisprudence* (London: Deacon, 1836).

67. Taylor, *Medical Jurisprudence*, iii.

68. Ibid., v.

69. Mohr, *Doctors and the Law*, 72.

70. Ibid., 74–75.

71. Quoted in Forbes, *Surgeons at the Bailey*, 10.

72. Mohr, *Doctors and the Law*, 14, 95, 15–28.

73. Ibid., 243.

74. Ibid., 27.

75. Ibid., 20, 26–28.

76. Beck, *Medical Jurisprudence*, 56, citing *Edinburgh Medical and Surgical Journal* 13 (1817): 491; italics are Beck's.

77. Ibid., 54.

78. Ibid., 68.

79. William R. Wilde, "Medico-Legal Observations upon the Case of Amos Greenwood, Tried at the Liverpool Assizes, December, 1857, for the Wilful [*sic*] Murder of Mary Johnson, and Sentenced to Penal Servitude for Life," *Dublin Quarterly Journal of Medical Science* 27 (Feb. 1859): 51, 69–82.

80. Ryan, *Manual of Medical Jurisprudence*, 162; *Oxford Dictionary of National Biography*, s.v. "Ryan, Michael (1800–1840)" (by G. Le G. Norgate, rev. Stephanie J. Snow), www.oxforddnb.com/view/article/24388 (accessed July 21, 2007).

81. Ryan, *Manual of Medical Jurisprudence*, 161.

82. Ibid., 161, 163; J. S. Forsyth, *A Synopsis of Modern Medical Jurisprudence, Anatomically, Physiologically, and Forensically, Illustrated; for the Faculty of Medicine, Magistrates, Lawyers, Coroners, and Jurymen* (London: W. Benning, 1829), 497–98.

83. Ryan, *Manual of Medical Jurisprudence*, 163.

84. Ibid., 164.

85. Ibid., 164–68.

86. Ibid., 167.

87. *Oxford Dictionary of National Biography*, s.v. "Cooper, Sir Astley Paston, first baronet (1768–1841)" (by W. F. Bynum), www.oxforddnb.com/view/article/6211 (accessed July 21, 2007).

88. "The Late Sir Astley Cooper," *Times* (London), Oct. 23, 1847, col. C, 6; "The Late Sir Astley Cooper; a Statue," *Times* (London), Sept. 20, 1844, col. A, 5; "Life of Sir Astley Cooper," *Times* (London), Apr. 14, 1843, col. F, 5; "Funeral of Sir Astley Cooper," *Times* (London), Feb. 22, 1841, col. F, 5.

89. From a Correspondent, "Memoir of Sir Astley Cooper," *Times* (London), Feb. 15, 1841, col. A, 6.

90. Doctors were still repeating Cooper's admonition in 1944, e.g., William J. Robinson, ed., "False Accusations of Rape," *Sexual Truths versus Sexual Lies, Misconceptions and Exaggerations* (New York: Eugenics Publishing Co., 1944), 381–82.

91. Wilde, " Medico-Legal Observations," 51, 53.

92. E.g., trial of Thomas Mercer, *OBP*, t16940830-9 (Aug. 30, 1694) (no penetration, guilty of assault but not rape; fine and pillory); trial of Jacob Whitlock, *OBP*, t16961014-10 (Oct. 14, 1696) (guilty, death); trial of Julian Brown, *OBP*, t17351015-28 (Oct. 15, 1735) (not guilty); trial of Francis Moulcer, *OBP*, t17441017-25 (Oct. 17, 1744) (guilty, death); trial of Christopher Larkin, *OBP*, t17510703-21 (July 3, 1751) (not guilty); trial of John Birmingham, *OBP*, t17530502-35 (May 2, 1753) (no penetration, not guilty of rape; held on new charge of assault with intent to commit rape); trial of John Grimes, *OBP*, t17540530-1 (May 30, 1754) (no penetration, not guilty of rape; held on new charge of assault with intent to commit rape); trial of Charles Brown, *OBP*, t17670603-52 (June 3, 1767) (no penetration, not guilty of rape; held on new charge of assault with intent to commit rape); trial of David Scott, *OBP*, t17960914-12 (Sept. 14, 1796) (guilty, death).

93. Wilde, "Medico-Legal Observations," 61. See also Thomas Weeden Cooke, Esq., "Foreign Intelligence, Practical Medicine, &c.: On Some Distressing Sequelae of the Diseases of Infancy—Purulent Discharges from the Aural, Nasal, and Vaginal Passages," *Ohio Medical and Surgical Journal* 3 (Nov. 1, 1850): 159.

94. Wilde, "Medico-Legal Observations," 75.

95. Ryan, *Manual of Medical Jurisprudence*, 163.

96. Y. P. Talmi et al., "Scrofula Revisited," *Journal of Laryngology and Otology* 102 (Apr. 1988): 387.

97. See *Oxford English Dictionary*, 2nd ed. (Oxford: Clarendon Press, 1989), *OED Online* (Oxford University Press), s.v. "scrofula," http://dictionary.oed.com/cgi/entry/50216968?single=1&query_type=word&queryword=scrofula&first=1&max_to_show=10; s.v. "scrofulous," http://dictionary.oed.com/cgi/entry/

50216976?single=1&query_type=word&queryword=scrofulous&first=1&max_to_show=10 (accessed July 22, 2007).

98. William Nisbet, *An Inquiry into the History, Nature, Causes and Different Modes of Treatment Hitherto Pursued, in the Cure of Scrophula and Cancer* (Edinburgh: printed by Alex. Chapman and Co., for Thomas Kay, London; and James Watson and Co., Edinburgh, 1795), 3.

99. Quoted in Francis Wharton and Moreton Stillé, *A Treatise on Medical Jurisprudence* (Philadelphia: Kay and Brother, 1855), 334.

100. See Joseph E. Hawkins, "Sketches of Otohistory, Part 2: Origins of Otology in the British Isles: Wilde and Toynbee," *Audiology & Neuro-Otology* 9 (2004): 129–31.

101. Ibid., 131–32; Times Correspondent, "Ireland," *Times* (London), Dec. 19, 1864, 6; *Oxford Dictionary of National Biography*, s.v. "Wilde, Sir William Robert Wills (1815–1876)" (by James McGeachie), www.oxforddnb.com/view/article/29403 (accessed July 22, 2007).

102. Wharton and Stillé, *Medical Jurisprudence*, 334; Wilde, "Medico-Legal Observations," 51; Spongberg, *Feminizing Venereal Disease*, 117–20.

103. Quoted in Wharton and Stillé, *Medical Jurisprudence*, 334.

104. Taylor, *Medical Jurisprudence*, 458.

105. Ibid., 462.

106. Ibid., 460; Wilde, "Medico-Legal Observations," 70–84.

107. Taylor, *Medical Jurisprudence*, 460; Forbes, *Surgeons at the Bailey*, vi, 91.

108. Taylor, *Medical Jurisprudence*, 460.

109. Burke Ryan, "On the Communicability of Gonorrhoea, in Reference to Medical Jurisprudence," *London Medical Gazette* 47 (1851): 744.

110. Ibid., 745.

111. Ibid.

112. Ibid.

113. Ibid.

114. Sydney Smith, ed., *Taylor's Principles and Practice of Medical Jurisprudence*, 9th ed. (London: J. and A. Churchill, 1934), 2:119; L. Emmett Holt Jr. and Rustin McIntosh, eds., *Holt's Diseases of Infancy and Childhood: A Textbook for the Use of Students and Practitioners*, 11th ed. (New York: D. Appleton-Century Co., 1940), 821.

115. Wharton and Stillé, *Medical Jurisprudence*, i. Moreton Stillé died before the book's publication, after which Dr. Alfred Stillé, who was active in advocating the formation of the American Medial Association and later worked as a consultant at the Women's Hospital of Pennsylvania, coauthored subsequent editions. Mohr, *Doctors and the Law*, 38, 110, 225–26; Wells, *Out of the Dead House*, 75.

116. Wharton and Stillé, *Medical Jurisprudence*, 326.

117. Johann Ludwig Casper, *Practiches Handbuch der Gerichtlichen Medicin: Nach Eigenen Erfahrungen* (Berlin: Thanatologischer Theil, 1857); Johann Ludwig Casper and George William Balfour, *A Handbook of the Practice of Forensic Medicine: Based upon Personal Experience* (London: New Sydenham Society, 1861–65); W. Keil et al., "Textbooks on Legal Medicine in the German-Speaking Countries," *Forensic Science International* 144 (2004): 289–302, 292.

118. Wharton and Stillé, *Medical Jurisprudence*, 326.

119. Ibid., 327–28, 326–27.

120. Ibid., 327.

121. Ibid., 328.

122. Ibid., 330.

123. Ibid., 326.

124. Ibid.

125. Ibid., 330.

126. See Charles E. Rosenberg, "Sexuality, Class and Role in Nineteenth-Century America," *American Quarterly* 25 (May 1973): 131, 143.

127. "Gonorrhoea in a Young Girl," *Medical and Surgical Reporter* 22 (Apr. 30, 1870): 377.

128. See, e.g., Theophilus Parvin, "Communications: Catarrhal Vulvo-Vaginitis in Children," *Medical and Surgical Reporter* 57 (Dec. 24, 1887): 827; "Etiology of Infectious Vulvo-Vaginitis of Childhood," *Medical and Surgical Reporter* 54 (Feb. 27, 1886): 279; "Shocking Depravity," *Boston Daily Globe*, Sept. 4, 1879, 4; and "Weekly Police Record," *NPG*, Dec. 5, 1846, 99. See also Sharon Block, *Rape and Sexual Assault in Early America* (Chapel Hill: University of North Carolina: 2006), 103, 104; Stephen Robertson, *Crimes against Children: Sexual Violence and Legal Culture in New York City, 1880–1960* (Chapel Hill: University of North Carolina Press, 2005), 42–43; Diane Miller Sommerville, *Rape and Race in the Nineteenth-Century South* (Chapel Hill: University of North Carolina, 2004), 64.

129. Jonathan Gillis, "Taking a Medical History in Childhood Illness: Representations of Parents in Pediatric Texts Since 1850," *Bulletin of the History of Medicine* 79 (2005): 393–429, 399, 411; Wells, *Out of the Dead House*, 22.

130. Jacob D. Wheeler, ed., "The People v. Hugh Flinn," Nov. 1822, *Reports of Criminal Law Cases Decided at the City-Hall of the City of New York*, vol. 1 (New York: Banks Gould and Co., 1854), 74. See also "Shocking Depravity," *Boston Daily Globe*, Sept. 4, 1897, 4.

131. Durkee, *Treatise*, 172. See also "Etiology of Infectious Vulvo-Vaginitis of Childhood," *Medical and Surgical Reporte*r 54 (Feb. 27, 1886): 279.

132. "Etiology of Infectious Vulvo-Vaginitis," 279.

133. I. E. Atkinson, "Report of Six Cases of Contagious Vulvitis in Children," *American Journal of Medical Science* 75 (Apr. 1878): 444.

134. See Eugene F. Cordell, "Atkinson, Isaac Edmonton," *Medical Annals of Maryland* (Baltimore: Press of Cordell and Wilkins, 1903), 308; *Centuries of Leadership: Deans of the University of Maryland School of Medicine* (Baltimore: University of Maryland, [ca. 2000]), 22.

135. Atkinson, "Report of Six Cases," 444, 446.

136. Ibid., 445.

137. Morris, "Gonorrhoeae in Women," 373.

138. Ibid., 374.

139. Ibid., 376.

140. Ibid., 375.

141. Ibid., 376. See also Nayan Shah, *Contagious Divides: Epidemics and Race in San Francisco's Chinatown* (Berkeley: University of California Press, 2001), 90; "Rectal Gonorrhea," *Medical and Surgical Reporter* 68 (Aug. 6, 1892): 229; and Hunter, *Treatise on the Venereal Disease*, 64.

142. E.g., Randolph Winslow, "Report of an Epidemic of Gonorrhea Contracted from Rectal Coition," *Medical News* 49 (Aug. 14, 1886): 7; "From the Essex Gazette, of Nov. 28," *New London (CT) Gazette*, Dec. 15, 1769, 3.

143. Morris, "Gonorrhoeae in Women," 379.

144. E.g., Andrew F. Currier, "Original Articles: Vulvo-Vaginitis in Children," *Medical News* 55 (July 6, 1889): 3; Henry C. Coe, "Gynecology," *American Journal of the Medical Sciences* 86 (Oct. 1888): 430; "Etiology of Vulvo-Vaginitis in Children," *Medical and Surgical Reporter* 59 (Sept. 15, 1888): 340; "Medical Progress," *Medical News* 47 (Nov. 7, 1885): 511.

145. Jerome Walker, "Reports, with Comments, of Twenty-One Cases of Indecent Assault and Rape upon Children," pts. 1 and 2, *Archives of Pediatrics* 3 (May 1886): 269; (June 1886): 321.

146. "Dr. Jerome Walker," *New York Times*, June 21, 1924, 13; "Free Public Lectures," *Brooklyn Eagle*, Mar. 30, 1902, 32; "Pure Air in Schools, a Paper by Dr. Jerome Walker Before the Medical Association," *Brooklyn Eagle*, Nov. 15, 1894, 9; "Food Products," *Brooklyn Eagle*, Oct. 25, 1889, 6; "American Public Health Association," *Brooklyn Eagle*, Sept. 1889, 1; "Educational," *Brooklyn Eagle*, Apr. 6, 1887, 1; "Prevention of Cruelty to Children," *Brooklyn Eagle*, Jan. 6, 1884, 3; "Little Ones," *Brooklyn Eagle*, Dec. 14, 1880, 4.

147. Historical Archives Advisory Committee, "Committee Report: American Pediatrics: Milestones at the Millennium," *Pediatrics* 107 (2001): 1482–91, 1484.

148. Walker, "Twenty-One Cases," 330–31.

149. Ibid., 331.

150. Ibid., 331, 330.
151. Ibid., 331.
152. Ibid., 339–40.
153. Ibid., 275–76, 339–40. See also Parvin, "Communications," 827.
154. Walker, "Twenty-One Cases," 273–74.
155. *Moore v. Ohio*, 17 Ohio 521 (1867).
156. Walker, "Twenty-One Cases," 279.
157. Ibid., 270.
158. Ibid., 272.
159. Ibid., 329.
160. "A Brute," *Los Angeles Times*, May 2, 1888, 8.
161. "The Butti Brute," *Los Angeles Times*, May 3, 1888.

Chapter Three: Gonorrhea and Incest Break Out

1. "American Medical Association," *Medical News* 62 (June 10, 1893): 23.

2. Henry C. Coe, ed., "Gynecology," *American Journal of the Medical Sciences* 98 (Aug. 1889): 214 (reporting on F. Spaeth, "The Recognition of Vulvo-Vaginitis in Childhood," *Müncher med. Wochenschrift* [Munich Medical Weekly], May 28, 1889); Henry C. Coe, ed., "Gynecology: The Etiology of Vulvo-Vaginitis in Children," *American Journal of the Medical Sciences* 96 (Oct. 1888): 430 (reporting on Pott, *Archiv für Gynekologie* [Archives of Gynecology], Bd. Xxxii. Heft 3); Henry C. Coe, "Gynecology: Etiology of Vulvo-Vaginitis in Children," *Medical and Surgical Reporter* 59 (Sept. 15, 1888), 340 (reporting on Dr. Pott's comments at the Second Congress of the German Gynecological Society, Halle, May 26, 1888, *Deutsche med. Wochenschrift* [German Medical Weekly], July 26, 1888).

3. Charles O'Donovan, "Gonorrhoea in Children: A Frequent Occurrence amongst the Negroes of Baltimore," *Archives of Pediatrics* 16 (1899): 25.

4. "The Gonococcus in the Diagnosis of Vulvitis," *Medical and Surgical Reporter* 64 (Jan. 24, 1891): 109.

5. John Lovett Morse, "Five Cases of Gonorrhoeae in Little Girls," *Archives of Pediatrics* 11 (1894): 596; Elliston J. Morris Jr. and Samuel M. Wilson, ed., "Current Literature Reviewed, 'Five Cases of Gonorrhoea in Little Girls,' " *Medical and Surgical Reporter* 71 (Sept. 22, 1894): 410.

6. L. Emmett Holt, *Diseases of Infancy and Childhood: For the Use of Students and Practitioners of Medicine* (New York: D. Appleton and Co., 1897), 641. By 1905 the book was in its third edition.

7. Joseph Louis Baer, "Epidemic Gonorrheal Vulvo-Vaginitis in Young Girls," *Journal of Infectious Diseases* 1 (Mar. 19, 1904): 313.

8. L. Emmett Holt, "Gonococcus Infections in Children, with Especial Reference to Their Prevalence in Institutions and Means of Prevention," *New York Medical Journal and Philadelphia Medical Journal* 81 (Mar. 18, 1905): 521, 525, 591.

9. Ibid., 534.

10. Flora Pollack, "The Acquired Venereal Infections in Children," *Johns Hopkins Hospital Bulletin* 20 (May 1909): 142, 147.

11. Clara P. Seippel, "Venereal Diseases in Children," *Illinois Medical Journal* 22 (July 1912): 50, 52; Jane Addams, *A New Conscience and an Ancient Evil* (New York: Macmillan, 1912), 184.

12. Kathleen Wehrbein, *A Survey of the Incidence, Distribution and Facilities for Treatment of Vulvo-Vaginitis in New York City, with Concomitant Sociological Data: Report of the Committee on Vaginitis* (New York: E. B. Treat and Co., 1927), 250.

13. Stephen A. Yesko, "Gonorrheal Vulvovaginitis in the Young." *American Journal of Diseases of Children* 33 (Apr. 1927): 633.

14. New York City Department of Health, "Clinic Does Research on Vaginitis," *JSH* 15 (June 1929): 368, 369.

15. In 1911 California became the first state to pass a law to include venereal disease as a reportable disease. Allan M. Brandt, *No Magic Bullet: A Social History of Venereal Disease in the United States since 1880* (New York: Oxford University Press, 1987), 42; Holt, "Gonococcus Infections in Children," 593.

16. Board of Health, *Weekly Bulletin of the Dept. of Health of the City of New York*, n.s., 2 (Aug. 23, 1913): 2; Hermann M. Biggs, "Venereal Disease: The Attitude of the Department of Health in Relation Thereto," *New York Medical Journal and Philadelphia Medical Journal and the Medical News* 97 (May 17, 1913): 1009; "Health Notes and Comment," *Monthly Bulletin of the Dept. of Health of the City of New York* 2 (Feb. 1912): 31; Minutes, Section on Public Health, NYAM, Venereal Disease file 1909–14, Dec. 1909, 2, NYAM.

17. N. A. Nelson, "Gonorrhea Vulvovaginitis: A Statement of the Problem," *NEJM* 207 (July 21, 1932): 135, 136; Walter Clarke, "Summary of a Social Hygiene Survey," *JSH* 27 (Feb. 1931): 65, 78. The Massachusetts Board of Health did not mention venereal disease in its publications until 1912 and did not require reporting of venereal diseases until December 1917. Robert N. Wilson, "The Eradication of Social Diseases in Large Cities," *Monthly Bulletin of the State Board of Health of Massachusetts* 7 (Oct. 1912): 360; "The Venereal Disease Campaign in Massachusetts," *Commonhealth* [Bulletin of the Massachusetts Department of Health] 5 (Jan. 1918): 3.

18. See Holt, *Diseases of Infancy and Childhood*, 40–41, and Jonathan Gillis, "Taking a Medical History in Childhood Illness: Representations of Parents in Pe-

diatric Texts since 1850," *Bulletin of the History of Medicine* 79 (2005): 393–429, 396–97.

19. See Thomas Percival, *Observations on the State of Population in Manchester, and Other Adjacent Places* (Manchester: written for the Royal Society and printed for the author, 1773), 3; Stanford T. Shulman, "History of Pediatric Infectious Diseases," *Pediatric Research* 55 (2003): 165–76, 167–68; Kathleen W. Jones, "Sentiment and Science: The Late Nineteenth Century Pediatrician as Mother's Advisor," *Journal of Social History* 17 (Autumn 1993): 79, 82–83; "Saving More Babies: Dr. L. Emmett Holt Discusses Work of Visiting Nurse and Infant Mortality," *NYT*, Jan. 25, 1920, 85; L. Emmett Holt, "Infant Mortality, Ancient and Modern, an Historical Sketch," *Archives of Pediatrics* 30 (1913): 885, www.neonatology.org/classics/holt.html (accessed Aug. 6, 2007); and L. Emmett Holt, "Scope and Limitations of Hospitals for Infants," *Transactions of the American Pediatric Society* 10 (1898): 147.

20. Evelynn Maxine Hammonds, *Childhood's Deadly Scourge: The Campaign to Control Diphtheria in New York City, 1880–1930* (Baltimore: Johns Hopkins University Press, 1999), 8, 73.

21. Herman B. Sheffield, "Contribution to the Study of Infectious Vulvo-Vaginitis in Children, with Remarks upon Purulent Ophthalmia, and a Report of Sixty-Five Cases," *American Medico-Surgical Bulletin* 9 (May 30, 1896): 726.

22. W. M. Leszynsky, "Leucorrhoea as the Cause of a Recent Epidemic of Purulent Ophthalmia in One of Our City Charitable Institutions," *New York Medical Journal* 43 (Mar. 27, 1886): 352.

23. Sheffield, "Contribution," 727.

24. "Gonorrheal Infection from Baths," *Medical and Surgical Reporter* 54 (Mar. 27, 1886): 414.

25. Sheffield, "Contribution," 731.

26. Herman B. Sheffield, "Vulvovaginitis in Children," *New York Medical Journal* 76 (Aug. 4, 1900): 189, 190–91; "Medical Progress: Vulvovaginitis in Children," *Medical News* 77 (Aug. 11, 1900): 213, 214.

27. Sheffield, "Vulvovaginitis in Children," 191. See "Therapeutic Notes: Vulvovaginitis in Children," *Medical News* 72 (Feb. 12, 1898): 209 (reporting on Comby, *Journ. de Méd. de Paris* [Paris Medical Journal], Oct. 3, 1897); "Gonorrhoea Occurring in Little Girls," *Medical and Surgical Reporter* 69 (Oct. 21, 1893): 645 (reporting on Cassell, *Berlin Klin. Wocken.* [Berlin Clinical Weekly]); Coe, "Gynecology," *American Journal of the Medical Sciences* 98 (Aug. 1889): 214; Coe, "Gynecology," *American Journal of the Medical Sciences* 96 (Oct. 1888): 430; Coe, "Gynecology," *Medical and Surgical Reporter*, 340. Sheffield published his own pediatrics textbook and translated Eugen Graetzer, *Practical*

Pediatrics: A Manual of the Medical and Surgical Diseases of Infancy and Childhood (Philadelphia: F. A. Davis, 1905), from German to English.

28. Sheffield, "Vulvovaginitis in Children," 191.

29. Ibid., 191. See also W. D. Trenwith, "Gonococcus Vaginitis in Little Girls," *New York Medical Journal* 83 (Feb. 3, 1906): 240. In his column on health in an African American newspaper, Dr. A. Wilberforce Williams reported that women sometimes asked him what their husbands mean when they say they have "strained" themselves and would have to sleep alone for two or three weeks. Williams mocked the euphemism and said that "strain" is "common to all men and boys who violate the SIXTH COMMANDMENT," and he warned the women about contracting gonorrhea. Wilberforce Williams, "Keep Healthy," *Chicago Defender*, Feb. 14, 1914, 4; Wilberforce Williams, "The Venereal Drive—Gonorrhea," *Chicago Defender*, May 25, 1918, 16. See also Lucien M. Brown, "Keeping Fit," *New York Amsterdam News*, Aug. 27, 1930, 10; Lucien M. Brown, "Keeping Fit," *New York Amsterdam News*, May 18, 1932, 9; and John R. West, "Gonorrhea Has Preventives," *New York Amsterdam News*, July 24, 1937, 13.

30. Sheffield, "Vulvovaginitis in Children," 191.

31. Ibid., 192.

32. E.g., Holt, "Gonococcus Infections in Children," 521; A. C. Cotton, "An Epidemic of Vulvovaginitis among Children," *Archives of Pediatrics* 22 (Feb. 1905): 106; Sara Welt-Kakels, "Vulvovaginitis in Little Girls: A Clinical Study of 190 Cases," *New York Medical Journal and Philadelphia Medical Journal* 80 (Oct. 8, 1904): 689.

33. Robert N. Wilson, "The Social Evil in University Life: A Talk with the Students of the University of Pennsylvania," *Medical News* 84 (Jan. 16, 1904): 97, 98; Marcus Rosenwasser, "A Practical Consideration of Gonorrhea in Women," *Medical and Surgical Reporter* 75 (Oct. 10, 1896): 472.

34. Reuel B. Kimball, "Gonorrhoea in Infants, with a Report of Eight Cases of Pyaemia," *New York Medical Record* 64 (Nov. 14, 1903): 761. See also Baer, "Epidemic Gonorrheal Vulvo-Vaginitis," 313; Holt, "Gonococcus Infections in Children," 521; A. C. Cotton, "Epidemic," 106; and Welt-Kakels, "Vulvovaginitis in Little Girls," 689.

35. Guy L. Hunner and Norman MacL. Harris, "Acute General Gonorrheal Peritonitis," *Bulletin of the Johns Hopkins Hospital* 13 (June 1902): 121, 127; Albert Neisser, "Gonorrhea: Its Dangers to Society," *Medical News* 76 (Jan. 20, 1900): 85, 86; "A Fatal Case of Gonorrhea," *Medical Times and Register* 36 (Aug. 13, 1898): 93 (discussing Ghon and Schlagenhaufer, "A Fatal Case of Gonorrhea," *Wien. klin. Woch.* [Vienna Clinical Weekly] 24 [1898]); L. Emmett Holt, *Diseases of Infancy and Childhood: For the Use of Students and Practitioners of Medicine* (1897; repr., Birmingham, AL: Gryphon Editions, 1980), 642–43; Fran-

cis Huber, "Acute Supperative Peritonitis Following Vulvo-Vaginal Catarrh," *Boston Medical and Surgical Journal* 121 (1889): 413.

36. Hunner and Harris, "Gonorrheal Peritonitis," 129. Their review of the medical literature found that the first articles linking gonorrhea and peritonitis were published in Europe in 1892 and in America three years later.

37. Ibid., 128–29.

38. Kimball, "Gonorrhoea in Infants," 761. The hospital had not been testing girls before admission, but Kimball estimated that only one-tenth of the infected girls had been admitted with the disease and the rest infected during their stay.

39. Ibid.

40. Welt-Kakels, "Vulvovaginitis in Little Girls," 741.

41. Isaac A. Abt, "Gonorrhea in Children," *JAMA* 31 (Dec. 17, 1898): 1474; "Medical Matters in Chicago," *Medical News* 73 (Dec. 17, 1898): 806.

42. Abt, "Gonorrhea in Children," 1474; Michael Reese Hospital and Medical Center, "Our Hospital: Mission and History," www.michaelreesehospital.com/Hospital.htm (accessed Aug. 5, 2007); Arthur F. Abt, "Obituary: Isaac Arthur Abt, 1867–1955," *Pediatrics* 18 (Aug. 1956): 327; Council of the Chicago Medical Society, *History of Medicine and Surgery and Physicians and Surgeons of Chicago* (Chicago: Biographical Publishing Corp., 1922), 363. By 1923 Abt was a nationally recognized authority on pediatrics and editor of the textbook *Pediatrics* (Philadelphia: Saunders, [ca. 1923–26]) and medical yearbook *Pediatrics*.

43. Abt, "Gonorrhea in Children," 1474–75.

44. E.g., Holt, "Gonococcus Infections in Children," 525.

45. Kimball, "Gonorrhoea in Infants," 761; "The Social World," *NYT*, Dec. 8, 1893, 5.

46. See John G. Clark, "A Critical Summary of Recent Literature on Gonorrhoea in Women," *American Journal of the Medical Sciences* 119 (Jan. 1900): 73, 76. On the meager finances of hospitals, see Rosemary Stevens, *In Sickness and in Wealth: American Hospitals in the Twentieth Century* (New York: Basic Books, 1989), chap. 2.

47. E.g., Baer, "Epidemic Gonorrheal Vulvo-Vaginitis," 313; Holt, "Gonococcus Infections in Children," 523–25; Kimball, "Gonorrhoea in Infants," 761.

48. Henry Koplik, "Prophylactic Measures to Prevent the Spread of Vulvovaginitis in Hospital Services," *Archives of Pediatrics* 10 (1903): 735.

49. Howard Markel, "Henry Koplik, MD, the Good Samaritan Dispensary of New York City, and the Description of Koplik's Spots," *Archives of Pediatrics and Adolescent Medicine* 150 (May 1996): 535.

50. Donald Gribetz, "Pediatrics at Mount Sinai," *Mount Sinai Journal of Medicine* 64 (Nov. 1977): 392–98, 393.

51. Ibid., 392–94.

52. Markel, "Koplik," 538.

53. Koplik, "Prophylactic Measures," 739; Holt, "Gonococcus Infections in Children," 593.

54. Holt, "Gonococcus Infections in Children," 521.

55. Shulman, "Pediatric Infectious Diseases," 167; Historical Archives Advisory Committee, "Committee Report: American Pediatrics: Milestones at the Millennium," *Pediatrics* 107 (June 2001): 1482, 1484; Jones, "Sentiment and Science," 79; Holt, *The Care and Feeding of Children: A Catechism for the Use of Mothers and Children's Nurses* (New York: D. Appleton and Co., 1894).

56. "Dr. Holt's Death Evokes Tributes," *NYT*, Mar. 13, 1924, 15.

57. "Work of Babies' Hospital: Report Shows How It Is Dependent on the Charitable," *NYT*, Mar. 27, 1904, 20; Holt, "Scope and Limitations," 147; "Caring for the Babies: What Their Hospital Has Accomplished in a Year," *NYT*, Feb. 23, 1890, 9.

58. Holt, "Gonococcus Infections in Children," 523.

59. Ibid., 590; Koplik, "Prophylactic Measures," 739. On the use of isolation wards in this era, see Jeanne Kisacky, "Restructuring Isolation: Hospital Architecture, Medicine, and Disease Prevention," *Bulletin of the History of Medicine* 79 (2005): 1, 35–43.

60. Holt, "Gonococcus Infections in Children," 590.

61. Ibid.

62. Ibid., 591.

63. Ibid., 521.

64. Stephen Robertson, *Crimes against Children: Sexual Violence and Legal Culture in New York City, 1880–1960* (Chapel Hill: University of North Carolina Press, 2005), 26, 43–44, 50–52.

65. J. Clifton Edgar and J. C. Johnston, "Rape," in *Medical Jurisprudence: Forensic Medicine and Toxicology*, ed. R. A. Witthaus and Tracy C. Becker (New York: William Wood and Co., 1894), 2:415, 467.

66. P. M. Dunn, "J. Clifton Edgar (1859–1939) of New York and His Obstetric Text," *Archives of Diseases in Childhood Fetal and Neonatal Edition* 90 (2005): 441; "What Is Doing in Society," *NYT*, May 28, 1899, 16; "What Is Doing in Society," *NYT*, May 7, 1899, 19; "What Is Doing in Society," *NYT*, Apr. 27, 1899, 7; "What Is Doing in Society," *NYT*, Apr. 26, 1899, 6; "What Is Doing in Society," *NYT*, Apr. 23, 1899, 16; "What Is Doing in Society," *NYT*, Apr. 19, 1899, 7.

67. See Oskar Schaeffer, *Anatomical Atlas of Obstetrics, with Special Reference to Diagnosis and Treatment*, trans. and ed. J. Clifton Edgar (Philadelphia: Saunders, 1901); James Clifton Edgar, *Practice of Obstetrics: Designed for the Use of Students and Practitioners of Medicine* (Philadelphia: Blakiston, 1903);

Carolyn Conant Van Blarcom, *Getting Ready to Be a Mother: A Little Book of Information and Advice for the Young Woman Who Is Looking Forward to Motherhood*, intro. J. Clifton Edgar and Frederick W. Rice (New York: Macmillan, 1922).

68. Edgar and Johnston, "Rape," 467.

69. Ibid.

70. Ibid., 442–43. See Henry C. Coe, "Gynecology: Gonorrhoea in Women from a Medico-Legal Standpoint," *American Journal of the Medical Sciences* 112 (July 1896): 117, 118–19; "Foreign Correspondence: Letter from Berlin," *Medical and Surgical Reporter* 63 (Nov. 1, 1890): 509, 510.

71. See "Matters We Ought to Know," review of *How the Other Half Lives*, by Jacob Riis, *NYT*, Jan. 4, 1891, 19.

72. Jacob Riis, *How the Other Half Lives: Studies among the Tenements of New York* (1890; Dover Publications, 1971). See Robert W. de Forest, "A Brief History of the Housing Movement in America," *Annals of the American Academy of Political and Social Science* 51 (Jan. 1914): 8, 8–10, 18.

73. See "The University Settlement Society," *NYT*, Feb. 1, 1895, 4; "The Tenement Houses," *NYT*, Nov. 14, 1894, 4; and "Family Life in American Cities," *NYT*, May 8, 1892, 4.

74. Edgar and Johnston, "Rape," 443. Dr. Antonio Stella raised similar arguments in his defense of Italian immigrants in 1908. "The Effects of Urban Congestion on Italian Women and Children," *Medical News* 73 (May 2, 1908): 722.

75. Edgar and Johnston, "Rape," 469.

76. Ibid., 482.

77. Ibid., 461, 467, 473–76.

78. W. Travis Gibb, "Indecent Assault upon Children," in *A System of Legal Medicine*, ed. Allan McLane Hamilton and Lawrence Godkin (New York: E. B. Treat, 1894), 649, 656, 654.

79. "Gibb Pavilion Opened: $24,000 Operating Structure Built by the City Dedicated," *NYT*, Feb. 26, 1909, 2; "Operation on W. P. Harrison," *NYT*, June 21, 1900, 2.

80. Gibb, "Indecent Assault," 649–50. In 1916 Gibb claimed he had examined more than twenty-five hundred girls. Robertson, *Crimes against Children*, 68, 262 n. 54.

81. Gibb, "Indecent Assault," 655–66.

82. Abraham L. Wolbarst, "Gonorrhea in Boys," *JAMA* 33 (Sept. 28, 1901): 827, 828. See Abraham Leo Wolbarst, *Gonorrhea in the Male: A Practical Guide to Its Treatment* (New York: International Journal of Surgery, 1911); Abraham L. Wolbarst, trans. and ed., *A Treatise on Cystoscopy and Urethroscopy* (St. Louis: C. V. Mosby, 1918).

83. J. E. C., "To Relieve Distress: A Proposed Fund for Dispensary Doctors to Disburse," *NYT*, Mar. 8, 1889, 5.

84. Riis, *How the Other Half Lives*, 85, 88. See also "New York's Own Russia," *NYT*, Feb. 23, 1892, 10.

85. "Healing the Sick Poor: Report of the Year's Work of the Eastern Dispensary," *NYT*, Mar. 20, 1885, 8.

86. Riis, *How the Other Half Lives*, 88.

87. Wolbarst, "Gonorrhea in Boys," 828.

88. Ibid., 827; "Society Proceedings, AMA, Section on Diseases of Children," *Medical News* 79 (July 6, 1901): 31, 35.

89. See Wolbarst, *Treatise on Cystoscopy*, and Wolbarst, *Gonorrhea in the Male*.

90. Wolbarst, "Gonorrhea in Boys," 828.

91. Ibid.

92. In addition to having privileges at Presbyterian Hospital, Cotton was chief of Cook County Hospital's infectious disease wards and taught at Rush Medical School, where he held chairs in pediatrics and the diseases of children. He also served as president of the Illinois and Chicago medical societies, the Chicago Pediatric Society, and the Chicago Examiner Association. "Death of Dr. Alfred C. Cotton," *Rush Alumni Association Bulletin* (July 1916): 4; Council of the Chicago Medical Society, *History of Medicine*, 139–41.

93. Wolbarst, "Gonorrhea in Boys," 830.

94. Ibid., 829.

95. Council of the Chicago Medical Society, *History of Medicine*, 363; "Dr. J. C. Cook Passes Away: Chicago Physician, Well Known for His Study of Children's Diseases, Is Dead," *Chicago Daily Tribune*, Mar. 22, 1908, A2; "Opening of La Rabida: Jackson Park Sanitarium to Care for Sick Children," *Chicago Daily Tribune*, June 5, 1898, 13; La Rabida Children's Hospital, "History of La Rabida," www.larabida.org/managex/index.asp?x=305&y=309&articlesource=309 (accessed Aug. 5, 2007).

96. Wolbarst, "Gonorrhea in Boys," 830.

97. Ibid.

98. Welt-Kakels, "Vulvovaginitis in Little Girls," 741.

99. Welt-Kakels treated girls at Mount Sinai from 1889 until her retirement in 1922. "Dr. Sara Welt," *NYT*, Dec. 28, 1943, 18; "A Great Hebrew Hospital," *NYT*, Feb. 21, 1892, 10.

100. Welt-Kakels, "Vulvovaginitis in Little Girls," 740.

101. Ibid., 740, 692

102. Ibid., 824.

103. Edward B. Morse, "Mount Sinai Hospital," *NYT*, Nov. 28, 1897, 10; "A Great Hebrew Hospital," *NYT*, Feb. 21, 1892, 10.

104. Trenwith, "Gonococcus Vaginitis in Little Girls," 240. Trenwith was on the staff at New York Hospital (currently New York–Cornell Medical Center) and drew from his own cases as well as those at the Children's Department of the Vanderbilt Clinic.

105. Jerome Walker, "Reports, with Comments, of Twenty-One Cases of Indecent Assault and Rape upon Children," pts. 1 and 2, *Archives of Pediatrics* 3 (May 1886): 269, 274; (June 1886): 321, 330–31.

106. John Higham, *Strangers in the Land: Patterns of American Nativism, 1860–1925*, 2nd ed. (New Brunswick, NJ: Rutgers University Press, 1988), 87–96. Italians and Jews were also lynched in this period, although in relatively few numbers.

107. Richard von Krafft-Ebing, *Psychopathia Sexualis*, trans. Charles G. Chaddock, 7th ed., rev. and enl. (Philadelphia: F. A. Davis Co., 1893); Harry Oosterhuis, *Stepchildren of Nature: Krafft-Ebing, Psychiatry, and the Making of Sexual Identity* (Chicago: University of Chicago Press, 2000), 47, 88–89.

108. Abraham L. Wolbarst, "Sexual Perversions: Their Medical and Social Implications," *Medical Journal and Record* 134 (July 1, 1931): 5, 7; Oosterhuis, *Stepchildren of Nature*, 46–47, 275–77. See, e.g., G. Frank Lydston, "A Lecture on Sexual Perversion, Satyriasis and Nymphomania," in *Addresses and Essays* (Louisville, KY: Renz and Henry, 1892), 243, 244.

109. Krafft-Ebing, *Psychopathia Sexualis*, 402.

110. Edgar and Johnston, "Rape," 462.

111. Lisa Duggan, *Sapphic Slashers: Sex, Violence, and American Modernity* (Durham, NC: Duke University Press, 2000), 3–4. On the private sexual misadventures, including father-daughter incest, of Gilded Age New Yorker Stanford White in this period, which became a public scandal at his death, see Suzannah Lessard, *The Architect of Desire: Beauty and Danger in the Stanford White Family* (New York: Delta Publishing, 1996), 10, 14–16, 197–213, 218–21, 277–301.

112. Gail Bederman, *Manliness and Civilization: A Cultural History of Gender and Race in the United States, 1880–1917* (Chicago: University of Chicago Press, 1995), 25–31, 35–36. See also Martin Summers, *Manliness and Its Discontents: The Black Middle Class and the Transformation of Masculinity, 1900–1930* (Chapel Hill: University of North Carolina Press, 2004), 1–15; Michele Mitchell, *Righteous Propagation: African Americans and the Politics of Racial Destiny after Reconstruction* (Chapel Hill: University of North Carolina Press, 2004), 10–12; Matthew Frye Jacobson, *Barbarian Virtues: The United States Encounters Foreign Peoples at Home and Abroad* (New York: Hill and Wang, 2000), 105–27,

139–54; Oosterhuis, *Stepchildren of Nature*, 51–55; Duggan, *Sapphic Slashers*, 3–4, 14–15, 19–27; Siobhan B. Somerville, *Queering the Color Line: Race and the Invention of Homosexuality in American Culture* (Durham, NC: Duke University Press, 2000), 3–13, 15–38; and George M. Fredrickson, *The Black Image in the White Mind: The Debate on Afro-American Character and Destiny, 1817–1914* (New York: Harper and Row, 1971), 228–82.

113. Amy L. Fairchild, *Science at the Borders: Immigrant Medical Inspection and the Shaping of the Modern Industrial Labor Force* (Baltimore: Johns Hopkins University Press, 2003), 31.

114. Edgar and Johnston, "Rape," 421. Georgia, Alabama, Texas, and Maryland included death among the penalties for statutory rape (420).

115. Fredrickson, *Black Image*, 276. See also Mitchell, *Righteous Propagation*, 79–84; Peter Bardaglio, *Reconstructing the Household: Families, Sex, and the Law in the Nineteenth-Century South* (Chapel Hill: University of North Carolina Press, 1995), 217–18; cf. Hunter McGuire and G. Frank Lydstron, "Sexual Crimes among the Southern Negroes—Scientifically Considered," *Virginia Medical Monthly* 20 (May 1893): 1.

116. Fredrickson, *Black Image*, 279; John S. Haller Jr. *Outcasts from Evolution: Scientific Attitudes of Racial Inferiority, 1859–1900* (Urbana: University of Illinois Press, 1971), 47–60.

117. Raymond Patterson, "Negro a Child in Mind," *Washington Post*, July 24, 1903, 5.

118. See Bederman, *Manliness and Civilization*, 23–31.

119. Edgar and Johnston, "Rape," 461.

120. See Winfried Schleiner, "Infection and Cure through Women: Renaissance Constructions of Syphilis," *Journal of Medieval and Renaissance Studies* 24 (Fall 1994): 499–517. Thanks to Laura McGough for bringing this article to my attention.

121. Trial of David Scott (Sept. 1796), *OBP*, t17960914-12.

122. Ibid.

123. William R. Wilde, "History of the Recent Epidemic of Infantile Leucorrhea, with an Account of Five Cases of Alleged Felonious Assaults Recently Tried in Dublin," *Medical News* 11 (Oct. 1853): 153, 154.

124. Giuseppe Pitrè, *Sicilian Folk Medicine*, trans. Phyllis H. Williams (Lawrence, KS: Coronado Press, 1971), 302; originally published as *Medicina Popolare Siciliana* (Turin: Carlo Clausen, 1896).

125. Gibb, "Indecent Assault," 649; Walker, "Reports," pt. 1, 274.

126. "Robley Dunglison," *Centuries of Leadership: Deans of the University of Maryland School of Medicine* (Baltimore: University of Maryland, School of Medicine, 2000), 13; Joby Topper, "Robley Dunglison, M.D., 1798–1869," Claude

Moore Health Sciences Library, University of Virginia, www.healthsystem.virginia.edu/internet/library/historical/uva_hospital/dunglison/ (accessed Aug. 5, 2007).

127. University of Virginia School of Medicine, "History: School of Medicine," *Factbook*, www.healthsystem.virginia.edu/internet/about/factbook/ch1.cfm#history (accessed Aug. 5, 2007).

128. "Robley Dunglison," *Centuries of Leadership*, 13.

129. Robley Dunglison, *Syllabus of the Lectures on Medical Jurisprudence and on the Treatment of Poisoning and Suspended Animation, Delivered in the University of Virginia* (Charlottesville: University of Virginia, 1827), 121. See also James C. Mohr, *Doctors and the Law: Medical Jurisprudence in Nineteenth-Century America* (Baltimore: Johns Hopkins University Press, 1993), 73.

130. Francis Wharton and Moreton Stillé, *A Treatise on Medical Jurisprudence* (Philadelphia: Kay and Bro., 1855), 327–28 (citing Johann Ludwig Caspar, *Pract. Hadbuch der Gerichtlichen Medicin* [Practical Handbook of Forensic Medicine], vol. 3 [Berlin, 1857], 103); William R. Wilde, "Medico-Legal Observations upon the Case of Amos Greenwood, Tried at the Liverpool Assizes, December, 1857, for the Wilful [*sic*] Murder of Mary Johnson, and Sentenced to Penal Servitude for Life," *Dublin Quarterly Journal of Medical Science* (Feb. 1859): 51, 110. See Mary Spongberg, *Feminizing Venereal Disease* (New York: New York University Press, 1997), 110. Spongberg also found a source indicating that certain Liverpool brothels specialized in providing virgins for the "cure" (211 n. 25).

131. W. Bathurst Woodman and Charles Meymott Tidy, *Forensic Medicine and Toxicology* (Philadelphia: Lindsay and Blakiston, 1877), 639. See also Taylor, *Principles and Practice of Medical Jurisprudence*, ed. Thomas Stevenson, 3rd ed. (Philadelphia: Lea, 1883), 2:427.

132. Jean Labbé, "Ambroise Tardieu: The Man and His Work on Child Maltreatment a Century before Kempe," *Child Abuse & Neglect* 29 (2005): 311, 314–15.

133. Walker, "Reports," 274.

134. Krafft-Ebing, *Psychopathic Sexualis*, 561.

135. Elbridge Gerry, "Must We Have the Cat-o'-Nine Tails?" *North American Review* 160 (Mar. 1895): 318, 321.

136. See Nora Ellen Groce and Reshma Trasi, "Rape of Individuals with Disability: AIDS and the Folk Belief of Virgin Cleansing," *Lancet* 363 (May 22, 2004): 1663–64; B. L. Meel, "The Myth of Child Rape as a Cure for HIV/AIDS in Transkei: A Case Report," *Medicine, Science, and the Law* 43 (2003): 85, 86–87; "Focus on the Virgin Myth and HIV/AIDS," *UN Integrated Regional Information Networks* (Apr. 25, 2002), www.aegis.com/news/irin/2002/IR020406.html (accessed Aug. 5, 2007); Gavin du Venage, "Rape of Children Surges in

South Africa: Minors Account for About 40% of Attack Victim [*sic*]," *San Francisco Chronicle*, Feb. 12, 2002, www.aegis.com/news/sc/2002/SC020203.html (accessed Aug. 5, 2007); Roger Davidson, "'This Pernicious Delusion': Law, Medicine, and Child Sexual Abuse in Early-Twentieth-Century Scotland," *Journal of the History of Sexuality* 10 (Jan. 2001): 62; Suzanne Daley, "Screening Girls for Abstinence in South Africa," *NYT*, Aug. 17, 1999, A3; Dean E. Murphy, "Africa's Silent Shame: Rape Is Spreading AIDS to Young Children. The Trend Is Fueled by Unscrupulous 'Healers' Who Suggest That Having Sex with a Virgin Can Cure the Disease," *LAT*, Aug. 16, 1998, pt. A, Foreign Desk, 1; A. S. Bada, "Traditional Belief: An Aid to the Spread of Sexually Transmitted Diseases," *Tropical Doctor* 27 (Oct. 1997): 253; Ellen R. Wald et al., "Gonorrheal Disease among Children in a University Hospital," *Sexually Transmitted Diseases* 7 (Apr.–June 1980): 41, 43.

137. See John Parascandola, "The Introduction of Antibiotics into Therapeutics," in *Sickness and Health in America: Readings in the History of Medicine and Public Health*, ed. Judith Walzer Leavitt and Ronald L. Numbers, 3rd ed., rev. (Madison: University of Wisconsin Press, 1997), 102, 102–7; Brandt, *No Magic Bullet*, 12.

138. See Phyllis H. Williams, *South Italian Folkways in Europe and America: A Handbook for Social Workers, Visiting Nurses, School Teachers, and Physicians* (New Haven: Yale University Press, 1938), 176.

139. Ferdinand C. Valentine and Terry M. Townsend, "Some Forensic Problems Concerning Venereal Diseases," *Albany Law Journal* 67 (Apr. 1905): 99, 105.

140. Flora Pollack, "A Report of the Women's Venereal Department of the Johns Hopkins Hospital Dispensary," *Maryland Medical Journal* 49 (Aug. 1906): 289, 290. See also Charles D. Lockwood, "Venereal Diseases in Children, Their Causes and Prevention," *Bulletin of the American Academy of Medicine* 11 (1910): 478, 479–10; Richard F. Woods, "Gonorrhoeal Vulvovaginitis in Children," *American Journal of the Medical Sciences* 125 (Feb. 1903): 311; and Wolbarst, "Gonorrhea in Boys," 827, 830 (comment of Dr. John C. Cook).

141. Jane Eliot Sewell, *Medicine in Maryland: The Practice and the Profession, 1799–1999* (Baltimore: Johns Hopkins University Press, 1999), 110.

142. Pollack, "Report of the Women's Venereal Department of the Johns Hopkins Hospital Dispensary," 289, 290.

143. W. Travis Gibb, "Criminal Aspect of Venereal Diseases in Children: Based upon the Personal Examination of Over 900 Children, the Alleged Victims of Rape, Sodomy, Indecent Assault, Etc.," *Transactions of the American Society of Sanitary and Moral Prophylaxis* 2 (1908): 25, 30

144. Ibid., 30.

145. Ibid., 25, 30.

146. Ibid., 30.

147. Gibb, "Criminal Aspect," 25.

148. Ibid., 25–26. "Swarthy" and "kinky-haired" southern Italians were grouped with "non-white" men at the turn of the century, in both Italy and America. Robert Orsi, "The Religious Boundaries of an Inbetween People: Street *Feste* and the Problem of the Dark-Skinned Other in Italian Harlem, 1920–1990," *American Quarterly* 44 (Sept. 1992): 313, 313–19.

149. Gibb, "Criminal Aspect," 25, 30.

150. Ibid., 30.

151. Pollack, "Acquired Venereal Infections," 144.

152. Ibid., 142.

153. Ibid., 142–43.

154. Ibid., 143, 145. The dispensary saw an average of thirty new cases each year. There were twenty other hospitals and dispensaries in Baltimore, and Pollack's estimate included patients seen in private practice.

155. Pollack, "Acquired Venereal Infections," 143–44. Pollack referred to page 561 of the 1906 translation, which states that "in Europe an idea is still prevalent that intercourse with children heals venereal disease." Richard von Krafft-Ebing, *Psychopathic Sexualis: With Especial Reference to the Antipathic Sexual Instinct, a Medico-Forensic Study*, trans. F. J. Rebman, 12th ed. (New York: Medical Art Agency, 1906), 561.

156. Pollack, "Acquired Venereal Infections," 148.

157. Ibid., 143.

158. Ibid., 148. See also W. P. Northrup, "General Gonococcal Peritonitis in Young Girls under Puberty," *Archives of Pediatrics* 20 (Dec. 1903): 388, 390, 392.

159. Pollack, "Acquired Venereal Infections," 144–47. Pollack had seen the girl who died of peritonitis about whom Hunner and Harris had written in 1902, and she said that there had been no fatalities since.

160. Ibid., 145.

161. Ibid., 144–45.

162. Ibid., 147.

163. Ibid.

164. See American Academy of Pediatrics, "Statement," *Pediatrics* 101 (Jan. 1998): 134.

165. E. D. Barringer, "Gonorrheal Vulvo-Vaginitis in Children," in *Diseases of the Genito-Urinary Organs: Considered from a Medical and Surgical Standpoint Including a Description of Gonorrhea in the Female and Conditions Peculiar to the Female Urinary Organs*, ed. Edward L. Keyes Jr. (New York: D. Appleton and Co., 1910), 122.

166. Emily Dunning Barringer, *Bowery to Bellevue: The Story of New York's First Woman Ambulance Surgeon* (New York: Norton, 1950), 184. See National Library of Medicine Online Exhibit, *Changing the Face of Medicine*, s.v. "Dr. Emily Dunning Barringer," Feb. 2004, www.nlm.nih.gov/changingthefaceofmedicine/ (accessed Aug. 5, 2007), and Ellen S. More, *Restoring the Balance: Women Physicians and the Profession of Medicine, 1850–1995* (Cambridge: Harvard University Press, 1999), 109, 183. John Sturges directed a film based on Barringer's memoir, *Girl in White* (1952), starring June Allyson.

167. Barringer, "Gonorrheal Vulvo-Vaginitis," 122. See also George P. Dale, "Moral Prophylaxis: Gonorrhoea," *American Journal of Nursing* 11 (July 1911): 782, 783.

168. Seippel, "Venereal Diseases in Children," 52.

169. Vice Commission of Chicago, *The Social Evil in Chicago: A Study of Existing Conditions* (Chicago: Gunthrop-Warren Printing Co., 1911), 240, 293–94.

170. Addams, *New Conscience*, 185.

171. William Healy, *Individual Delinquent: A Text-Book of Diagnosis and Prognosis for All Concerned in Understanding Offenders* (Boston: Little, Brown, 1915), 233.

172. Gurney Williams, "Rape in Children and in Young Girls: Based on the Personal Investigation of Several Hundred Cases of Rape and of Over Fourteen Thousand Vaginal Examinations," pts. 1 and 2, *International Clinics*, 23rd ser., 2 (1913): 245, 254–50; vol. 3, 245, 255.

173. Ibid., pt. 2, 252–53. See also Gilbert H. Stewart, *Legal Medicine* (Indianapolis: Bobbs-Merrill Co., 1910), 131. An attorney and professor of medical jurisprudence, Stewart claimed the superstition was widespread and encouraged, in part, by "old women who profess to some medical knowledge."

174. Frederick J. Taussig, "The Prevention and Treatment of Vulvovaginitis in Children," *American Journal of Medical Sciences* 148 (Oct. 1914): 480, 482–83.

175. E.g., Stephen A. Yesko, "Gonorrheal Vulvovaginitis in the Young," *American Journal of Diseases of Children* 33 (Apr. 1927): 633, 635–36; George Robertson Livermore and Edward Armin Schumann, *Gonorrhea and Kindred Affections* (New York: D. Appleton and Co., 1929), 204; Walter M. Brunet and Robert L. Dickinson, "Gonorrhea in the Female," *VDI* 10 (Apr. 20, 1929): 149, 164.

176. John Lovett Morse, *Clinical Pediatrics* (Philadelphia: Saunders, 1926), 667.

177. Matthew Frye Jacobson, *Whiteness of a Different Color: European Immigrants and the Alchemy of Race* (Cambridge: Harvard University Press, 2001), 7–8, 68–90; Fairchild, *Science at the Borders*, 215–20, 256.

178. B. H. Regenburg and R. A. Durfee, "Venereal Disease and the Patient," *JSH* 20 (Nov. 1934): 369, 376–77.

179. Sydney Smith, ed., *Taylor's Principles and Practice of Medical Jurisprudence*, 9th ed. (London: J. and A. Churchill, 1934), 2:111.

180. Havelock Ellis, *Studies in the Psychology of Sex*, vol. 4: *Sex in Relation to Society* (1906; repr., New York: Random House, 1936), 337.

181. John B. West, "Gonorrhea Has Preventives," *New York Amsterdam News*, July 24, 1937, 13; "Social Diseases to Be Revealed," *New York Amsterdam News*, June 26, 1937, 2; "Central Harlem Health Center's Model Home to Replace 'House Where Mme. Walker Lived,'" *New York Amsterdam News*, Apr. 24, 1937, 24.

182. Williams, *South Italian Folkways*, 167–68. Williams claimed to have worked with more than five hundred Italian families in eleven years.

183. Nels A. Nelson and Gladys L. Crain, *Syphilis, Gonorrhea and the Public Health* (New York: Macmillan, 1938), 141, 143.

184. Ibid., 143. See Howard Markel, "Academic Pediatrics: The View from New York City a Century Ago," *Academic Medicine* 71 (Feb. 1996): 146.

185. See Stella, "Effects of Urban Congestion," 723, and Thomas A. Guglielmo, *White on Arrival: Italians, Race, Color, and Power in Chicago, 1890–1945* (New York: Oxford University Press, 2003), 21–23, 77–79.

Chapter Four: Protecting Fathers, Blaming Mothers

1. David J. Pivar, *Purity Crusade: Sexual Morality and Social Control, 1868–1900* (Westport, CT: Greenwood Press, 1973), 259, 171–72, 238–39; Linda Gordon, *The Moral Property of Women: A History of Birth Control Politics in America* (Urbana: University of Illinois Press, 2002), 72–75.

2. Pivar, *Purity Crusade*, 6–7, 131–71; Allan M. Brandt, *No Magic Bullet: A Social History of Venereal Disease in the United States since 1880* (New York: Oxford University Press, 1987), 7; John C. Burnham, "The Progressive Era Revolution in American Attitudes toward Sex," *Journal of American History* 59 (Mar. 1973): 885, 886.

3. Burnham, "Progressive Era Revolution," 890.

4. Gordon, *Moral Property of Women*, 114.

5. Burnham, "Progressive Era Revolution," 886; Mark Thomas Connelly, *The Response to Prostitution in the Progressive Era* (Chapel Hill: University of North Carolina Press, 1980), 67–90, 155–57, 179–95. Brandt's *No Magic Bullet* is the most complete history of twentieth-century social hygiene.

6. Brandt, *No Magic Bullet*, 14.

7. Prince A. Morrow, "Report of the Committee of Seven of the Medical Society of the County of New York on the Prophylaxis of Venereal Disease in New York City," *New York Medical News* 74 (Dec. 21, 1901): 961; cf. Richard C.

Cabot, "Observations Regarding the Relative Frequency of the Different Diseases Prevalent in Boston and Its Vicinity," *Boston Medical and Surgical Journal* 165 (Aug. 3, 1911): 155; Prince A. Morrow, "The Frequency of Venereal Diseases: A Reply to Dr. Cabot," *Boston Medical and Surgical Journal* 165 (Oct. 5, 1911): 520. See also Brandt, *No Magic Bullet*, 12–13.

8. Prince A. Morrow, *Social Diseases and Marriage: Social Prophylaxis* (New York: Lea Brothers and Co., 1904); Prince A. Morrow, "The Relations of Social Diseases to the Family," *American Journal of Sociology* 14 (Mar. 1909): 622.

9. Morrow, *Social Diseases*, 25; Burnham, "Progressive Era Revolution," 892–904. See C.-E. A. Winslow, *The Evolution and Significance of the Modern Public Health Campaign* (New Haven: Yale University Press, 1923), 55, 57–59; Prince A. Morrow et al., "Symposium on Venereal Diseases," *JAMA* 47 (Oct. 20, 1906): 1244–58.

10. Prince A. Morrow, "Education within the Medical Profession," pamphlet reprinted from the *Medical News* 85 (June 24, 1905), box 1, folder 7, American Social Hygiene Collection, Rare Book and History of Medicine Collection, Malloch Rare Book Room, NYAM; Prince A. Morrow, "Publicity as a Factor in Venereal Prophylaxis," *JAMA* 47 (Oct. 20, 1906): 1244.

11. Morrow, *Social Diseases*, v, 19–21, Brandt, *No Magic Bullet*, 25–27; Pivar, *Purity Crusade*, 242–44; Burnham, "Progressive Era Revolution," 890–92.

12. Morrow, "Education within the Medical Profession," 10–14; Morrow, *Social Diseases*, 30–33, 90–110; Morrow, "Report of the Committee," 962–63.

13. Morrow, *Social Diseases*, 19–24; Morrow, "Education within the Medical Profession," 10–11. See also Albert H. Burr, "The Guarantee of Safety in the Marriage Contract," *JAMA* 47 (Dec. 8, 1906): 1887; George Gray Ward, "Common Causes of Gynecological Disease," *Medical News* 86 (May 13, 1905): 876; and Albert Neisser, "Gonorrhea: Its Dangers to Society," *Medical News* 76 (Jan. 20, 1900): 85, 88–89.

14. Brandt, *No Magic Bullet*, 23.

15. Morrow, *Social Diseases*, 23, 30, 340–44.

16. Brandt, *No Magic Bullet*, 15–17, 24–25; Gordon, *Moral Property of Women*, chap. 6. See, e.g., Morrow, *Social Diseases*, 108–110; Morrow, "Education within the Medical Profession," 9; Edith Houghton Hooker, "Race Suicide," *Life's Clinic: A Series of Sketches Written from between the Lines of Some Medical Case Histories* (New York: Association Press, 1916, 1917), 14–20; "Eliot Blames Man for the Social Evil," *NYT*, May 23, 1913, 8; Sophonisba P. Breckenridge and Edith Abbott, *The Delinquent Child and the Home* (1912; repr., New York: Arno Press, 1970), 106; and Egbert H. Grandin, "Race Suicide from the Gynecological Standpoint," *Medical News* 85 (July 9, 1904): 51, 52. On "race suicide," see Amy L. Fairchild, *Science at the Borders: Immigrant Medical Inspec-*

tion and the Shaping of the Modern Industrial Labor Force (Baltimore: Johns Hopkins University Press, 2003), 8–9, 193–204, and Miriam King and Steven Ruggles, "American Immigration, Fertility, and Race Suicide at the Turn of the Century," *Journal of Interdisciplinary History* 20 (Winter 1990): 347, 365–66.

17. Brandt, *No Magic Bullet*, 24–25; Committee on Education, American Public Health Association, "Report of Committee on Education of the Public as to the Communicability and Prevention of Gonorrhoea and Syphilis," *American Journal of Public Health* 2 (Mar. 1912): 194, 197–200; Prince A. Morrow, "A Plea for the Organization of a 'Society of Sanitary and Moral Prophylaxis,'" *Medical News* 84 (June 4, 1904): 1073.

18. Morrow, *Social Diseases*, 119, 116.

19. Prince A. Morrow, *Wood's Pocket Manuals, Venereal Memoranda: A Manual for the Student and Practitioner* (New York: William Wood and Co., 1885), 15.

20. Morrow, *Social Diseases*, 117–18.

21. Ibid., 117.

22. J. Clifton Edgar, "Gonococcus Infection as a Cause of Blindness, Vulvovaginitis and Arthritis," *JAMA* 49 (Aug. 3, 1907): 411, 413–14; see also Alice Hamilton, "Gonorrheal Vulvo-Vaginitis in Children: With Special Reference to an Epidemic Occurring in Scarlet-Fever Wards," *Journal of Infectious Diseases* 5 (Mar. 1908): 133, 144.

23. Edgar, "Gonococcus Infection," 413–14; Andrew F. Currier, "Vulvo-Vaginitis in Children," *Medical News* 55 (July 6, 1889): 3.

24. Edgar, "Gonococcus Infection," 414.

25. Alfred Cleveland Cotton, *Medical Diseases of Infancy and Childhood* (Philadelphia: Lippincott, 1906), 386–87.

26. Chicago Society of Social Hygiene, *For the Protection of Wives and Children from Venereal Contamination* (Chicago: Chicago Society of Social Hygiene, 1907).

27. "Dr. Rachelle Yarros: Physician, Scientist, Humanitarian," *Medical Women's Journal* 38 (Jan. 1931): 10; "Urges Women to Join in Hygiene Campaign," *Washington Post*, Nov. 24, 1921, 11; Rachel S. Yarros, "Experiences of a Lecturer," *JSH* 5 (Apr. 1919): 205; "Dr. Rachelle S. Yarros Given Important Post by State of Illinois," *Atlanta Constitution*, Apr. 6, 1919, B11; Mary Ritter Beard, *Women's Work in Municipalities* (New York: Ayers Publishing, 1915), 127; "Crowds Hear Talks on Sex," *Chicago Daily Tribune*, June 16, 1914, 3; "Sex Hygiene Lectures Announced by Mrs. Young," *Chicago Daily Tribune*, Oct. 14, 1913, 10; "Dr. Rachel Yarros Speaks on Social Hygiene to Large Audience," *Lexington (KY) Herald*, Apr. 18, 1913, [1].

28. Chicago Society of Social Hygiene, *For the Protection of Wives*, 3.

29. See Fairchild, *Science at the Borders*, 80–81; Nancy Tomes, "Spreading the Germ Theory: Sanitary Science and Home Economics, 1880–1930," in *Women and Health in America: Historical Readings*, ed. Judith Walzer Leavitt (Madison: University of Wisconsin Press, 1999), 596; John Duffy, "Social Impact of Disease in the Late Nineteenth Century," in *Sickness and Health in America: Readings in the History of Medicine and Public Health*, ed. Judith Walzer Leavitt and Ronald L. Numbers, 3rd ed., rev. (Madison: University of Wisconsin Press, 1997), 418, 424; and Suellen Hoy, *Chasing Dirt: The American Pursuit of Cleanliness* (New York: Oxford University Press, 1995), 102–4.

30. See Hoy, *Chasing Dirt*, 87–121; Nancy Tomes, *The Gospel of Germs: Men, Women, and the Microbe in American Life* (Cambridge: Harvard University Press, 1998), 185–95; and Alan M. Kraut, *Silent Travelers: Germs, Genes, and the Immigrant Menace* (New York: Basic Books, 1994). Naomi Rogers documented medical rhetoric about supposed connections between cleanliness, citizenship, class, and polio infection in the first half of the twentieth century. Naomi Rogers, *Dirt and Disease: Polio before FDR* (New Brunswick, NJ: Rutgers University Press, 1992).

31. Hamilton, "Gonorrheal Vulvo-Vaginitis," 134. See also L. Emmett Holt, "Gonococcus Infections in Children, with Especial Reference to Their Prevalence in Institutions and Means of Prevention," *New York Medical Journal and Philadelphia Medical Journal* 81 (Mar. 18, 1905): 521, and Sara Welt-Kakels, "Vulvovaginitis in Little Girls: A Clinical Study of 190 Cases," *New York Medical Journal and Philadelphia Medical Journal* 80 (Oct. 8, 1904): 689, 740.

32. Hamilton, "Gonorrheal Vulvo-Vaginitis," 144.

33. See Alice Hamilton, *Exploring the Dangerous Trades: The Autobiography of Alice Hamilton, M.D.* (Boston: Little, Brown, 1943), 99–100, and Barbara Sicherman, *Alice Hamilton: A Life in Letters* (Cambridge: Harvard University Press, 1984), 145, 144–46.

34. Sicherman, *Alice Hamilton*, 145, 144–46.

35. Hamilton, *Exploring the Dangerous Trades*, 98.

36. Sicherman, *Alice Hamilton*, 146.

37. Hamilton, *Exploring the Dangerous Trades*, 99–100.

38. Palmer Findley, *Gonorrhoea in Women* (St. Louis: C. V. Mosby, 1908), 20.

39. Ibid., 21, 70.

40. Ibid., 70.

41. Alice Hamilton, "Venereal Diseases in Institutions for Women and Girls," *Proceedings of the National Conference of Charities and Corrections* 37 (1910): 53, 54.

42. Ibid., 54.

43. Ibid., 54–55.

44. Lavinia L. Dock, *Hygiene and Morality: A Manual for Nurses and Oth-*

ers, Giving an Outline of the Medical, Social, and Legal Aspects of the Venereal Diseases (1910; New York: G. P. Putnam's Sons, 1912). See also Charlotte A. Aikens, *Clinical Studies for Nurses: A Text-Book for Second and Third Year Pupil Nurses and a Hand-Book for All Who Are Engaged in Caring for the Sick* (Philadelphia: Saunders, 1911); cf. Abraham Wolbarst, "On the Occurrence of Syphilis and Gonorrhea in Children by Direct Infection," *American Medicine* 7 (Sept. 1912): 493, 494.

45. Dock, *Hygiene and Morality*, iii.

46. Ibid., 48.

47. Ibid., 147.

48. Ibid., 138–46, 148–49.

49. Mary Burr, "Some Statistics of Criminal Assault upon Young Girls" (1909), in Dock, *Hygiene and Morality*, app. C, 178, 182.

50. Hamilton, "Gonorrheal Vulvo-Vaginitis," 142.

51. E. D. Barringer, "Gonorrheal Vulvo-Vaginitis in Children," in *Diseases of the Genito-Urinary Organs*, ed. Edward L. Keyes Jr. (New York: D. Appleton and Co., 1910), 122.

On scrofula and gonorrhea vulvovaginitis: Hamilton, "Gonorrheal Vulvo-Vaginitis," 145, and Antonio Stella, "The Effects of Urban Congestion on Italian Women and Children," *Medical News* 18 (May 2, 1908): 722, 728.

On public health in this period: Evelynn Maxine Hammonds, *Childhood's Deadly Scourge: The Campaign to Control Diphtheria in New York City, 1880–1930* (Baltimore: Johns Hopkins University Press, 1999), 9–11, and John Duffy, *The Sanitarians: A History of American Public Health* (Urbana: University of Illinois Press, 1990), 193–220.

52. Barringer, "Gonorrheal Vulvo-Vaginitis," 122; Reuel B. Kimball, "Gonorrhoea in Infants, with a Report of Eight Cases of Pyaemia," *Medical Record* 64 (Nov. 14, 1903): 761.

53. Barringer, "Gonorrheal Vulvo-Vaginitis," 122.

54. "Aiding Medical Progress: Dedication of Two Gifts to Science; Opening Exercises of the Sloane Maternity Hospital and Vanderbilt Clinic," *NYT*, Dec. 30, 1887, 9.

55. "Correspondence: Letter from New York," *Medical News* 59 (July 18, 1891): 81.

56. See, e.g., "Dispensaries and Physicians: Medical Men Say the Public Abuses Many Free Privileges Extended to Them," *NYT*, Apr. 4, 1909, 10.

57. "Rich Gifts to Columbia: The Vanderbilts, Sloanes, and Others Subscribe $1,050,000: All Given for New Buildings," *NYT*, Jan. 8, 1895, 1.

58. B. Wallace Hamilton, "Gonococcus Vulvovaginitis in Children: With Results of Vaccine Treatment in Out-Patients," *JAMA* 54 (Apr. 9, 1910): 1196.

59. Ibid., 1196.

60. Ibid., 1197.

61. Charles-Edward Amory Winslow and Pauline Brooks Williamson included instructions on how to douche a girl infected with gonorrhea in *The Laws of Health and How to Teach Them* (New York: Charles E. Merrill Co., 1926), 70.

62. Jessie D. Hodder, "Sex Problems," *Third Annual Report of the Social Service Department of the Massachusetts General Hospital*, Oct. 1, 1907, to Oct. 1, 1908 (Boston: Fort Hill Press, [ca. 1908]), 26, 31. Available at Massachusetts General Hospital Archives and Special Collections, Social Service Department Records, box 2, MGH. MGH social work case files are not open for public inspection. After a lengthy discussion about my research interests, on November 5, 1999, MGH archivist Jeffrey Mifflin examined every female social work case from 1905 to 1940 in the files of the Social Service Department. He found no cases in which a female under 21 years of age was being treated for gonorrhea, vaginitis, or vulvovaginitis.

63. Hodder, "Sex Problems," 31.

64. Ibid.

65. Ibid., 31. The Schlesinger Library at Harvard University holds Hodder's papers, in which I could find no mention of the issue of infected girls.

66. Hodder, *Fourth Annual Report of the Social Service Department of the Massachusetts General Hospital*, Oct. 1, 1908, to Jan. 1, 1910 (Boston: Fort Hill Press, [ca. 1910]), 25, 26–27, Social Service Department Records, box 2, MGH. Reports for subsequent years did not provide the ages of infected girls.

67. Richard C. Cabot, M.D., "Problems of Sex," *Fifth Annual Report of the Social Service Department of the Massachusetts General Hospital*, Jan. 1, 1910, to Jan. 1, 1911 (Boston: Fort Hill Press, [ca. 1911]), 16, 16–17, Social Service Department Records, box 2, MGH. On Cabot and hospital social work, see Daniel J. Walkowitz, *Working with Class: Social Workers and the Politics of Middle-Class Identity* (Chapel Hill: University of North Carolina Press, 1999), 44–46.

68. Fritz B. Talbot and Richard M. Smith, "Report of the Social Work in the Children's Medical Department," *Sixth Annual Report of the Social Service Department of the Massachusetts General Hospital*, Jan. 1, 1911, to Jan. 1, 1912 (Boston: Fort Hill Press, [ca. 1912]), 13, 14, Social Service Department Records, box 2, MGH.

69. Nathaniel W. Faxon, *Massachusetts General Hospital, 1935–1955* (Cambridge: Harvard University Press, 1959), 264–65.

70. Talbot and Smith, "Report," 16–17.

71. Richard M. Smith, "Vulvovaginitis in Children," *American Journal of Diseases of Children* 6 (Nov. 1913): 355, 357.

72. Ibid., 357.

73. Ibid.

74. Ibid., 356–57.

75. Ibid., 358.

76. Smith, "Vulvovaginitis," 358. See also B. K. Rachford, "Epidemic Vaginitis in Children," *American Journal of the Medical Sciences* 153 (Feb. 1917): 207, 210–11.

77. Smith, "Vulvovaginitis," 358–59.

78. Edith Rogers Spaulding, "Vulvovaginitis in Children," *American Journal of Diseases of Children* 5 (Mar. 1913): 248.

79. Ibid., 253.

80. Ibid., 266.

81. Clara P. Seippel, "Venereal Diseases in Children," *Illinois Medical Journal* 22 (July 1912): 50, 55. The home had room for fifteen girls. Vice Commission of Chicago, *Social Evil in Chicago: A Study of Existing Conditions* (Chicago: Gunthrop-Warren Printing Co., 1911), 278.

82. "Dr. Clara Seippel Heads Medical Women's Club," *Chicago Daily Tribune*, June 14, 1917, 8, col. 4; Inez Travers, "News of the Chicago Women's Clubs," *Chicago Daily Tribune*, Nov. 22, 1914, D4; Inez Travers, "Work of the Anti Cruelty Society Heavy at This Season," *Chicago Daily Tribune*, June 14, 1914, F4; "Final Poll Talk for Women: Almost 800 Sign List as Pioneer Election Officials of Sex," *Chicago Daily Tribune*, Feb. 1, 1914, 3.

83. Byron Cummings to Dr. Clara Seippel Webster, Sept. 6, 1927; "Clara Webster, Pioneer Woman Physician Dies," *Arizona Daily Star* (Tucson), Apr. 7, 1965; Clara Seippel, University of Arizona application for employment, Feb. 20, 1929, all in "Clara Seippel Webster, 1877–1965," University of Arizona Biographical File, University of Arizona Library Special Collections, Main Library, University of Arizona, Tucson, Arizona.

84. "Frances Home Group Takes Up Child Program," *Chicago Daily Tribune*, June 4, 1942, 18; "Home for Young Venereals Shut after 32 Years," *Chicago Daily Tribune*, July 1, 1941, 3; "Midway School Buys Residence on Kimbark Ave.," *Chicago Daily Tribune*, May 21, 1925, 26; "Frances Juvenile Home in Need of Larger Quarters," *Chicago Daily Tribune*, Dec. 12, 1920, F5; "Patten Gives $10,000 to Aid Frances Juvenile Home," *Chicago Daily Tribune*, Aug. 19, 1911, 4

85. "Patten Aids Frances Home," *Chicago Daily Tribune*, Feb. 5, 1911, 3.

86. Ibid., 3; Seippel, "Venereal Diseases in Children," 53.

87. Seippel, "Venereal Diseases in Children," 53–54.

88. W. A. Evans, M.D., *Some Information about the So-Called Venereal Diseases, Most of Which Appeared in the Chicago Tribune* (1913), NYAM / Venereal Disease—Section on Public Health file, Venereal Diseases 1909–14 folder,

NYAM; [City of Chicago], *Chicago's Care of Her Children*, Children's Charity Work: Pamphlet Box (n.p., [ca. 1912–16]), Widener Library, Harvard University, Cambridge, MA. In 1917 the Cincinnati Hospital and Children's Clinic treated seventy-five girls, most of whom had transferred from orphanages or "children's homes." Rachford, "Epidemic Vaginitis," 207.

89. Seippel, "Venereal Diseases in Children," 52.

90. Ibid.

91. Ibid.

92. Ibid., 51.

93. Ibid.

94. Jane Addams, "Charity and Social Justice," *Survey* 24 (June 11, 1910): 441, 447.

95. Anne Meis Knupfer, "'To Become Good, Self-Supporting Women': The State Industrial School for Delinquent Girls at Geneva, Illinois, 1900–1935," *Journal of the History of Sexuality* 9 (Oct. 2000): 420, 423, 426; Louise Morrow and Olga Bridgman, "Gonorrhea in Girls: Treatment of Three Hundred Cases," *JAMA* 58 (May 25, 1912): 1564. On female delinquency in the Progressive Era, see Anne Meis Knupfer, *Reform and Resistance: Gender, Delinquency, and America's First Juvenile Court* (New York: Routledge, 2001); Janis Appier, *Policing Women: The Sexual Politics of Law Enforcement and the LAPD* (Philadelphia: Temple University Press, 1998); Mary E. Odem, *Delinquent Daughters: Protecting and Policing Adolescent Female Sexuality in the United States, 1885–1920* (Chapel Hill: University of North Carolina Press, 1995); Steven Schlossman and Stephanie Wallach, "The Crime of Precocious Sexuality: Female Juvenile Delinquency and the Progressive Era," in *Women and the Law: A Social Historical Perspective*, ed. D. Kelly Weisberg (Cambridge, MA: Schenkman Publishing Co., 1982), 1:45.

96. Vice Commission of Chicago, *Social Evil*, 174–76. See also Maude E. Miner, *Slavery of Prostitution: A Plea for Emancipation* (New York: Macmillan, 1916), 39, 55.

97. Elizabeth A. Sullivan and Edith R. Spaulding, "The Extent and Significance of Gonorrhea in a Reformatory for Women," *JAMA* 66 (Jan. 8, 1916): 95, 101.

98. Jane Addams, *A New Conscience and an Ancient Evil* (New York: Macmillan, 1912), 109.

99. Arthur W. Towne, "A Community Program for Protective Work with Girls," *JSH* 6 (Jan. 1920): 57, 66.

100. W. I. Thomas, *The Unadjusted Girl: With Cases and Standpoint for Behavioral Analysis* (Boston: Little, Brown, 1923), 99–102. At the Girls' Industrial School in Cleveland in the early 1920s, one-third of the girls reported having been sexually assaulted by their fathers, and one-quarter were infected with venereal disease. Howard Whipple Green, *An Analysis of 377 Records of Girls Commit-*

ted to the Girls Industrial School by the Juvenile Court at Cleveland during a Six Year Period, 1920–25 (Cleveland: Cleveland Social Hygiene Assn., 1929).

101. Charles A. Fife, J. Claxton Gittings, and Howard Childs Carpenter, "Report of the Committee of the American Pediatric Society on Vaginitis in Childhood," *Transactions of the American Pediatric Society* 26 (1915): 331, 334.

102. Ibid., 334–35.

103. George W. Goler, *Syphilis, Gonorrhea, and Gonorrhoeal Opthalmia: The Curse of Man, the Scourge of Woman, the Blight of Children* (Rochester, NY: Health Bureau, 1916), 2, Venereal Diseases 1915–17 folder, NYAM.

104. Goler, *Syphilis, Gonorrhea*, 4. See also Thurman B. Rice, *The Venereal Diseases* (Chicago: AMA, 1933), 8.

105. See Edith Abbott, *The Tenements of Chicago, 1908–1935* (Chicago: University of Chicago Press, 1936), 211–13.

106. Walter M. Brunet and Robert L. Dickinson, "Gonorrhea in the Female," *VDI* 10 (Apr. 20, 1929): 149, 166.

107. Gladys Crain, "Facts about Gonorrhea," *Commonhealth* [Bulletin of the Massachusetts Department of Health] 20 (Apr.–June 1933): 84, reprinted from *Public Health Nursing* (Jan. 1933).

108. Ibid., 86. By 1938 Brunet's and Crane's view had been compressed into the statement "But, particularly in poor communities, where there is a considerable lack of personal hygiene, with bad housing and crowded sleeping quarters, much child gonorrhea can be found." Nadine B. Geitz, "Nurses' Number," *JSH* 24 (Apr. 1938): 161, 171; see also Nadine B. Geitz, *Social Hygiene Nursing Techniques: A Manual of Procedure in the Diagnosis, Treatment and Public Health Control of Syphilis and Gonorrhea*, rev. ed. (New York: ASHA, 1943), 10, 53.

109. Brandt, *No Magic Bullet*, 36–37.

110. William Snow, *The American Social Hygiene Association: Some Notes on the Historical Background, Development and Future Opportunities of the National Voluntary Organization for Social Hygiene in the United States* (New York: ASHA, June 1946), pamphlet, box 1, folder 1, ASHA Collection; *Milestones in Venereal Disease Control* (New York: ASHA, 1949), pamphlet, box 1, folder 1, ASHA Collection; Minutes, American Vigilance Association, Feb. 17, 1913, box 2, folder 6, ASHA Collection; pamphlet, New York Social Hygiene Society, Inc., formerly the American Society for Sanitary and Moral Prophylaxis, box 125, folder 18, ASHA Collection. The organization changed its name in 1960. Press Release, Jan. 18, 1960, box 18, folder 6, ASHA Collection; David Klaassen and Kay Faminio, *American Social Hygiene Association*, brochure (Research Triangle Park, NC: ASHA, 1994), 3–5, 17. My thanks to David Klaassen, archivist of the Social Welfare History Archives, for giving me a copy of the brochure. See also Brandt, *No Magic Bullet*, 38.

111. Brandt, *No Magic Bullet*, 38; Dunham to Topping, memorandum, Oct. 26, 1931, ser. 3, box 7, folder 164, Bureau of Social Hygiene Project and Research Files [1918–40], Rockefeller Archive Center of Rockefeller University, Sleepy Hollow, New York (Wilmington, DE: Scholarly Resources, 1979; microfilm, Lamont Library, Harvard University, Cambridge, Massachusetts); cf. Klaassen and Faminio, *American Social Hygiene Association*, 4–5.

112. Roster of officers and directors, 1913–57 (ca. 1957), box 18, folder 7, ASHA Collection.

113. Charles W. Eliot, "The American Social Hygiene Association," *JSH* 1 (Dec. 1914): 1, 2.

114. William F. Snow, "Progress, 1900–1915," *JSH* 2 (Jan. 1916): 37, 41.

115. E.g., Maurice A. Bigelow, "Youth and Morals," *JSH* 14 (Jan. 1928): 1; Thomas Parran, "Social Hygiene and Public Health," *JSH* 13 (Jan. 1927): 17; "Children and the Motion Picture Problem," *JSH* 10 (May 1924): 309; Karl S. Lashley and John B. Watson, "A Psychological Study of Motion Pictures in Relation to Venereal Disease Campaigns," *JSH* 7 (Apr. 1921): 181; Brandt, *No Magic Bullet*, 159.

116. Charles W. Eliot, "Public Opinion and Sex Hygiene," *JSH* 12 (1926): 451, 455–56. Eliot read his paper at the Fourth International Congress on School Hygiene and the Annual Meeting of the Federation, Buffalo, New York, 1913. Box 2, folder 3, ASHA Collection; see also Brandt, *No Magic Bullet*, 126–29, 136–37.

117. The Massachusetts Society for Social Health, Articles of Incorporation, June 15, 1915, box 4, folder 30, MSSH Records; Massachusetts Society for Social Health board of directors to old subscribers, letter dated Feb. 14, 1928, box 1, folder 1, MSSH Records; "Introducing the Bulletin," *Bulletin of the MSSH*, Mar. 1931, 1; Maida Herman Solomon, transcript of oral history, interviewed by Barbara Miller Solomon, Anne Evans, Eva S. Mosely, and Elizabeth Gould Herrera, vol. 2, 1977, 5-254–59; 6-271–79, OH-24, Arthur and Elizabeth Schlesinger Library on the History of Women in America, Radcliffe Institute for Advanced Study, Harvard University, Cambridge, Massachusetts.

118. E.g., Max J. Exner, "The Sex Factor in Character Training," *JSH* 10 (Oct. 1924): 385.

119. E.g., Donald R. Hooker, "In Defense of Radicalism," *JSH* 3 (Apr. 1917): 157; "That Word 'SEX,'" *JSH* 14 (Nov. 1928): 449.

120. See Julian B. Carter's provocative discussion of this shift in *Hearts of Whiteness: Normal Sexuality and Race in America, 1880–1940* (Durham, NC: Duke University Press, 2007), 118–40. See also Gordon, *Moral Property of Women*, 123–38.

121. "And What of the Family? Three Extracts from Addresses Given in the Conference on Family Life in America Today, Buffalo, October 2–5, 1927," *JSH*

14 (Feb. 1928): 90; Edward L. Keyes, "Social Hygiene and the National Association," *JSH* 11 (Dec. 1925): 513, 518–19.

122. See George B. Mangold, "Social Work and Social Hygiene," *JSH* 10 (Jan. 1924): 12, 15.

123. Michele Mitchell, *Righteous Propagation: African Americans and the Politics of Racial Destiny after Reconstruction* (Chapel Hill: University of North Carolina, 2004), 80–81, 84–86, 101–3; Christina Simmons, "African Americans and Sexual Victorianism in the Social Hygiene Movement, 1910–40," *Journal of the History of Sexuality* 4 (July 1993): 51. See Franklin O. Nichols, "Social Hygiene and the Negro," *JSH* 15 (Oct. 1929): 408, 410.

124. E.g., Dr. A. Wilberforce Williams, "How to Keep Well," *Chicago Defender*, Aug. 21, 1920, 12, and Aug. 28, 1920, 12; Dr. A. Wilberforce Williams, "The Venereal Drive: Gonorrhea in Women, the Black Peril," *Chicago Defender*, June 15, 1918, 16; Dr. A. Wilberforce Williams, "Keep Healthy," *Chicago Defender*, Nov. 1, 1913, 4; Dr. A. Wilberforce Williams, "Keeping Healthy," *Chicago Defender*, Oct. 25, 1913, 4.

125. Mary Church Terrell, "Tells Astounding Facts at World's Purity Federation," *Chicago Defender*, Nov. 12, 1927, 4. See also Dr. A. Wilberforce Williams, "Talks on Preventative Measures, First Aid Remedies, Hygienics and Sanitation," *Chicago Defender*, Aug. 30, 1919, 20.

126. R. A. Adams, "Arrows," *New York Amsterdam News*, Dec. 12, 1927, 16.

127. Aubrey Bowser, "Superstition and Syphilis," review of Roark Bradford, *This Side of Jordan*, *New York Amsterdam News*, Apr. 10, 1929, 20.

128. Mary Swain Routzahn, "The Aims of Social Hygiene Publicity," *JSH* 15 (Feb. 1929): 65, 66.

129. E.g., Anna Garlin Spencer, "The Social Education of Women in Relation to Family Life," *JSH* 15 (May 1929): 257; Ira S. Wile, "Sex as Biological Social Factor," *JSH* 15 (May 1929): 277.

130. "The Conference," *JSH* 11 (Jan. 1925): 54, 55.

131. Katharine Bement Davis, "Study of the Sex Life of the Normal Married Woman," pts. 1–3, *JSH* 8 (Apr. 1922): 173; 9 (Jan. 1923): 1; 9 (Mar. 1923): 129. Davis published her book-length study in 1929. Katharine Bement Davis, *Factors in the Sex Life of Twenty-Two Hundred Women* (New York: Harper and Brothers, 1929).

132. E.g., Walter M. Brunet, "The Great Imitator," *JSH* 14 (Oct. 1928): 400, 401–2; Bascom Johnson, "Civic Housecleaning," *JSH* 14 (Nov. 1928): 464, 466–69.

133. E.g., B. H. Regenburg and R. A. Durfee, "Venereal Disease and the Patient," *JSH* 20 (Nov. 1934): 369; A. J. Lanza, "Venereal Diseases and the Family," *JSH* 11 (Oct. 1925): 396.

134. Edith M. Baker, "Social Case Work in Hospital and Clinic." *JSH* 13 (Nov. 1927): 477, 491.

135. Haven Emerson, "A Family Service Ward," *JSH* 9 (Mar. 1923): 239, 240.

136. See Anna Garlin Spencer, *Woman's Share in Social Culture* (New York: Mitchell Kennerley, 1912), 254–74, and William L. O'Neill, "Divorce in the Progressive Era," in *The American Family in Social-Historical Perspective*, ed. Michael Gordon (New York: St. Martin's Press, 1973), 251–66.

137. E.g., Baker, "Social Case Work," 477.

138. Tomes, "Spreading the Germ Theory," 605.

139. E.g., Baker, "Social Case Work," 491.

140. Frederick J. Taussig, "The Contagion of Gonorrhoea among Little Girls," *JSH* 1 (1915): 415, 422.

141. Bertha C. Lovell, "Some Problems in Social Hygiene in a Clinic for Women's Diseases," *JSH* 2 (1916): 501.

142. Ibid., 505.

143. Baker, "Social Case Work," 479, 490.

144. Ibid., 490.

145. Rebecca Jo Plant, "The Repeal of Mother Love: Momism and the Reconstruction of Motherhood in Philip Wylie's America" (Ph.D. diss., Johns Hopkins University, 2001), 176–78.

146. Ruth Schwartz Cowan, *More Work for Mother: The Ironies of Household Technology From the Open Hearth to the Microwave* (New York: Basic Books, 1983), 164–66.

147. Ibid., 160–63.

148. Ibid., 121–22, 173–78; Ruth Schwartz Cowan, "The 'Industrial Revolution' in the Home: Household Technology and Social Change in the 20th Century," *Technology and Culture* 17 (Jan. 1976): 1, 13.

149. Cowan, *More Work for Mother*, 172.

150. Ibid., 177–78; Cowan, "Industrial Revolution," 14–15.

151. Cowan, *More Work for Mother*, 120–227; 178–79; Cowan, "Industrial Revolution," 4–5.

152. Peter N. Stearns, *Anxious Parents: A History of Modern Childrearing in America* (New York: New York University Press, 2003), chap. 2, esp. 17–21.

153. Ibid., 28–31.

154. Ibid., 36–37.

155. Ibid., 43.

156. See Cowan, *More Work for Mother*, 120.

157. Baker, "Social Case Work," 490; Cowan, "Industrial Revolution," 21–22.

158. Stephen A. Yesko, "Gonorrheal Vulvovaginitis in the Young," *American Journal of Diseases of Children* 33 (Apr. 1927): 630, 642.

159. Taussig, "Contagion of Gonorrhoea," 421.

160. B. Wallace Hamilton commented, "In numerous instances direct evidence that father, mother, and all female children were suffering from a gonococcus infection at the same time has been obtained." Hamilton, "Gonococcus Vulvovaginitis," 88. See also Stearns, *Anxious Parents*, 46–47.

161. Lovell, "Some Problems," 505.

162. See CDC, "Sexually Transmitted Diseases Treatment Guidelines, 2006," *MMWR* 55 (Aug. 4, 2006): 1, 83.

163. Edna Pearson Wagner, "Social Aspects of Gonorrheal Vaginitis As Shown by a Study of Fifty Cases of the Children's Bureau in New Orleans" (master's thesis, Tulane University, 1940), 51–52.

164. B. Wallace Hamilton, "Gonococcus Vulvovaginitis in Children," *VDI* 8 (Mar. 20, 1927): 84, 88.

165. Edward L. Keyes, "Special Articles: Report of a Conference of Venereal Disease Control Officers and Clinicians, Held in New York January 20, 1927," *VDI* 8 (Mar. 20, 1927): 71.

166. Yesko, "Gonorrheal Vulvovaginitis," 635.

167. "Home Care of Vaginitis Cases," *JHS* 20 (Mar. 1934): 166, 167, excerpt from an article by Constance Jacobs published in *Public Health Nursing* (Jan. 1934).

168. Wagner, "Social Aspects of Gonorrheal Vaginitis," 66–67.

169. Baker, "Social Case Work," 490.

170. Wagner, "Social Aspects of Gonorrheal Vaginitis," 84, 80–84.

171. Ibid., 39.

172. Crain, "Facts about Gonorrhea," 84.

173. Wagner, "Social Aspects of Gonorrheal Vaginitis," 73, 44.

174. Ibid., 59–60.

175. Ibid., 29–30.

176. Ibid., 55.

177. Wagner, "Social Aspects of Gonorrheal Vaginitis," 45.

178. See, e.g., Dock, *Hygiene and Morality*, 148–49.

179. Odem, *Delinquent Daughters*, 58–62.

180. See, e.g., Spencer, *Woman's Share in Social Culture*, 254–74; Gail Bederman, *Manliness and Civilization: A Cultural History of Gender and Race in the United States, 1880–1917* (Chicago: University of Chicago Press, 1995), 135–67.

181. Jane Addams, *The Spirit of Youth and the City Streets* (Urbana: University of Illinois Press, 1972), 33, 34–47. See William L. O'Neill, "Divorce in the Progressive Era," in Gordon, *American Family*, 251–66.

Chapter Five: Incest Disappears from View

1. Paul Popenoe and Walter Clarke, "The Forum," *JSH* 17 (May 1931): 224, 225.

2. "Biography: Publicity Material in Connection with Dr. Paul Popenoe," undated manuscript, box 174, folder 18, Paul Bowman Popenoe Papers, 1874–1991, Collection Number 04681, American Heritage Center, University of Wyoming, Laramie, Wyoming. Between 1927 and 1930, Popenoe published portions of the study in various journals, including the *Journal of Social Hygiene*. E.g., Paul Popenoe, "Eugenic Sterilization in California: I, The Insane," *JSH* 13 (May 1927): 257; "Eugenic Sterilization in California: II, Attitude of Patients toward the Operations," *JSH* 14 (May 1928): 280. In 1929 the Human Betterment Foundation published a compilation of Popenoe's studies, coauthored with the foundation's organizer. E. S. Gosney and Paul Bowman Popenoe, *Sterilization for Human Betterment: A Summary of Results of 6,000 Operations* (New York: Macmillan, 1929). See also Alexandra Minna Stern, *Eugenic Nation: Faults and Frontiers of Better Breeding in Modern America* (Berkeley: University of California Press, 2005), 105–9; Mike Anton, "Forced Sterilization Once Seen as Path to a Better World," *LAT*, July 16, 2003, col. 1.

3. Popenoe to Ruth Jarvis Mackay-Scott, Apr. 9, 1929, box 1, folder 5, Mackay-Scott Papers, 1928–41; Human Betterment Foundation, "Human Betterment Today," pamphlet (ca. 1938), folder 6, Mackay-Scott Papers; "Popenoe Goods," box 2, and Case Histories, 1934–58, American Institute for Family Relations, box 159, folders 1–33, both in Paul Bowman Popenoe Papers (see n. 2 above). See also Stern, *Eugenic Nation*, 155–67.

4. Included in letters to birth control advocate Mary Ware Dennett and the Voluntary Parenthood League were requests by wives or fiancées for information about venereal disease. E.g., [Council Bluffs] to Ware Dennett, Apr. 22, 1928, and Ware Dennett to [Council Bluffs], Apr. 30, 1928, folder 371, Mary Ware Dennett, 1872–1947, Papers, 1874–1944, Arthur and Elizabeth Schlesinger Library on the History of Women in America, Radcliffe Institute for Advanced Study, Harvard University, Cambridge, Massachusetts.

5. See Betty Hannah Hoffman, "The Man Who Saves Marriages," *Ladies Home Journal*, Sept. 1960, 71; Joan Didion, "Marriage à la Mode," *National Review*, Aug. 13, 1960, 90. Occidental College in Los Angeles bestowed an honorary degree on Popenoe in 1929. David Popenoe, "Remembering My Father: An Intellectual Portrait of 'The Man Who Saved Marriages,'" typescript (1991), 2–3, 6. My thanks to David Popenoe for sending me his essay and for telling me that he had deposited his father's papers at the American Heritage Center at the University of Wyoming, Laramie.

6. Popenoe and Clarke, "The Forum," *JSH* 17 (May 1931): 225.

7. See Walter Clarke, "Social Hygiene in Settlement Work," *JSH* 2 (July 1916): 383, Biographical Data, box 48, folder 2, and box 78, folder 13, ASHA Collection.

8. Popenoe and Clarke, "The Forum," 225–26.

9. Ibid., 226.

10. Nancy Tomes, *The Gospel of Germs: Men, Women, and the Microbe in American Life* (Cambridge: Harvard University Press, 1998), 85.

11. Robert W. de Forest, "A Brief History of the Housing Movement in America," *Annals of the American Academy of Political and Social Science* 51 (Jan. 1914): 8, 15.

12. Nancy Tomes, "Spreading the Germ Theory: Sanitary Science and Home Economics, 1880–1930," in *Women and Health in America: Historical Readings*, ed. Judith Walzer Leavitt (Madison: University of Wisconsin Press, 1999), 596, 603.

13. Ruth Schwartz Cowan, *More Work for Mothers: The Ironies of Household Technology From the Open Hearth to the Microwave* (New York: Basic Books, 1983), 88–89.

14. Tomes, "Spreading the Germ Theory," 603; Tomes, *Gospel of Germs*, 84–87, 157–66.

15. Frederick J. Taussig, "The Prevention and Treatment of Vulvovaginitis in Children," *American Journal of Medical Sciences* 148 (Oct. 1914): 480, 483.

16. Frederick J. Taussig, "The Contagion of Gonorrhoea among Little Girls," *JSH* 1 (1915): 415, 416.

17. Ibid., 415.

18. Ibid., 416–17.

19. Taussig, "Prevention and Treatment," 490.

20. Taussig, "Contagion of Gonorrhoea," 417.

21. Ibid.

22. Ibid., 418.

23. Ibid., 417.

24. Ibid., 421.

25. Taussig, "Prevention and Treatment," 483–84.

26. Taussig, "Contagion of Gonorrhoea," 421.

27. Evangeline Hall Morris, *Public Health Nursing in Syphilis and Gonorrhea* (Philadelphia: Saunders, 1946), 98, 106–7.

28. Clara P. Seippel, "Venereal Diseases in Children," *Illinois Medical Journal* 22 (July 1912): 50, 52.

29. Bertha C. Lovell, "Some Problems in Social Hygiene in a Clinic for Women's Diseases," *Journal of Social Hygiene* 2 (1916): 501, 505.

30. See Suellen Hoy, *Chasing Dirt: The American Pursuit of Cleanliness* (New York: Oxford University Press, 1995), 126–28; John Duffy, *A History of Public Health in New York City, 1866–1966*, vol. 2 (New York: Russell Sage Foundation, 1974), 217–18; Edith Abbott, *The Tenements of Chicago, 1908–1935* (Chicago: University of Chicago Press, 1936), 205–22, 211–13; Board of Education, "Sanitary Conditions in Lavatories," *School Health News* 1 (Dec. 1915): 8; Tenement House Committee, Committee on Housing, Charity Organization Society of New York City, *Housing Reform in New York City: Report of the Tenement House Committee of the Charity Organization Society of the City of New York, 1911, 1912, 1913* (New York: M. B. Brown, 1914), 2, 32–33; New York Association for Improving the Condition of the Poor, Department of Social Welfare, and Bureau of Public Health and Hygiene, *Comfort Stations in New York City: A Social Sanitary and Economic Survey*, publication no. 80 (New York: AICP, [ca. 1910]).

31. Richard Meckel, "Going to School, Getting Sick: The Social and Medical Construction of School Diseases in the Late Nineteenth Century," in *The Formative Years: Children's Health in the United States, 1880–2000*, ed. Alexandra Minna Stern and Howard Markel (Ann Arbor: University of Michigan Press, 2002), 185–88, 195–98, 201–3; Duffy, *History of Public Health*, 217–18; Hoy, *Chasing Dirt*, 126–28; H. D. Chapin, "The Unsanitary Conditions of the Primary Schools of the City of New York," *Sanitarian* 28 (1892): 331.

32. Duffy, *History of Public Health*, 217–18.

33. Stephen Woolworth, " 'The Warring Boards': Sanitary Regulation and the Control of Infectious Disease in the Seattle Public Schools, 1892–1900," *Pacific Northwest Quarterly* 96 (Winter 2004/5): 14.

34. See "School House Sanitation," *School Health News* 1 (Dec. 1915): 2, Dept. of Health Collection. The *News* was published monthly by the Bureau of Public Health Education, Department of Health, and distributed to the city's public school teachers. The COS took an active role in laws regulating tenement privies, boasting that seven thousand "vile" privies were replaced with toilets between 1911 and 1913. Tenement House Committee, Committee on Housing, Charity Organization Society of New York City, *Housing Reform*, 2, 32–33. See also Leonard Porter Ayres, *School Building and Equipment* (Cleveland: Survey Committee of the Cleveland Foundation, 1916), 63–66, and Helen C. Putnam, *School Janitors, Mothers and Health* (Easton, PA: American Academy of Medicine Press, 1913), 77–80.

35. J. H. Berkowitz, *Sanitary School Surveys as a Health Protective Measure* (New York: New York Association for Improving the Condition of the Poor, 1916), reprinted from *Modern Hospital* 6 (Mar. 1916). See also New York Association for Improving the Condition of the Poor, Department of Social Welfare, and Bureau of Public Health and Hygiene, *Comfort Stations*.

36. William A. Link, "Privies, Progressivism, and Public Schools: Health Re-

form and Education in the Rural South, 1909–1920," *Journal of Southern History* 54 (Nov. 1988): 623, 626.

37. Ibid., 627, 631.

38. Ibid., 631–32, 636–39.

39. Board of Education, "Sanitary Conditions in Lavatories," *School Health News* 1 (Dec. 1915): 8, Dept. of Health Collection. The board recommended improved ventilation and disinfection to eliminate "obnoxious toilets."

40. "Venereal Disease," *School Health News* 2 (Apr. 1916): 6, Dept. of Health Collection. According to one physician, by 1910 Boston, Chicago, and Philadelphia all had measures to exclude infected girls from school. Charles Lockwood, "Venereal Diseases in Children, Their Causes and Prevention," *Bulletin of the American Academy of Medicine* 11 (Feb. 1910): 478, 482.

41. Bureau of Public Health Education, Department of Health, "Routine Procedure in Guarding the Schools against Syphilis and Gonorrhoea," *School Health News* 7 (Feb. 1921): 1, 2, Dept. of Health Collection.

42. Wehrbein Draft, 4–5.

43. Ibid., 5.

44. Ibid., 6.

45. C.-E. A. Winslow and Savel Zimand, *Health under the "El": The Story of the Bellevue-Yorkville Health Demonstration in Mid-Town New York* (New York: Harper and Brothers, 1937), 151–52.

46. Sub-Committee on Venereal Disease of the Committee on Cooperation (Charity Organization Society of New York City), "Report Concerning the Venereal Disease Problem in the Boroughs of Manhattan and Bronx, City of New York," typescript, Oct. 17, 1921, box 163, COS Papers. On the COS and the professionalization of social work, see Daniel J. Walkowitz, *Working with Class: Social Workers and the Politics of Middle-Class Identity* (Chapel Hill: University of North Carolina Press, 1999), 32–34.

47. Charity Organization Society of New York City, "Survey of Gonorrheal Vaginitis in Children," *JSH* 13 (Mar. 1927): 144.

48. Minutes, Committee on Venereal Disease, June 1, 1927, box 109, COS Papers.

49. Snow to Affeld, Nov. 18, 1925, box 163, COS Papers.

50. Murray P. Horwood, "A Tuberculosis Survey of Philadelphia," *American Journal of Public Health* 14 (Feb. 1924): 128, 13–31.

51. Minutes, Committee on Venereal Disease, Feb. 19, 1926, box 109, COS Papers.

52. Louis Chargin, "Venereal Diseases: Their Administrative Control As Developed in the City of New York," *Department of Health of the City of New York Reprint Series*, no. 33 (Aug. 1915), 19.

53. See Michael M. Davis and Mary C. Jarrett, *A Health Inventory of New York City: A Study of the Volume and Distribution of Health Services in the Five Boroughs* (New York: Welfare Council of New York City, 1929), 132. Hospitals reported 28.1 percent of the cases, physicians only 11.6 percent.

54. The U.S. Public Health Service collected data on reported venereal disease infections from all the states. It separated infections by gender but not age, so it is not possible to tell how many infected females were girls. Mark J. White, "Venereal Disease Clinics Statistics, July 1, 1923–Dec. 31, 1923," *VDI* 5 (Mar. 20, 1924): 119, 121.

55. Affeld to District Secretaries, memorandum, May 6, 1926, box 109, COS Papers; Minutes, Committee on Venereal Disease, May 14, 1926, box 109, COS Papers.

56. Kathleen Wehrbein, *A Survey of the Incidence, Distribution and Facilities for Treatment of Vulvo-Vaginitis in New York with Concomitant Sociological Data* (New York: E. B. Treat and Co., 1927); reprint from *Archives of Pediatrics* 8 (Apr. 1927); [Committee on Venereal Disease], "Survey of Gonorrhea Vaginitis in Children," June 2, 1926, box 109, COS Papers; Charity Organization Society of New York City, "Survey of Gonorrheal Vaginitis," 144, 146.

57. Charity Organization Society of New York City, "Survey of Gonorrheal Vaginitis," 147.

58. Ibid., 146.

59. Minutes of the Committee on Venereal Disease, Mar. 21, 1927, box 109, COS Papers. The Committee on Vaginitis met from approximately June 1926 to March 1927. See Minutes, Committee on Vaginitis, Oct. 6, 1926–Feb. 15, 1927, Bliss Papers; Minutes, Committee on Venereal Disease, Mar. 21, 1927, box 109, COS Papers.

60. Minutes, Committee on Vaginitis, Oct. 6, 1926, Bliss Papers.

61. A 1929 study of medical facilities in New York City concluded that it was "nearly impossible to obtain any data from most venereal disease clinics and hospitals, such as the amount of service given to venereal disease, statistical data, administrative control and follow-up data or procedures." Davis and Jarrett, *Health Inventory*, 145–46.

62. Wehrbein, *Survey*, 248 (all citations are to the Treat edition).

63. Ibid., 249. Dr. Brunet recommended that the published report omit the names of the hospitals but they are identified in the draft of the report. See Wehrbein Draft, 6.

64. Wehrbein, *Survey*, 250; Wehrbein Draft, 3. Girls also received treatment at women's clinics, but Bellevue and Vanderbilt were the only clinics that set aside specific hours devoted to the treatment of girls. Vanderbilt began treating girls in 1907. E. H. Lewinski-Corwin, "Venereal Disease Clinics: Section of Report on

New York Dispensaries by the Public Health Committee of the New York Academy of Medicine," *JSH* 6 (July 1920): 337, 348, Venereal Diseases, 1915–17 folder, NYAM.

65. Wehrbein, *Survey*, 249; Wehrbein Draft, 3.

66. Wehrbein, *Survey*, 254.

67. Wehrbein Draft, 8.

68. Wehrbein, *Survey*, 256, 265, 262.

69. Ibid., 253, 261–62.

70. Ibid., 262.

71. Stephen A. Yesko, "Gonorrheal Vulvovaginitis in the Young," *American Journal of Diseases of Children* 33 (Apr. 1927): 633; cf. Frederick Joseph Taussig, *Diseases of the Vulva*, Gynecological and Obstetrical Monographs (New York: D. Appleton and Co., 1923), 66.

72. Wehrbein Draft, 265; Minutes, Committee on Vaginitis, Dec. 8, 1926, Feb. 8, 1927, Feb. 15, 1927, Bliss Papers; Minutes, Committee on Venereal Disease, Mar. 21, 1927, box 109, COS Papers.

73. [Wehrbein] to Committee on Vaginitis, memorandum attached to Minutes, Committee on Vaginitis, Feb. 15, 1927, Bliss Papers.

74. "Vulvo-Vaginitis Clinic in the Bellevue-Yorkville Health Center," *JSH* 13 (Nov. 1927): 498, 499; Winslow and Zimand, *Health under the "El,"* 150, 151; Duffy, *History of Public Health*, 323–25.

75. Minutes, Committee on Venereal Disease, Nov. 19, 1926, box 109, COS Papers; Minutes, Committee on Vaginitis, Feb. 8, 1927, Bliss Papers; Minutes, Committee on Venereal Disease, Feb. 25, 1927, 3, box 163, COS Papers; Minutes, Committee on Venereal Disease, Mar. 21, 1927, box 109, COS Papers. On Lillian Wald and the Henry Street Settlement, see Karen Buhler-Wilkerson, *No Place Like Home: A History of Nursing and Home Care in the United States* (Baltimore: Johns Hopkins University Press, 2001), chap. 5.

76. See also "Stated Objectives of the Demonstration," 1923–24, box 62, folders 371 and 378, COS Papers. See also "The Bellevue-Yorkville Health Demonstration," pamphlet, ca. Nov. 1929, box 62, folder 371-8, COS Papers.

77. Affeld to Committee on Venereal Disease, memorandum, Apr. 28, 1927, attaching Walter S. Brunet, "Research," Apr. 21, 1927, box 109, COS Papers; see also New York City Department of Health, "Clinic Does Research on Vaginitis," *JSH* 15 (June 1929): 368, 369.

78. Minutes, Committee on Venereal Disease, June 1, 1927, 1, box 109; COS Papers; "Dr. Brunet Directs Vaginitis Study," *JSH* 14 (Oct. 1928): 437.

79. Pamphlet, "The Bellevue-Yorkville Health Demonstration," Nov. 1929, box 62, folder 371-8 [1926–30], COS Papers.

80. Winslow and Zimand, *Health under the "El,"* 150, 151; "Report of Nurs-

ing Activities, Bellevue-Yorkville Health Demonstration, 1927," ca. 1928, 10–11, box 64, folder 371-8J, COS Papers.

81. Walter M. Brunet et al., "Cervico-Vaginitis of Gonococcal Origin in Children: Report of a Project of the Bellevue-Yorkville Health Demonstration of New York City," *Hospital Social Service Magazine*, Supp. 1 (Mar. 1933): 1, 75; "Weekly Bulletin," New York City Department of Health, Dec. 1, 1928, excerpted in "Clinic Does Research on Vaginitis," *JSH* 15 (Oct. 1929): 368, 370.

82. "Sees Hospitals Run for Tammany Jobs," *NYT*, Nov. 1, 1929, 7; "Women Find Perils in City's Hospitals," *NYT*, Mar. 12, 1925, 11; "Assail Conditions in 2 City Hospitals," *NYT*, Dec. 6, 1924, 15.

83. Winslow and Zimand, *Health under the "El,"* 151–52.

84. Social Hygiene Educational Campaign, Department of Health, New York City: *Information for Women*, #6A (New York: Department of Health, 1930), Health Pamphlet Collection, box 130, #113541-3, NYAM.

85. Minutes, Board of Managers, June 11, 1931, Apr. 14, 1932, Dec. 12, 1932, box 63, folder 371-8b, COS Papers.

86. "Report of Nursing Activities," 10–11, COS Papers.

87. Brunet et al., "Cervico-Vaginitis of Gonococcal Origin," 93–98.

88. Ibid., 1–4, 72.

89. Ibid.

90. Ibid., 72.

91. Ibid., 74.

92. Brunet et al., "Cervico-Vaginitis of Gonococcal Origin," 6.

93. Ibid., 75.

94. Ibid., 84; cf. A. K. Paine, "Vulvovaginitis," *Boston Medical and Surgical Journal* 185 (Dec. 22, 1921): 750; A. K. Paine, "Gonorrhea in the Female," *Commonhealth* [Bulletin of the Massachusetts Department of Health] 15 (July–Sept. 1928): 51, 54.

95. Brunet et al., "Cervico-Vaginitis of Gonococcal Origin," 7.

96. Ibid., 6, 75, 79.

97. Ibid., 7.

98. Department of Health, City of New York, *Important Information: Special Instructions for Vaginitis Cases* (New York: Department of Health, 1929), Dept. of Health Collection; New York City Health Department, *Instructions to Parents Regarding Vaginitis*, Official Project #65-97-415, "Treatment of Social Diseases" (New York: Works Progress Administration for the City of New York, 1937), Government Documents Collection, NYAM. Pamphlets were also printed in Italian and other European languages.

99. N. A. Nelson, "Gonorrhea Vulvovaginitis: A Statement of the Problem," *NEJM* 207 (July 21, 1932): 135, 136.

100. John H. Stokes, *To-day's World Problem in Disease Prevention*, Department of Health, Ottawa, Canada (Ottawa: F. A. Acland, 1923), 39.

101. See George Stevens, *Public Health Aspects of Gonorrhea in the Female*, bulletin no. 20 (Los Angeles: Board of Health Commissioners, Mar. 1937), 4, Oviatt; Ella Oppenheimer and Ray H. Everett, "School Exclusions for Gonorrheal Infections in Washington, D.C.," *American Journal of Public Health* 24 (May 1934): 529; S. H. Rubin, "The Point of View of the School Physician," *NEJM* 207 (July 21, 1932): 142; Walter Clarke, "Summary of a Social Hygiene Survey," *JSH* 27 (Feb. 1931): 65, 78; Bureau of Public Health Education, Department of Health, "Routine Procedure in Guarding the Schools against Syphilis and Gonorrhoea," *School Health News* 7 (Feb. 1921): 1, 2; "Venereal Disease," *School Health News* 2 (Apr. 1916): 6; cf. Nelson, "Gonorrhea Vulvovaginitis," 135.

102. N. A. Nelson, "Some Diseases We Don't Talk About—and Why Not: The Prevalence of Gonorrhea and Syphilis," *Commonhealth* 20 (Apr.–June 1933): 57, 58 (text of radio broadcast, WBZ, Dec. 24, 1930); see also N. A. Nelson, "Gonorrhea and Syphilis: A Description of the Two Diseases," *Commonhealth* 20 (Apr.–June 1933): 59 (text of radio broadcast, WEEI, Aug. 12, 1932).

103. Nelson, "Gonorrhea Vulvovaginitis," 135.

104. See Edward T. Devine, "Address: The Neisserian Medical Society of Massachusetts," *NEJM* 203 (July 21, 1932): 132.

105. Ibid., 137.

106. Ibid., 137; Wehrbein Draft, 4–5.

107. Rubin, "Point of View," 142.

108. Ibid., 143.

109. Clarke, "Social Hygiene Survey," 78.

110. Ibid., 84.

111. Ella Oppenheimer and Ray H. Everett, "School Exclusions for Gonorrheal Infections in Washington, D.C.," *JSH* 20 (Mar. 1934): 129; *American Journal of Public Health* 24 (May 1934): 529; first published as typewritten report (Washington, DC, 1932), Countway Library, Boston, Massachusetts.

112. See B. K. Rachford, "Epidemic Vaginitis in Children," *American Journal of the Medical Sciences* 153 (Feb. 1917): 207, 211.

113. Paul R. Stalnaker, "Gonococcal Vulvovaginitis before Puberty," *Texas State Journal of Medicine* 29 (Oct. 1933): 395; cf. Rachford, "Epidemic," 207.

114. "Editorial, How Educate Infected Children?" *JSH* 20 (Mar. 1934): 156; Walter Clarke, "Report of a Survey of Medical Aspects of Social Hygiene in San Francisco, California, 1931," draft typescript, 81, box 98, folder 4, ASHA Collection.

115. N. A. Nelson, "The Darkness of Ignorance and the Fog of Prudery," *Commonhealth* 22 (Oct.–Dec. 1935): 226, 227 (date of broadcast and station not provided).

116. Ibid., 227–28.

117. N. A. Nelson, "Gonorrhea and Syphilis in the Public Schools," *Commonhealth* 22 (Apr.–June 1935): 118–19; see also Nels A. Nelson, "Gonorrhea and Syphilis in the Public Schools," *Bulletin of the MSSH* 5 (Mar. 1935): 1, box 8, folder 83, MSSH Records; Nels A. Nelson, "Epidemiology of Gonorrhea and Syphilis," *Commonhealth* 20 (1933): 1. Nelson changed the division's name from "venereal" to "genito-infectious" diseases and remained the director of the division until leaving for Johns Hopkins in 1941.

118. Nelson, "Gonorrhea and Syphilis in the Public Schools," *Commonhealth*, 119.

119. Stevens, *Public Health Aspects*, 4.

120. John Duffy, *The Sanitarians: A History of American Public Health* (Urbana: University of Illinois Press, 1990), 266.

121. N. A. Nelson and P. S. Pelouze, "The Control of Gonorrhea," *JSH* 23 (Jan. 1937): 35.

122. Ibid., 43.

123. Nels A. Nelson, *Gonorrhea,* pamphlet (Boston: MSSH, 1936), 4.

124. "Gonorrhea: How Does a Person Catch Gonorrhea?" American Social Hygiene Association pamphlet no. 940 (1935), box 173, folder 4, ASHA Collection.

125. Edna Pearson Wagner, "Social Aspects of Gonorrhea Vaginitis as Shown by a Study of Fifty Cases of the Children's Bureau in New Orleans" (master's thesis, Tulane University, 1940), 41, 29–30.

126. Social Hygiene Committee, New York Tuberculosis and Health Association, *Survey of Female Gonorrhea Clinics, New York City* (New York: Social Hygiene Committee, New York Tuberculosis and Health Association, 1936), 4. See also American Social Hygiene Association, "Survey of New York City Infections and Hospitals," 1935, 93–94, 124, box 101, folder 3, ASHA Collection.

127. Walter Clarke, "Syphilis and Gonococcal Infections in Children," (paper presented to the Section on Pediatrics of the NYAM, Oct. 8, 1936). The American Social Hygiene Association published the paper as a pamphlet (American Social Hygiene Association pamphlet no. A-35, box 173, folder 7, ASHA Collection), and the paper was also reprinted in *Preventative Medicine* 5 (Mar. 1937).

128. Albert G. Bower and Edith B. Pilant, *Communicable Diseases for Nurses* (Philadelphia: Saunders, 1937), 274–76.

129. John B. West, "Gonorrhea Has Preventives," *New York Amsterdam News*, July 24, 1937, 13.

130. Reuel A. Benson and Arthur Steer, "Vaginitis in Children: A Review of the Literature," *American Journal of Diseases of Children* 53 (Mar. 1937): 806; see also J. L. Reichert et al., "Infection of the Lower Part of the Genital Tract in Girls," *American Journal of Diseases of Children* 54 (Sept. 1937): 459.

131. Benson and Steer, "Vaginitis in Children," 810–11.

132. Ibid., 822.

133. Ibid.

134. Ibid.

135. Nels A. Nelson and Gladys L. Crain, *Syphilis, Gonorrhea and the Public Health* (New York: Macmillan, 1938), 142–43.

136. Ibid., 144.

137. Nelson, "Gonorrhea Vulvovaginitis," 135–40, 136.

138. Ibid., 136.

139. Ernest B. Howard, "Communicable Diseases," *Commonhealth* 26 (Jan.–Mar. 1939): 18.

140. Reuel A. Benson and Arthur Steer, "Vaginitis Clinic," *JSH* 25 (Jan. 1939): 21, 23.

Chapter Six: Incest in the Twentieth Century

1. "Reporter's Transcript," *People v. McAfee*, Court of Appeal records, 2Crim1444, 1927, 8–38, Sacramento.

2. Ibid., 39–50.

3. Ibid., 139–45.

4. Ibid., 149–52, 309, 355–58.

5. Ibid., 159–62, 198, *People v. McAfee*, 82 Cal. App. 389 (2d Dist. 1927).

6. E.g., "Appellants Opening Brief," *People v. Hewitt*, Court of Appeal records, 2Crim1860, 1929, 19, Sacramento.

7. *People v. Guiterez*, 14 P. 838 (3d Dist. 1932); *People v. Little*, 122 Cal. App. 275 (3d Dist. 1932); "Direct Testimony of Fay McCullom," *People v. McCullom*, Court of Appeal records, 4Crim216, 1931, 33–36, 42, Sacramento; *People v. Domenighini*, 81 Cal. App. 484 (1st Dist. 1927); *People v. Oliver*, 214 P. 272 (3d Dist. 1923); *People v. Burrows*, 27 Cal. App. 428, 430 (3d Dist. 1915); *People v. Costa*, 24 Cal. App. 739 (2d Dist. 1914); *People v. Fong Chung*, 5 Cal. App. 587 (1st Dist. 1907); *People v. Ah Lean*, 7 Cal. App. 626 (1st Dist. 1908).

8. *In Re Manchester*, 33 Cal. 2d 740 (1949). See also *People v. Brown*, 114 Cal. App. 2d 7, (2d Dist. 1952).

9. See, e.g., *In Re Manchester*, 33 Cal. 2d 740 (1949); "Trial Record," *In Re Manchester*, Supreme Court of California Records, Crim4892, 1949, Sacramento.

10. See, e.g., *People v. Stratton*, 141 Cal. 604 (1904).

11. E.g., *People v. Deatrick*, 30 Cal. App. 507 (2d Dist. 1916), (25 years).

12. See, e.g., "Trial Record," *People v. Wilson*, Court of Appeal records, 1Crim1328, 1926; *People v. Hewitt*, 1929; *People v. Meraviglia*, Court of Appeal

records, 3Crim854, 1925; *People v. Rankins*, Court of Appeal records, 2Crim3809, 1944; *People v. Hall*, Court of Appeal records, 3Crim1598, 1938, all in Sacramento.

13. *People v. Rankins*, 66 Cal. App. 2d 956, 957 (2d Dist. 1944).

14. See, e.g., Edith R. Spaulding and William Healy, "Inheritance as a Factor in Criminality: A Study of a Thousand Cases of Young Repeated Offenders," *Journal of the American Institute of Criminal Law and Criminology* 4 (Mar. 1914): 837–58; cf. Clarence Darrow, *Crime: Its Causes and Treatment* (New York: Thomas Y. Crowell Co., 1922), 233–34, 243–49. See also Matthew Frye Jacobson, *Barbarian Virtues: The United States Encounters Foreign Peoples at Home and Abroad* (New York: Hill and Wang, 2000), 154–72.

15. Draft of speech, A-127, folder 232, box 24, Dummer Papers.

16. See Kathleen W. Jones, *Taming the Troublesome Child: American Families, Child Guidance, and the Limits of Psychiatric Authority* (Cambridge: Harvard University Press, 1999), 66–68; Janis Appier, *Policing Women: The Sexual Politics of Law Enforcement and the LAPD* (Philadelphia: Temple University Press, 1998), 20–21, 78–79.

17. E.g., *People v. Jackson*, 63 Cal. App. 2d 586 (3d Dist. 1944); *People v. De Coe*, 77 P.2d 883 (2d Dist. 1938); *People v. Johnson*, 115 Cal. App. 704 (2d Dist. 1931); *People v. Young*, 261 P. 530 (3d Dist. 1927); *People v. King*, 56 Cal. App. 484 (1st Dist. 1922); *People v. Whitney*, 34 Cal. App. 737 (3d Dist. 1917); "Brutes Are Bound Over," *Fort Wayne (IN) News and Sentinel*, Oct. 19, 1918, 2; "Enormity of Charge Gets Man New Trial," *Atlanta Constitution*, Mar. 3, 1913, 8.

18. *People v. Hobday*, 131 Cal. App. 626, 627 (2d Dist. 1933). See also *Strand v. State*, 36 Wyo. 78, 82–83 (1927); cf. John Henry Wigmore, *A Supplement 1923–1933 to the Second Edition (1923) of A Treatise on the System of Evidence in Trials at Common Law* (Boston: Little, Brown, 1934), 393 n. 1.

19. "Reporter's Transcript on Appeal," *People v. Hobday*, Court of Appeal record, 2Crim2282, 1933, 144–45, 168–70, Sacramento.

20. Ibid., 627.

21. *People v. Roberts*, 50 Cal. App. 2d 558, 564 (2d Dist. 1942).

22. See, e.g., Lauretta Bender and Abram Blau, "Reaction of Children to Sexual Relations with Adults," *Journal of Orthopsychiatry* 7 (Oct. 1937): 500; Estelle B. Freedman, " 'Uncontrolled Desires': The Response to the Sexual Psychopath, 1920–1960," *Journal of American History* 74 (June 1987): 83. See also Elizabeth Lunbeck, " 'A New Generation of Women': Progressive Psychiatrists and the Hypersexual Female," in *Women and Health in America*, ed. Judith Walzer Leavitt, 2nd ed. (Madison: University of Wisconsin Press, 1999), 229–49; John C. Burnham, "Psychiatry, Psychology, and the Progressive Movement," *Paths into Amer-*

ican Culture: Psychology, Medicine, and Morals (Philadelphia: Temple University Press, 1988), 187–94; and Judith Lewis Herman and Lisa Hirschman, *Father-Daughter Incest* (Cambridge: Harvard University Press, 1981), 36–42.

23. Jones, *Taming the Troublesome Child*, 41–43.

24. William Healy and Mary Tenney Healy, *Pathological Lying, Accusation, and Swindling: A Study in Forensic Psychology* (Boston: Little, Brown, 1926), 182–87 (originally published in *Criminal Science Monographs* 1 [Sept. 1915]); William Healy, *The Individual Delinquent: A Text-book of Diagnosis and Prognosis for All Concerned in Understanding Offenders* (Boston: Little, Brown, 1915, 1927), 736–39. See also Herman and Hirschman, *Father-Daughter Incest*, 11.

25. Healy and Healy, *Pathological Lying*, 185.

26. Ibid., 182, 187.

27. Jones, *Taming the Troublesome Child*, 1–3, 57–61; Alice Boardman Smuts, *Science in the Service of Children, 1893–1935* (New Haven: Yale University Press, 2006), 3–6; Judge Baker Children's Center, "Our History" (2007), www.jbcc.harvard.edu/about/history.htm (accessed Aug. 12, 2007).

28. Jones, *Taming the Troublesome Child*, 9–10, 62–90.

29. See ibid., 56–57.

30. Wigmore, *Supplement*, 379–95; John Henry Wigmore, *A Treatise on the System of Evidence in Trials at Common Law*, 3rd ed. (Boston: Little, Brown, 1940), 3:459.

31. Wigmore, *Supplement*, 384.

32. Ralph W. Webster, *Legal Medicine and Toxology* (Philadelphia: Saunders, 1930), 258, 263.

33. Sydney Smith, ed., *Taylor's Principles and Practice of Medical Jurisprudence*, 9th ed. (London: J. and A. Churchill, 1934), 2:120, 118.

34. "Girl Accuses Step-Father of Incest; 6 of Rape," *Chicago Defender* (national ed.), Apr. 29, 1939, 2; "18-Year-Old Girl Fires on Stepfather," *Chicago Defender* (national ed.), May 12, 1934, 13; "Girls Accuse Father with Grave Crime," *Chicago Defender* (national ed.), July 12, 1929, 2; "Father Charged with Rape by His Daughter," *New York Amsterdam News*, Mar. 23, 1927, 1; "Around the Hub," *Chicago Defender* (national ed.), Dec. 16, 1922, 18; "Charged with Incest," *Chicago Defender* (national ed.), Apr. 2, 1921, 2.

35. "Incest Charges Leveled against Her Love Mate," *New York Amsterdam News*, Mar. 25, 1939, 11; "Peekskill," *New York Amsterdam News*, June 3, 1939, 11.

36. "Rivers Criticized for Reprieving White Man," *New York Amsterdam News*, Dec. 17, 1939, 11.

37. *Absent Mother*: "Gray Hair Saved Perrett Going to Noose, Says Juror,"

Fort Worth Star-Telegram, Apr. 29, 1913, 5; "Bridges Found Guilty," *Morning World-Herald* (Omaha), Mar. 28, 1907, 3; "Regan Is Held for Trial," *Grand Forks (ND) Daily Herald*, Mar. 9, 1907, 6; "Nubs of News," *Grand Forks (ND) Daily Herald*, Oct. 5, 1905, 4; "Brutal Father Is in Jail," *Los Angeles Herald*, Dec. 25, 1902, 3; "Girl Kills Her Father," *Los Angeles Herald*, June 13, 1902, 2.

Pregnancy: "Autoist Sentenced Following Fatality," *Hartford (CT) Courant*, Sept. 17, 1925, 14; "But Two Cases in Willimantic Court," *Hartford (CT) Courant*, May 16, 1919, 19; "Ex-Preacher in Jail Facing Incest Charge," *Fort Worth Star-Telegram*, June 20, 1913, 2; "Enormity of Charge Gets Man New Trial," *Atlanta Constitution*, Mar. 3, 1913, 8; "Farmer Faces a Serious Charge," *Grand Forks (ND) Daily Herald*, Mar. 13, 1912, 4; "Pleads Guilty to Awful Crime" *Grand Forks (ND) Daily Herald*, June 29, 1911, 7; "Bomb Thrown at Barcelona, 18-Year Sentence on Charge of Incest," *Atlanta Constitution*, Sept. 29, 1909, 4; "Naset Is Bound Over," *Grand Forks (ND) Daily Herald*, July 6, 1906, 3; "Los Angeles County; Its Cities, Towns, Hamlets and Suburban Places, Unspeakable Crimes Charged to Father," *LAT*, Apr. 20, 1902, 11; "New Trial Refused, Senak Will Suffer for An Unnatural Offense," *Wilkes-Barre (PA) Times*, Jan. 3, 1900, 5.

Violence and Murder: "Convict Ex-Deputy of Slaying Doctor," *Chicago Daily Tribune*, Aug. 13, 1938, 1; "Father, Daughter, Bound Over to Superior Court," *Hartford (CT) Courant*, Aug. 9, 1934, 9; "Jail Father for Incest," *California Eagle* (Los Angeles), Mar. 10, 1933, 1; "Prison for Woman Who Would Shield Husband's Incest," *California Eagle* (Los Angeles), Dec. 19, 1930; "Boy Confesses Seven Murders, Murder of Twenty Years Ago in Court," *LAT*, May 6, 1928, 14; "Unwritten Law," *LAT*, Sept. 26, 1922, I14; "Committed to Jail on Serious Charge," *Charlotte (NC) Observer*, Nov. 26, 1918, 14; "Would Kill His Daughter," *Charlotte (NC) Daily Observer*, July 24, 1907, [1]; "Unspeakable Crimes Charged to Father," *LAT*, Apr. 20, 1902, 11; "Murder and Suicide and House on Fire," *LAT*, Jan. 11, 1901, 3.

Suicide: "Attempts Suicide," *Grand Forks (ND) Daily Herald*, Apr. 27, 1911, 5; "Jail Is Charged by Mob," *Atlanta Constitution*, May 9, 1901, 6; "Shot Three, Slew Himself," *LAT*, Dec. 17, 1900, I11; "He Preferred Death," *Morning World Herald* (Omaha), Apr. 23, 1900, 8.

Friends: "Father of Young Girl to Answer," *LAT*, Oct. 29, 1924, A11; "San Luis Obispo," *LAT*, Oct. 28, 1901, 10.

Long Duration, Multiple Daughters: "Trial Record," *People v. Hall*, 1938, Sacramento; *People v. Pribnow*, 61 Cal. App. 252 (1st Dist. 1923); "Four Draw Sentences to State Prison," *Hartford (CT) Courant*, Sept. 23, 1936, 24; "Jail Father for Incest," *California Eagle* (Los Angeles), Mar. 10, 1933, 1; "Trial Record," *People v. Wilson* 1926, Sacramento; "Trial Record," *People v. Jones*, Court of Appeal records, 2Crim1258, 1926, Sacramento; "Brutes Are Bound Over,"

Fort Wayne (IN) News and Sentinel, Oct. 19, 1918, 2; "Plaza Farmer on Trial Charged with Crime Against Girl," *Grand Forks (ND) Daily Herald*, Mar. 5, 1914, 5; "Albert Haef Pleads Guilty," *Grand Forks (ND) Daily Herald*, July 9, 1913, 7; "Thirty-Five Years for Negro," *Dallas Morning News*, July 3, 1908, 3; "Negro Charged with Incest" *Charlotte (NC) Daily Observer*, July 20, 1907, 3; "Naset Is Bound Over," *Grand Forks (ND) Daily Herald*, July 6, 1906, 3; "Nubs of News," *Grand Forks (ND) Daily Herald*, Oct. 5, 1905, 4; "Unspeakable Crimes Charged to Father," *LAT*, Apr. 20, 1902, 11; "Jail Is Charged By Mob," *Atlanta Constitution*, May 9, 1901, 6; "Riverside County's Late Happenings, Bad Jones Case," *LAT*, Nov. 15, 1900, I15; "A Horrible Crime," *Arizona Weekly Journal-Miner* (Prescott), June 13, 1900, [1]; "Charged with Incest," *Sunday World Herald* (Omaha), Jan. 28, 1900, 7.

38. *Court System*: "Has Hot Words for Jurymen," *Atlanta Constitution*, Apr. 29, 1903, 1; "Carl Andre Not Guilty," *Morning World Herald* (Omaha), Jan. 30, 1900, 5.

Abortion: "Peaster Banker Not Guilty of Incest," *Fort Worth Star-Telegram*, Oct. 23, 1913, 5; "Jail Is Charged by Mob," *Atlanta Constitution*, May 9, 1901, 6.

Child Protectors: "Offending the Little Ones," *NYT*, Feb. 3, 1913, 10; "10 Plead Guilty in Plainville on Sex Case Counts," *Hartford (CT) Courant*, Feb. 10, 1932, 6; "Farmer Faces a Serious Charge," *Grand Forks (ND) Daily Herald*, Mar. 13, 1912, 4; "Attempts Suicide," *Grand Forks (ND) Daily Herald*, Apr. 27, 1911, 5; "Unnatural Crime Charged," *Morning World-Herald* (Omaha), Aug. 4, 1906, 5.

Closed Courtroom: "10 Plead Guilty in Plainville on Sex Case Counts," *Hartford (CT) Courant*, Feb. 10, 1932, 6; "Father Charged with Incest," *Charlotte (NC) Daily Observer*, Oct. 31, 1908, 10; "Cleared of Incest Charge," *Atlanta Constitution*, Aug. 12, 1907, 10; "Charged with Incest," *Sunday World Herald* (Omaha), Jan. 28, 1900, 7.

Women: "Merchants Approve Vote of Meeting, Ashford Man Bound Over," *Hartford (CT) Courant*, Feb. 27, 1919, 14; "Farmer Gets 100 Days for Attacking Own Daughter," *Fort Worth Star-Telegram*, Oct. 10, 1913, 2; "Putnam Man Is Held for Incest," *Hartford (CT) Courant*, Sept. 11, 1913, 1; "Albert Haef Pleads Guilty," *Grand Forks (ND) Daily Herald*, July 9, 1913, 7; "Atlanta Sentence for Counterfeiter, Long Sentence for Marchorino," *Hartford (CT) Courant*, Dec. 4, 1912, 5; "Salisbury News of a Day," *Charlotte (NC) Daily Observer*, Sept. 5, 1911, 9; "Negro Charged with Incest," *Charlotte (NC) Daily Observer*, July 20, 1907, 3; "Full of Family Scandals," *Morning World-Herald* (Omaha), Mar. 1, 1907, 2; "Father of Girl Accused," *Grand Forks (ND) Daily Herald*, Jan. 27, 1907, 6; "Putnam," *Hartford (CT) Courant*, May 21, 1906, 16; "Girl Dies to Save Her Sinful Father," *LAT*, Sept. 22, 1903, 2; "Charged with Incest," *Morning World-Herald* (Omaha), July 15, 1902, 6; "Ingraham Is Held for Rape,"

Morning Olympian (Olympia, WA), Nov. 26, 1901, [3]; "Charges Multiply, J. H. Vance's 11 Year-Old Daughter Also Accused Him of Incest," *Morning World Herald* (Omaha), Jan. 29, 1900, 3.

Gonorrhea Infection: "Brutes Are Bound Over," *Fort Wayne (IN) News and Sentinel*, Oct. 19, 1918, 2; "Mrs. Richards Changes Front, Now Declares Her Husband Guiltless of Criminally Assaulting His Stepdaughter," *Sunday World-Herald* (Omaha), Apr. 20, 1902, 7.

39. "Charged with Awful Crime, White Man Jailed at Anderson, S.C., on Charge of Incest," *Atlanta Constitution*, Nov. 14, 1900, 2. See also "Father Charged with Incest," *Charlotte (NC) Daily Observer*, Oct. 31, 1908, 10.

40. "Revolting Story," *LAT*, July 4, 1909, 13.

41. "Trial Record," *People v. Wilson*, 1926, Sacramento; *People v. Deatrick*, 30 Cal. App. 507 (2d Dist. 1916); "Putnam Man Is Held For Incest," *Hartford (CT) Courant*, Sept. 11, 1913, 1; "Prefers Charges against the Cedar Rapids Police," *Sunday World-Herald* (Omaha), June 26, 1904, 2; "Too Many Wives, but No Wife," *LAT*, June 7, 1902, A2; "[Coast Record], Boy Poisoner Acquitted," *LAT*, Sept. 21, 1900, 15.

42. *People v. Meraviglia*, 73 Cal. App. 402 (3d Dist. 1925); *People v. Britt*, 62 Cal. App. 674 (2d Dist. 1923); *People v. Rabbit*, 64 Cal. App. 264 (2d Dist. 1923); *People v. Burrows*, 27 Cal. App. 428 (3d Dist. 1915); *People v. Lewis*, 18 Cal. App. 359 (3d Dist. 1912).

43. *People v. Lee*, 55 Cal. App. 2d 163 (2d Dist. 1942); *People v. Nobles*, 44 Cal. App. 2d 422 (2d Dist. 1941); *People v. Gump*, 61 P. 970 (1st Dist. 1936); *People v. Hewitt*, 101 Cal. App. 306 (2d Dist. 1929); *People v. Lopez*, 33 Cal. App. 530 (2d Dist. 1917); *People v. Letoile*, 31 Cal. App. 166 (2d Dist. 1916); *People v. Burrows*, 27 Cal. App. 428 (3d Dist. 1915).

44. "Father of 13 Not-Guilty of Incest Charge," *Belleville (IL) News-Democrat*, Dec. 2, 1916, 2; "Negroes Bound over Charges of Assault on Negro Actress," *Miami Herald*, Sept. 30, 1916, 8; "Woman Who Accused Father Enters Pleas," *Belleville (IL) News-Democrat*, Oct. 13, 1915, 2; "Daughter Says Charges against Father False," *Belleville (IL) News-Democrat*, Oct. 7, 1915, [1]; "Discharged in Justice Court," *Grand Forks (ND) Daily Herald*, May 18, 1911, [1]; "William Lampere Freed," *Boston Daily Globe*, Aug. 16, 1905, 3.

45. *People v. Hewitt*, 101 Cal. App. 306, 307–8 (2d Dist. 1929); "Appellant's Opening Brief," *People v. Hewitt*, 1929, 19, Sacramento.

46. "Direct Testimony of Harry C. T. Rankins," *People v. Rankins*, 1944, 216–17, Sacramento.

47. Ibid., 223–25.

48. "Autoist Sentenced Following Fatality," *Hartford (CT) Courant*, Sept. 17, 1925, 14.

49. New Associated Press, "Murder in a Saloon," *LAT*, Aug. 14, 1902, 3; "Brief Coast Dispatches," *LAT*, Aug. 16, 1902, 3.

50. "Three Men Held in Danielson's Girl's Case," *Hartford (CT) Courant*, Dec. 30, 1929, 11.

51. "10 Plead Guilty in Plainville on Sex Case Counts," *Hartford (CT) Courant*, Feb. 10, 1932, 6.

52. "Oroville Boy, Sisters Sent to Institution," *Placerville (CA) Republican*, Apr. 13, 1937, 1.

53. "To Have Hearing Tuesday," *Grand Forks (ND) Daily Herald*, Jan. 22, 1911, 6.

54. Darrow, *Crime*, 88–89.

55. Jacob A. Goldberg and Rosamond W. Goldberg, *Girls on City Streets: A Study of 1400 Cases of Rape* (New York: ASHA, 1935; repr., New York: Arno Press, 1974), 301, 186.

56. Ibid., 186.

57. Ibid., 186–87.

58. Ibid., 157.

59. Ibid., 188–214.

60. Goldberg and Goldberg, *Girls on City Streets*, 188, 291.

61. Ibid., 188.

62. Ibid., 186–87. The Goldbergs' rationale is strikingly close to Foucault's comments about class, psychoanalysis, and incest. Michel Foucault, *History of Sexuality*, vol. 1: *An Introduction*, trans. Robert Hurley (New York: Vintage Books, 1980), 109, 129.

63. See, e.g., "Social Hygiene at the New England Health Institute," *Bulletin of the Massachusetts Society for Social Hygiene* 1 (May 1931), MSSH Records. The Massachusetts Society for Social Health was renamed the Massachusetts Society for Social Hygiene after 1931.

64. Social Hygiene Committee, New York Tuberculosis and Health Association, *Survey of Female Gonorrhea Clinics in New York City* (New York: Social Hygiene Committee, New York Tuberculosis and Health Association, 1936). See also Report of the Executive Secretary, Massachusetts Society for Social Health, Executive Committee Reports, folder 13, box 2, Oct. 19, 1932, Nov. 15, 1933, MSSH Records; Report of the Executive Secretary, folder 14, box 2, Nov. 21, 1934, MSSH Records; Bertha Chace Lovell, "Do We Need Evening Pay Clinics for Venereal Diseases?" *Bulletin of the MSSH* 2 (Apr. 1919): 1.

65. Social Hygiene Committee, New York Tuberculosis and Health Association, *Survey*, 4. See also American Social Hygiene Association, "Survey of New York City Infections and Hospitals," 1935, 93–94, 124, box 101, folder 3, ASHA Collection.

66. Doctors who viewed dispensaries as business competition may have exaggerated the number of well-paid patients treated at dispensaries. Still, when philanthropists organized and opened a venereal disease clinic in Chicago in 1919, it charged $185 annually for treatment, compared with $525 charged by private physicians, a difference that any middle-class person would have found substantial and attractive. See Paul Starr, *The Transformation of American Medicine: The Rise of a Sovereign Professional and the Making of a Vast Industry* (New York: Basic Books, 1982), 182–84, 194.

67. Social Hygiene Committee, New York Tuberculosis and Health Association, *Survey*, 4.

68. Walter Clarke, "Syphilis and Gonococcal Infections in Children" (paper presented to the Section on Pediatrics of the NYAM, Oct. 8, 1936). The American Social Hygiene Association published the paper as a pamphlet (American Social Hygiene Association pamphlet no. A-35, box 173, folder 7, ASHA Collection), and the paper was also reprinted in *Preventative Medicine* 5 (Mar. 1937). See also Reuel A. Benson and Arthur Steer, "Vaginitis in Children: A Review of the Literature," *American Journal of Diseases of Children* 53 (Mar. 1937): 806, 822.

69. Lawrence R. Wharton, *Gynecology: With a Section on Female Urology* (Philadelphia: Saunders, 1943), 368.

70. George H. Bigelow, *Communicable Disease Control: Report of the Committee on Communicable Disease Control*, White House Conference on Child Health and Protection (New York: Century Co., 1931).

71. Between 1912 and 1925, Los Angeles incarcerated infected women arrested for prostitution in the Los Feliz Detention Hospital. See "Detention Home," "Detention Hospital," and "Los Feliz Hospital" files, Los Angeles City Archives, Los Angeles, California; Alida C. Bowler, "A Police Department's Social Hygiene Activities," *JSH* 15 (1929): 528, 532–34; Franklin Hichborn, "The Anti-Vice Movement in California, Part 2, Rehabilitation," *JSH* 6 (July 1920): 365, 369–71; "A Home for Street Women," *LAT*, May 5, 1918; "New Hospital Born of War," *LAT*, Jan. 27, 1918.

See *In Re Dillon and Adams*, 186 P. 170 (Cal. App. 2d 1919); *In Re Shepard*, 195 P. 1077 (Cal. App. 2d 1921); *In Re A. Arata*, 198 P. 814 (Cal. App. 2d 1921); *In re Dayton*, 199 P. 548 (Cal. App. 2d 1921); *In Re King*, 16 P. 2d 694 (Cal. App. 2d 1932).

On San Francisco, see Walter Clarke, "Report of a Survey of Medical Aspects of Social Hygiene in San Francisco," typescript, 1931, 67–78, 84–89, box 98, folder 4, 38, ASHA Collection; *In Re Travers*, 192 P. 454 (Cal. App. 1st 1920); Julius Rosenstirn, *Our Nation's Health Endangered by Poisonous Infection through the Social Malady: The Protective Work of the Municipal Clinic of San*

Francisco and Its Fight for Existence (San Francisco: Town Talk Press, 1913). California's quarantine policy for prostitutes remained in effect until 1975. *People v. Superior Court of Alameda County* et al., 19 Cal. 3d 338, 353–54 (1977).

Atlanta placed infected "prostitutes" in the city's venereal disease hospital, a building on the outskirts of town. Armed police managed the hospital, which had neither kitchen facilities nor an attending physician. Walter Clarke et al., "Atlanta Social Hygiene Survey," typescript, 1926, 15–16, box 98, folder 10, ASHA Collection.

On the nineteenth-century Social Evil Hospital in St. Louis, see John C. Burnham, "Medical Inspection of Prostitutes In America in the Nineteenth Century: The St. Louis Experiment and Its Sequel," *Bulletin of the History of Medicine* 45 (June 1971): 203, 206.

On Chicago, see Michael Willrich, *City of Courts: Socializing Justice in Progressive Era Chicago* (New York: Cambridge University Press, 2003), 201–7; Council of the Chicago Medical Society, *History of Medicine and Surgery and Physicians and Surgeons of Chicago* (Chicago: Biographical Publishing Corp., 1922), 333–34; "Health Rules Advocated by Dr. Bundesen," *Chicago Daily Tribune*, Apr. 15, 1922, 1.

See also Allan M. Brandt, *No Magic Bullet: A Social History of Venereal Disease in the United States since 1880* (New York: Oxford University Press, 1987), 44; John Duffy, *A History of Public Health in New York City, 1866–1966*, vol. 2 (New York: Russell Sage Foundation, 1974), 579–80, 583; Zelda Popkin, "Sociological Court Is Urged for Women," *NYT*, Nov. 25, 1934; Bruce Cobb, "The Women's Court in its Relation to Venereal Diseases," *JSH* 6 (Jan. 1920): 83 (New York City); "Medical Examination for Persons Arrested," *State* (Columbia, SC), Apr. 18, 1918, 12 (South Carolina).

72. Anastasia Sims, *The Power of Femininity in the New South: Women's Organizations and Politics in North Carolina, 1880–1930* (Columbia: University of South Carolina Press, 1997), 74–76.

73. Case Files, 1909–14, Booth Records.

74. Richard M. Smith, "Vulvovaginitis in Children," *American Journal of Diseases of Children* 6 (Nov. 1913): 355, 359.

75. Committee on Medical Care for Children, White House Conference on Child Health and Protection, *Hospitals and Child Health: Hospitals and Dispensaries, Convalescent Care, Medical Social Service* (New York: Century Co., 1932), 41. On the White House Conference, see Smuts, *Science in the Service of Children*, 242–46.

76. Elizabeth Alice Clement, *Love for Sale: Courting, Treating, and Prostitution in New York City, 1900–1945* (Chapel Hill: University of North Carolina Press, 2006), 121–25; Brandt, *No Magic Bullet*, 77–92.

77. Clement, *Love or Sale*, 117–18, 125–43; Brandt, *No Magic Bullet*, 59–61, 72–80; Appier, *Policing Women*, 43–56; "Boston Work of the Section on Women and Girls Law Enforcement Division Commission on Training Camp Activities," *Bulletin of the MSSH* 1 (Oct. 1918): 1, 3–4.

78. Mary Macey Dietzler, *Detention Houses and Reformatories as Protective Social Agencies in the Campaign of the U.S. Government against Venereal Diseases: A Report on Certain Detention Houses and Reformatories . . .* (Washington, DC: U.S. Interdepartmental Social Hygiene Board, June 1922), 10–26; U.S. Interdepartmental Social Hygiene Board, *Manual for the Various Agents of the U.S. Interdepartmental Social Hygiene Board* (Washington, DC: U.S. Government Printing Office, 1920), 1–16.

79. Clement, *Love for Sale*, 144–50, 170–76.

80. Dietzler, *Detention Houses*, 10–11, 15–16, 22–23, 34–35, 84–85, 224–27; C. E. Miner and Henrietta Additon, "Report on the Division of Protective Social Measures," in *Report of the U.S. Interdepartmental Social Hygiene Board for the Fiscal Year Ended June 30, 1921*, U.S. Interdepartmental Social Hygiene Board (Washington, DC: U.S. Government Printing Office, 1921), 30; Appier, *Policing Women*, 22, 43–45.

81. Dietzler, *Detention Houses*, 22–23; Brandt, *No Magic Bullet*, 84–90.

82. Dietzler, *Detention Houses*, 23. See also Wilbur A. Sawyer, "Venereal Disease Control in the Military Forces," *American Journal of Public Health* 9 (1919): 337, 339.

83. Thomas A. Storey, "Evaluation of Governmental Aid to Detention Houses and Reformatories," in Dietzler, *Detention Houses*, 3–9; Dietzler, *Detention Houses*, 69; U.S. Interdepartmental Social Hygiene Board, *Manual*, 16–19, 21–22, 75, 76; see also Raymond Fosdick to Ethel Sturges Dummer, Sept. 20, 1917, folder 357, and The War Department, Committee on Training Camp Activities, "Committee on Protective Work for Girls," undated pamphlet, Washington, DC, folder 357, both in Dummer Papers; Brandt, *No Magic Bullet*, 90–92. Some women objected to the policy because only women, and not men, were detained. See, e.g., Katharine C. Bushnell, *What's Going On: A Report of Investigations by Katharine C. Bushnell, M.D., Regarding Certain Social and Legal Abuses in California That Have Been in Part Aggravated and in Part Created by the Federal Social Hygiene Programme* (Oakland, CA: n.p., 1919).

84. U.S. Interdepartmental Social Hygiene Board, *Manual*, 16–23.

85. Dietzler, *Detention Houses*, 46, 56–57, 59, 92, 122–26, 130, 135–38, 149–53, 162–66, 170, 173–74, 213–14.

86. Valeria H. Parker, introduction to *Annual Report of the U.S. Interdepartmental Social Hygiene Board, 1922* (Washington, DC: U.S. Government Printing Office, 1922), 16–19; Sidney Morgan, Maude E. Stearns, and Mary E. Magee,

"General Analysis of the Answers Given in 15,010 Case Records of Women and Girls Who Came to the Attention of the Field Workers of the U.S. Interdepartmental Social Hygiene Board, and of Its Predecessors, the War Department and the Navy Department Commissions on Training-Camp Activities," app. F, *Report of the U.S. Interdepartmental Social Hygiene Board for the Fiscal Year Ended June 30, 1921*, 163–95; Miner and Additon, "Report on the Division of Protective Social Measures," 32–50; U.S. Interdepartmental Social Hygiene Board, *Manual*, 21–29, app. 9;

87. Dietzler, *Detention Houses*, 67.

88. Ibid., 103–4.

89. Ibid., 67.

90. Ibid.

91. Ibid., 68.

92. Dietzler, *Detention Houses*, 68.

93. See, e.g., George H. Preston, "The Cost of Feeblemindedness," *Atlanta Constitution*, May 29, 1922, 8. Preston was superintendent of the Georgia Training School for Mental Defectives.

94. Rhea C. Ackerman, "Annual Report of Juvenile Hall, 1930–1931," box 2, folder 1, 61–63, Ackerman Collection; "Annual Report, 1933–34," box 2, folder 2, 7, Ackerman Collection; Clarke, "Report of a Survey of Medical Aspects," 34–38.

95. "Juvenile Home Head Describes Bad Conditions," *Chicago Daily Tribune*, Feb. 9, 1937, 11; "Juvenile Home Scandal Bares Long Dissension," *Chicago Daily Tribune*, Jan. 26, 1937, 13; "Scandal Turns Scrutiny upon Juvenile Home," *Chicago Daily Tribune*, Jan. 21, 1937, 1–2; see Appier, *Policing Women*, 140–41.

96. "An Attic's Story Bares Evil at Juvenile Home," *Chicago Daily Tribune*, Jan. 31, 1937, 4; "Launch Cleanup of Conditions in Juvenile Home," *Chicago Daily Tribune*, Jan. 30, 1937, 7.

97. Case Files, 1909–14, Booth Records; Clarke, "Report of a Survey of Medical Aspects," 36, 124–25.

98. Michael M. Davis and Mary C. Jarrett, *A Health Inventory of New York City: A Study of the Volume and Distribution of Health Services in the Five Boroughs* (New York: Welfare Council of New York City, 1929), 145; Comments of Dr. A. A. Little Jr. in Paul R. Stalnaker, "Gonococcal Vulvovaginitis before Puberty," *Texas State Journal of Medicine* 29 (Oct. 1933): 395, 400. The Community Chest also provided funds to the Rosemary Cottage in Pasadena, a residential home for up to twenty adolescent girl. It did not offer medical services and only accepted infected girls who were victims of incest. Beverly Dye Blair, "Leisure Activities of Dependent, Adolescent Girls Living in a Group Care Situation" (master's thesis, University of Southern California, 1939), 9–10; Edwin A.

Cottrell, *Pasadena Social Agencies Survey* (Pasadena: Community Chest, 1940), 201.

99. "Direct Testimony of Mamie Rankins and Etta Gray," *People v. Rankins*, Feb. 28, 1944, Sacramento.

100. See "Annual Report, 1935–1936," box 2, folder 4, 37, Ackerman Collection.

101. See Ruth Bailey Stephens, "Practices in the Los Angeles City Schools in Dealing with Socially Maladjusted Girls" (master's thesis, University of Southern California, 1940), 32; C. R. Shaw and E. D. Myers, "The Juvenile Delinquent," in *Illinois Crime Survey*, ed. John Henry Wigmore (Chicago: Illinois Association for Criminal Justice in cooperation with the Chicago Crime Commission, 1929), 698–99. On the House of the Good Shepherd and infected and delinquent girls in Chicago in this period, see Anne Meis Knupfer, *Reform and Resistance: Gender, Delinquency, and American's First Juvenile Court* (New York: Routledge, 2001), 95–96, 157–76. On the Good Shepherd in Davenport, Iowa, in the 1890s, see Sharon E. Wood, *Freedom of the Streets: Work, Citizenship, and Sexuality in a Gilded Age City* (Chapel Hill: University of North Carolina Press, 2005), chap. 8.

102. See Pacific Protective Association, *Ruth Home* (pamphlet, [ca. 1936]), Special Collections, University of California, Los Angeles, California; Stephens, "Practices in the Los Angeles City Schools," 32.

103. Welfare Planning Council, Los Angeles Region, *Facilities and Services Available to Women and Children with Venereal Disease in Los Angeles County* (Los Angeles: Venereal Disease Council of the City and County of Los Angeles, Inc., and Welfare Council of Metropolitan L.A., Oct. 1949), 47. In 1943, community organizations formed the Venereal Disease Council, which was funded by the Community Chest and affiliated with the Los Angeles County Tuberculosis and Health Association. Proposed Kinsey Conference Planning Materials, box 83, folder 6, ASHA Collection; Walter Clarke, "Los Angeles County Social Hygiene Survey," 1947–48, typescript, box 98, folder 3, 128, ASHA Collection.

104. Welfare Planning Council, L.A. Region, *Facilities and Services*, 5. On Works Progress Administration funding of new construction for hospitals generally, see Rosemary Stevens, *In Sickness and in Wealth: American Hospitals in the Twentieth Century* (New York: Basic Books, 1989), 168–70.

105. Cottrell, *Pasadena Social Agencies*, 274.

106. By the 1940s additional trade school classes included dressmaking, beauty tech, typing, and laundry. Welfare Planning Council, L.A. Region, *Facilities and Services*, 48.

107. Ackerman notes on "Annual Report, 1930–31," box 2, folder 1, Ackerman Collection.

108. "Annual Report, 1930–31," box 2, folder 1, 61–63; "Annual Report, 1934–35," box 2, folder 3, 5, both in Ackerman Collection.

109. "Annual Report, 1933–34," box 2, folder 3, 13; "Annual Report, 1933–34," box 2, folder 2, 7, both in Ackerman Collection.

110. Annual Reports: "1938–39," box 2, folder 7; "1939–40," box 2, folder 8; "1940–41," box 2, folder 9; "1941–42," box 2, folder 10, all in Ackerman Collection; Clarke, "Los Angeles County Social Hygiene Survey," 128.

111. Welfare Planning Council, L. A. Region, *Facilities and Services*, 44.

112. See John Parascandola, "The Introduction of Antibiotics into Therapeutics," in *Sickness and Health in America: Readings in the History of Medicine and Public Health*, ed. Judith Walzer Leavitt and Ronald L. Numbers, 3rd ed., rev. (Madison: University of Wisconsin Press, 1997), 102, 102–7; Ruth Schwartz Cowan, *A Social History of American Technology* (New York: Oxford University Press, 1997), 310–18; Edward Shorter, *The Health Century* (New York: Doubleday, 1987), 37–46.

113. Welfare Planning Council, L. A. Region, *Facilities and Services*, ii, 2. From March 1947 to March 1949 the home treated only 264 girls, one-quarter of whom were pregnant. More than half were white girls under age 15 who were placed in the isolation unit and received no social services (4–5).

114. Brandt, *No Magic Bullet*, 125–26.

115. Ibid., 122; John Duffy, *The Sanitarians: A History of American Public Health* (Urbana: University of Illinois Press, 1990), 266. See also Suzanne Poirier, *Chicago's War on Syphilis, 1937–1940: The Times, "The Trib," and the Clap Doctor* (Urbana: University of Illinois Press, 1995), 19–20; Terra Ziporyn, *Disease in the Popular American Press: The Case of Diphtheria, Typhoid Fever, and Syphilis, 1870–1920* (New York: Greenwood Press, 1988), 113–42.

116. "Thomas, Parran, Jr.," *Virtual Office of the Surgeon General*, www.surgeongeneral.gov/library/history/bioparran.htm (accessed Aug. 15, 2007).

117. See "Shadow on the Land," *Chicago Daily Tribune*, Feb. 2, 1938, 12; John Parascandola, "VD at the Movies: PHS Films of the 1930s and 1940s," *Public Health Reports* 3 (Mar.–Apr. 1996): 173.

118. Thomas J. Parran, *Shadow on the Land: Syphilis* (New York: Reynal and Hitchcock, 1937); Thomas Parran Jr., "The Extent of the Problem of Gonorrhea and Syphilis in the United States," *JSH* 16 (Jan. 1930): 31; Brandt, *No Magic Bullet*, 136–37.

119. "Doctors Open Social Disease Fight Today," *Washington Post*, Dec. 28, 1936, 11. See Institute of Public Opinion, "Voters Favor Bold Fight on Venereal Diseases: Dr. Parran's Drive Approved by Majority of 9 to 1," *Washington Post*, Dec. 20, 1936, B1.

120. Brandt, *No Magic Bullet*, 138–42.

121. Ibid., 142–47.

122. Ibid., 147, 154; Duffy, *History of Public Health*, 584.

123. Edward L. Keyes Jr., "Gonorrhea—Stepchild of Medicine," *JSH* 25 (Oct. 1939): 226; Ernest B. Howard, "Communicable Diseases," *Commonhealth* 26 (Jan.–Mar. 1939): 18.

124. Keyes, "Gonorrhea," 226; Charles Walter Clarke, "Newer Research Findings for Dealing with Syphilis and Gonorrhea," *American Journal of Public Health* 29 (July 1939): 761, 762.

125. Alfred Cohn, Arthur Steer, and Eleanor L. Adler, "Further Observations on Gonococcal Vulvovaginitis," *Transactions of the American Neisserian Medical Society*, 6th annual session (New York, June 10–11, 1940), 24. Cohn and Steer were physicians with the New York City Department of Health; Adler was a pediatrician at New York University and the Children's Medical Service of Bellevue Hospitals.

126. "Gonococcal Vaginitis: A Preliminary Report on One Year's Work," *VDI* 21 (July 1940): 208, 211, 219.

127. Cohn, Steer, and Adler, "Further Observations on Gonococcal Vaginitis," 25.

128. L. Emmett Holt Jr. and Rustin McIntosh, ed., *Holt's Diseases of Infancy and Childhood: A Textbook for the Use of Students and Practitioners*, 11th ed. (New York: D. Appleton-Century Co., 1940), 821. My thanks to the late Dr. Willard Marmelzat of Los Angeles for bringing this book, his medical school textbook in pediatrics in 1942, to my attention.

129. Holt and McIntosh, *Holt's Diseases of Infancy*, 821–22.

130. Ibid.

131. Thomas A. Gonzalez, Morgan Vance, and Milton Helpern, *Legal Medicine and Toxicology* (New York: D. Appleton-Century Co., 1937), 359–61.

132. Ibid., 361.

133. Wharton, *Gynecology*, 368.

Epilogue

1. Harriet Hardy, transcript of oral history interview, Oct. 13–14, 1977, 32, Regina Morantz, interviewer, Oral History Project on Women in Medicine, Medical College of Pennsylvania (1978), Medical College of Pennsylvania Collection, OH-3, Arthur and Elizabeth Schlesinger Library on the History of Women in America, Radcliffe Institute for Advanced Study, Harvard University, Cambridge, Massachusetts.

2. Ibid., 27.

3. Aimee Zillmer and Ruth Larsen, "What She Thinks about It," *Journal of Social Hygiene* 28 (Nov. 1942): 464, 467.

4. Maurice Bigelow, "Some Dangerous Communicable Diseases: A Manual for Teachers and Students," pub. no. A531, 1943, box 175, folder 5, ASHA Collection.

5. Evangeline Hall Morris, *Public Health Nursing in Syphilis and Gonorrhea* (Philadelphia: Saunders, 1946), 98, 106–7.

6. Samuel D. Allison and June Johnson, *VD Manual for Teachers* (New York: Emerson Books, 1946), 79–80.

7. R. A. Vonderlehr, "Venereal Diseases," in *Successful Marriage: An Authoritative Guide to Problems Related to Marriage from the Beginning of Sexual Attraction to Matrimony and the Successful Rearing of a Family*, ed. Morris Fishbein and Ernest W. Burgess (Garden City, NY: Doubleday, 1948), 448, 449.

8. Nancy Tomes, *The Gospel of Germs: Men, Women, and the Microbe in American Life* (Cambridge: Harvard University Press, 1998), 253–54.

9. John Duffy, *A History of Public Health in New York City, 1866–1966*, vol. 2 (New York: Russell Sage Foundation, 1974), 586–87; David S. Folland et al., "Gonorrhea in Preadolescent Children: An Inquiry into Source of Infection and Mode of Transmission," *Pediatrics* 60 (Aug. 1977): 153.

10. E.g., Lawrence F. Nazarian, "Current Prevalence of Gonococcal Infections In Children," *Pediatrics* 39 (Mar. 1967): 372; Stephen L. Abbott, "Gonococcal Tonsillitis-Pharyngitis in a 5-Year-Old Girl," *Pediatrics* 52 (Aug. 1973): 287, 289.

11. See David L. Kerns, "Medical Assessment of Child Sexual Abuse," in *Sexually Abused Children and Their Families*, ed. Patricia Beezley Mrazek and C. Henry Kempe (New York: Pergamon, 1981), 129, 130.

12. Geraldine Branch and Ruth Paxton, "A Study of Gonococcal Infections among Infants and Children," *Public Health Reports* 80 (Apr. 1965): 347, 356.

13. Ibid., 351. See also Ida I. Nakashima and Gloria E. Zakus, "Incest: Review and Clinical Experience," *Pediatrics* 60 (Nov. 1977): 696; Ann Wolbert Burgess and Lynda Lytle Holmstrom, "Sexual Trauma of Children and Adolescents," *Nursing Clinics of North America* 10 (Sept. 1975): 551, 553.

14. See Richard P. Kluft, review of *The Battered Child*, 5th ed., *American Journal of Psychiatry* 156 (Nov. 1999): 1828; Vincent A. Fulginiti, "C. Henry Kempe," *Journal of Pediatrics* 126 (Jan. 1995): 152; Joseph Greengard, "The Battered-Child Syndrome," *American Journal of Nursing* 64 (June 1964): 98; "Battered-Child Syndrome," *Time*, July 2, 1962, www.time.com/time/magazine/article/0,9171,896393,00.html (accessed Mar. 28, 2008); C. Henry Kempe et al., "Battered Child Syndrome," *JAMA* 181 (July 7, 1962): 17.

15. "Child Abuse Prevention and Treatment Act," Public Law 93-273, 42 U.S.C. 5101 (1974).

16. Kirsten A. Lentsch and Charles F. Johnson, "Do Physicians Have Adequate Knowledge of Child Sexual Abuse? The Results of Two Surveys of Practicing Physicians, 1986 and 1996," *Child Maltreatment* 5 (Feb. 2000): 72, 76; Ruth S. Kempe and C. Henry Kempe, *The Common Secret: Sexual Abuse of Children and Adolescents* (New York: W. H. Freeman and Co., 1984), chap. 3; Suzanne M. Sgroi, " 'Kids with Clap': Gonorrhea as an Indicator of Child Sexual Assault," *Victimology* 2 (1977): 251, 255.

17. Karen S. Israel, Kathleen B. Rissing, and George F. Brooks, "Neonatal and Childhood Gonococcal Infections," *Clinical Obstetrics and Gynecology* 18 (Mar. 1975): 143, 145.

18. Hughes Evans, "The Discovery of Child Sexual Abuse in America," in *Formative Years: Children's Health in the United States, 1880–2000*, ed. Alexandra Minna Stern and Howard Markel (Ann Arbor: University of Michigan Press, 2002), 233–34; Philip Jenkins, *Moral Panic: Changing Concepts of the Child Molester in Modern America* (New Haven: Yale University Press, 1998), 119–20; Erna Olafson et al., "Modern History of Child Sexual Abuse Awareness: Cycles of Discovery and Suppression," *Child Abuse & Neglect* 17 (1993): 7, 7–24; Elizabeth Pleck, *Domestic Tyranny: The Making of Social Policy against Family Violence from Colonial Times to the Present* (New York: Oxford University Press, 1987), chap. 9; C. Henry Kempe, "Sexual Abuse: Another Hidden Pediatric Problem: The 1977 C. Anderson Aldrich Lecture," *Pediatrics* 62 (1978): 382.

19. Sgroi, "Kids with Clap," 251, 260; Hania W. Riis, letter to the editor, "Nonvenereal Transmission of Gonococcal Vaginitis," *Pediatrics* 61 (Jan. 1978): 144, 145; Folland et al., "Gonorrhea in Preadolescent Children," 153; Nakashima and Zakus, "Incest," 696; Suzanne M. Sgroi, "Pediatric Gonorrhea beyond Infancy," *Pediatrics Annals* 8 (May 1979): 73; Kempe, "Sexual Abuse," 382; "Sexual Molestation of Children: The Last Frontier of Child Abuse," *Child Today* 4 (1975): 18, 19.

20. Evans, "Discovery of Child Sexual Abuse," 249–50; Sgroi, "Kids with Clap," 251; Folland et al., "Gonorrhea in Preadolescent Children," 153; Nakashima and Zakus, "Incest," 696; Kempe, "Sexual Abuse," 382.

21. See James M. Meek, Asghar Askari, and A. Barry Belman, "Prepubertal Gonorrhea," *Journal of Urology* 122 (Oct. 1979): 532; R. C. Low, C. T. Cho, and B. A. Dudding, "Gonococcal Infections in Young Children," *Clinical Pediatrics* 16 (July 1977): 623; Israel, Rissing, and Brooks, "Neonatal and Childhood Gonococcal Infections," 143; and William B. Shore and Jerry A. Winkelstein, "Nonvenereal Transmission of Gonococcal Infections to Children," *Journal of Pediatrics* 79 (1971): 661; cf. Jan E. Paradise et al., "Vulvovaginitis in Premenar-

cheal Girls: Clinical Features and Diagnostic Evaluation," *Pediatrics* 70 (Aug. 1982): 193, 197, and Mary Ellen Rimsza and Elaine H. Niggemann, "Medical Evaluation of Sexually Abused Children: A Review of 311 Cases," *Pediatrics* 69 (Jan. 1982): 8, 13.

22. Sgroi, "Kids with Clap," 258.

23. James H. Gilbaugh and Peter C. Fuchs, "The Gonococcus and the Toilet Seat," *NEJM* 301 (July 12, 1979): 91; Michael F. Rein, letter to the editor, *NEJM* 301 (Dec. 13, 1979): 1347.

24. E.g., Shore and Winkelstein, "Nonvenereal Transmission," 66; Israel, Rissing, and Brooks, "Neonatal and Childhood Gonococcal Infections," 143; Low, Cho, and Dudding, "Gonococcal Infections," 625–26.

25. Branch and Paxton, "Study of Gonococcal Infections," 350–51; Nakashima and Zakus, "Incest," 698.

26. Ellen R. Wald et al., "Gonorrheal Disease among Children in a University Hospital," *Sexually Transmitted Diseases* 7 (Apr.–June 1980): 41, 43.

27. My thanks to Dr. Gertrude Finkelstein, Emerita Attending Staff, Cedars–Sinai Medical Center, Los Angeles, who shared this and many other insights into OB-GYN practice with me. See Kathleen Merkley, "Vulvovaginitis and Vaginal Discharge in the Pediatric Patient," *Journal of Emergency Medicine* 31 (Aug. 2005): 400; Angelo P. Giardino and Martin A. Finkel, "Evaluating Child Sexual Abuse," *Pediatric Annals* 34 (May 2005): 382; and Jan E. Paradise et al., "Influence of the History on Physicians' Interpretations of Girls' Genital Findings," *Pediatrics* 103 (May 1999): 980.

28. See Nakashima and Zakus, "Incest," 700.

29. Kempe, "Sexual Abuse," 383.

30. Nancy Kellogg and the Committee on Child Abuse and Neglect, "The Evaluation of Sexual Abuse in Children," *Pediatrics* 115 (Aug. 2005): 506; Kathi L. Makoroff et al., "Genital Examinations for Alleged Sexual Abuse of Prepubertal Girls: Findings by Pediatric Emergency Medicine Physicians Compared with Child Abuse Trained Physicians," *Child Abuse & Neglect* 26 (2002): 1235; Jan E. Paradise et al., "Assessments of Girls' Genital Findings and the Likelihood of Sexual Abuse: Agreement among Physicians Self-Rated as Skilled," *Archives of Pediatrics & Adolescent Medicine* 151 (Sept. 1997): 883; Carole Jenny, "Pediatrics and Child Sexual Abuse: Where We've Been and Where We're Going," *Pediatric Annals* 26 (May 1997): 284, 285–86; Adaora A. Adimora et al., eds., *Sexually Transmitted Diseases: Companion Handbook*, 2nd ed. (New York: McGraw-Hill, 1994), 379; Astrid Heger, "Helping the Medical Professional Make the Diagnosis of Sexual Abuse," in *Evaluation of the Sexually Abused Child: A Medical Textbook and Photographic Atlas*, ed. Astrid Heger and S. Jean Emans (New York: Oxford University Press, 1992), 1, 4.

31. David P. Ascher, "The Gonococcus and the Toilet Seat Revisited," *Pediatric Infectious Diseases Journal* 8 (Mar. 1989): 191. Dr. David L. Ingram responded by reiterating the high probability of incest and the importance of the patient history.

32. American Academy of Pediatrics, Committee on Child Abuse and Neglect, "Guidelines for the Evaluation of Sexual Abuse of Children," *Pediatrics* 87 (Feb. 1991): 254, 255. See also Richard D. Krugman, "Recognition of Sexual Abuse in Children," *Pediatrics Review* 8 (1986): 25, 26–28.

33. Kellogg and the Committee on Child Abuse and Neglect, "Evaluation of Sexual Abuse in Children," 506.

34. Ibid., 510–11; Heger, "Helping the Medical Professional," 6. See also Vincent J. Palusci, Ralph A. Hicks, and Frank E. Vandervort, "Experience and Reason: 'You Are Hereby Commanded to Appear': Pediatrician Subpoena and Court Appearances in Child Maltreatment," *Pediatrics* 107 (June 2001): 142.

35. S. Kirson Weinberg, *Incest Behavior* (New York: Citadel Press, 1955), 39–40.

36. E.g., Judson T. Landis, "Experiences of 500 Children with Adult Sexual Deviation," *Psychiatric Quarterly Supplement* 30 (1956): 91; Joseph Weiss et al., "A Study of Girl Sex Victims," *Psychiatric Quarterly* 29 (Jan. 1955): 1; Irving Kaufman, Alice L. Peck, and Consuelo K. Tagiuri, "The Family Constellation and Overt Incestuous Relations between Father and Daughter," *American Journal of Orthopsychiatry* 24 (Apr. 1954): 266; Benjamin Apfelberg, Carl Sugar, and Arnold Z. Pfeffer, "A Psychiatric Study of 250 Sex Offenders," *American Journal of Orthopsychiatry* 100 (May 1944): 762; Lauretta Bender and Abram Blau, "Reaction of Children to Sexual Relations with Adults," *Journal of Orthopsychiatry* 7 (Oct. 1937): 500.

See also Rachel Devlin, " 'Acting Out the Oedipal Wish': Father-Daughter Incest and the Sexuality of Adolescent Girls in the United States, 1941–1965," *Journal of Social History* 38 (2005): 609; Rachel Devlin, *Relative Intimacy: Fathers, Adolescent Daughters, and Postwar American Culture* (Chapel Hill: University of North Carolina Press, 2005), chap. 5; Olafson et al., "Modern History of Child Sexual Abuse Awareness," 7; Pleck, *Domestic Tyranny*, 156–60; and Judith Lewis Herman and Lisa Hirschman, *Father-Daughter Incest* (Cambridge: Harvard University Press, 1981), 12–14.

37. Paul Sloane and Eva Karpinski, "Effects of Incest on the Participants," *American Journal of Orthopsychiatry* 12 (1942): 666.

38. E.g., Georgene H. Seward, "Sex and the Social Order," in Fishbein and Burgess, *Successful Marriage*, 419, 423.

39. Alfred Kinsey et al., *Sexual Behavior in the Human Female* (Philadelphia: Saunders, 1953), 117; Herman and Hirschman, *Father-Daughter Incest*, 12–14.

40. Kinsey et al., *Sexual Behavior in the Human Female*, 121; Herman and Hirschman, *Father-Daughter Incest*, 17. See also Patricia Phelan, "Incest and Its Meaning: The Perspectives of Fathers and Daughters," *Child Abuse & Neglect* 19 (1995): 7, 22.

41. See Stephen Robertson, *Crimes against Children: Sexual Violence and Legal Culture in New York City, 1880–1960* (Chapel Hill: University of North Carolina Press, 2005), chaps. 7, 10; Pleck, *Domestic Tyranny*, chap. 8; Jenkins, *Moral Panic*, 49–106; Janis Appier, *Policing Women: The Sexual Politics of Law Enforcement and the LAPD* (Philadelphia: Temple University Press, 1998), 140–55; Estelle B. Freedman, "Uncontrolled Desires," in *True Stories from the Past*, ed. William Graebner (New York: McGraw-Hill, 1993), 83; and George Chauncey Jr., "The Postwar Sex Crime Panic," in Graebner, *True Stories from the Past*, 160.

42. Jennifer Terry, *An American Obsession: Science, Medicine, and Homosexuality in Modern Society* (Chicago: University of Chicago Press, 1999), 274.

43. See ibid., 274, and Jenkins, *Moral Panic*, 50–67.

44. Devlin, *Relative Intimacy*, 7; Devlin, "Acting Out the Oedipal Wish," 617.

45. Devlin, *Relative Intimacy*, 7.

46. Ibid., 41.

47. Ibid., 10

48. John H. Gagnon, "Female Child Victims of Sex Offenses," *Social Problems* 13 (1965): 176, 268, 247, 250. Erskine Caldwell's *Tobacco Road* became a best seller in 1932 and sold nearly four million copies over the next fifty years. The story concerns a rural Georgia family that is so poor that family members steal food from one another. A stage version ran for seven years on Broadway, and in 1941 John Ford directed a film version. Florence King, "The Man from Tobacco Road," review of *The Journey from Tobacco Road: A Biography* by Dan B. Miller, *New York Times Book Review*, Jan. 12, 1995, late ed., final, col. 1, 29; Edwin McDowell, "For Erskine Caldwell, Fifty Years of Successes," *NYT*, Dec. 1, 1982, C25; Jack Kirkland, "Tobacco Road: A Three-Act Play," in *S.R.O.: The Most Successful Plays of the American Stage*, ed. Bennett Cerf and Van H. Cartmell (Garden City, NY: Doubleday, 1944), 633; Erskine Caldwell, "Of 'Tobacco Road': Mr. Caldwell Defends the Character of Jeeter Lester and His Family," *NYT*, May 10, 1936, X1.

49. Margaret Randall, *This Is about Incest* (Ithaca, NY: Firebrand Books, 1987); Ellen Bass and Louise Thornton, ed., *I Never Told Anyone: Writings by Women Survivors of Child Sexual Abuse* (New York: Harper and Row, 1983); A. H. McNaron and Yarrow Morgan, ed., *Voices of the Night: Women Speaking about Incest* (Minneapolis: Cleis Press, 1982); Florence Rush, *Best Kept Secret:*

Sexual Abuse of Children (New York: McGraw-Hill, 1981); Charlotte Allen Vale, *Daddy's Girl* (New York: Wyndham Books, 1980); Louise Armstrong, *Kiss Daddy Goodnight: A Speak-Out on Incest* (New York: Pocket Books, 1978); Susan Forward and Craig Buck, *Betrayal of Innocence: Incest and Its Devastation* (New York: Penguin Books, 1978), 5; Karen C. Meiselman, *Incest: A Psychological Study of Causes and Effects, with Treatment Recommendations* (San Francisco: Jossey-Bass, 1978). Toni Morrison published *The Bluest Eye* in 1970.

50. E.g., Bettina F. Aptheker, *Intimate Politics: How I Grew Up Red, Fought for Free Speech, and Became a Feminist Rebel* (Emeryville, CA: Seal Press, 2006); Katherine Harrison, *The Kiss: A Memoir* (New York: Random House, 1997); Louis DeSalvo, *Vertigo: A Memoir* (New York: Dutton, 1996); Alison Rose, "Polished Sapphire," *New Yorker*, Dec. 25, 1995, 48; Barbara Graham and Christiana Ferrari, "Unlock the Secrets of Your Past: Early Childhood Memories," *Redbook*, Jan. 1993, 84; Renee Fredrickson, *Repressed Memories: A Journey to Recovery from Sexual Abuse* (New York: Simon and Schuster, 1992); Elizabeth Gleick, "Frances Lear," *People Weekly*, July 13, 1992, 91; Frances Lear, "Shari Karney Is a Zealous California Attorney Who Devotes the Full Force of Her Professional Energy to the Fight against Incest and Child Sexual Abuse," *Lear's*, Feb. 1992, 17; Louise M. Wisechild, ed., *She Who Was Lost Is Remembered: Healing from Incest through Creativity* (Seattle: Seal Press, 1991); Sue E. Blume, *Secret Survivors: Uncovering Incest and Its Aftereffects in Women* (New York: Ballantine Books, 1987); Eleanore Hill, *Family Secret: A Personal Account of Incest* (New York: Dell, 1985).

See Amy J. Curtis-Webber, "Not Just a Pretty Victim: The Incest Survivor and the Media," *Journal of Popular Culture* 28 (Spring 1995): 37, and Rosaria Champagne, "Oprah Winfrey's *Scared Silent* and the Spectatorship of Incest," *Discourse* 17 (Winter 1994–95): 123.

51. See Liz Kelly, *Surviving Sexual Violence* (Minneapolis: University of Minnesota Press, 1988), 138–58; Pleck, *Domestic Tyranny*, 182–86, 198–200; and Brenda J. Vander Mey and Ronald L. Neff, *Incest as Child Abuse: Research and Application* (New York: Praeger, 1986): 8–14.

52. E.g., Lenore Terr, *Unchained Memories: True Stories of Traumatic Memories, Lost and Found* (New York: Basic Books, 1994); Vicki Bell, *Interrogating Incest: Feminism, Foucault and the Law* (New York: Routledge, 1993); Elizabeth Waites, *Trauma and Survival: Post-Traumatic and Dissociative Disorders in Women* (New York: Norton, 1993); Diane Wood Middlebrook, *Anne Sexton: A Biography* (New York: Random House, 1991); Louis DeSalvo, *Virginia Woolf: The Impact of Childhood Sexual Abuse on Her Life and Work* (Boston: Beacon Press, 1989); Linda Gordon, *Heroes of Their Own Lives: The Politics and History of Family Violence, Boston 1880–1960* (New York: Penguin, 1988), 204–49;

Pleck, *Domestic Tyranny*; Diana E. H. Russell, *The Secret Trauma: Incest in the Lives of Girls and Women* (New York: Basic Books, 1986); Rush, *Best Kept Secret*; Sandra Butler, *Conspiracy of Silence: The Trauma of Incest* (San Francisco: New Glide Publications, 1978); Meiselman, *Incest.*

53. Boston Women's Health Collective, *Our Bodies, Ourselves* (Boston: New England Free Press, 1970).

54. Herman and Hirschman, *Father-Daughter Incest*, 8–9.

55. Judith Lewis Herman, "Father-Daughter Incest," *Signs* 2 (1977): 735.

56. Herman and Hirschman, *Father-Daughter Incest*, 7.

57. Ibid.

58. E.g., Russell, *Secret Trauma*; Alice Miller, *Thou Shalt Not Be Aware: Society's Betrayal of the Child*, trans. Hildegarde Hannum and Hunter Hannum (New York: Farrar, Straus and Giroux, 1984); Rush, *Best Kept Secret*; Butler, *Conspiracy of Silence*; Meiselman, *Incest.*

59. Jeffrey Moussaief Masson, *The Assault on Truth: Freud's Suppression of the Seduction Theory* (New York: Farrar, Straus and Giroux, 1984). See also Olafson et al., "Modern History of Child Sexual Abuse Awareness," 7, 10–12; Judith Lewis Herman, *Trauma and Recovery* (New York: Basic Books, 1992), 1–20; Pleck, *Domestic Tyranny*, 250–56; and Herman and Hirschman, *Father-Daughter Incest*, 9–11.

60. Journalist Janet Malcolm brought the issue into public view, first with a two-part *New Yorker* article and then with a book about the firestorm that Masson's allegations lit within his relatively small professional circle. Subsequent lawsuits between Malcolm and Masson that went to the U.S. Supreme Court kept the issue in the news for ten years. *Masson v. New Yorker Magazine, Inc.*, 501 U.S. 496 (1991); Janet Malcolm, *In the Freud Archives* (New York: Knopf, 1984); "Annals of Scholarship: Trouble in the Archives," *New Yorker* pt. 1, Dec. 5, 1983, 59; pt. 2, Dec. 12, 1983, 60. See also Ann Scott, "Feminism and the Seductiveness of the 'Real Event,'" *Feminist Review* 28 (Spring 1988): 88.

61. See Olafson et al., "Modern History of Child Sexual Abuse Awareness," 10–12.

62. See, e.g., Alice Miller, *Breaking Down the Wall of Silence: The Liberating Experience of Facing Painful Truth*, trans. Simon Worrall (New York: Meridian, 1993), 5–6.

63. E.g., Sapphire, *Push* (New York: Knopf, 1996); Sapphire, *American Dreams* (San Francisco: High Risk Books, 1994); Rachel Adelson, "Deadly Silence," *Village Voice*, Dec. 22, 1987, 33; Eleanore Hill, *Family Secret: A Personal Account of Incest* (New York: Dell, 1985).

64. Some critics castigated Walker for disseminating an unflattering representation of African American men in the mainstream media. Jacqueline Bobo,

"Sifting through the Controversy: Reading *The Color Purple*," *Callaloo* 12 (Spring 1989): 332.

65. *Something about Amelia*, ABC-TV, Jan. 9, 1984. On television and incest narratives, see Janet Walker, *Trauma Cinema: Documenting Incest and the Holocaust* (Berkeley: University of California Press, 2005), chap. 3.

66. Tom Shales, "Emmys for 'Amelia,' 'Hill Street,'" *Washington Post*, Sept. 24, 1984, final ed., Style, D1; Charlotte Sutton, "TV Ratings," *Washington Post*, final ed., Style, B9; John Carmody, TV Column, *Washington Post*, Jan. 11, 1984, final ed., Style, D12; Tom Shales, "ABC Tackles the Last Taboo: 'Amelia': A Grim and Powerful Look at Incest," *Washington Post*, Jan. 9, 1984, final ed., Style, B1; John J. O'Connor, "TV: 'Amelia' on ABC a Movie about Incest," *NYT*, late city final ed., sec. C, col. 1, 16; Harry F. Waters, "Toppling the Last Taboo," *Newsweek*, Jan. 9, 1984, 62; Richard Stengel, review of *Something about Amelia*, *Time*, Jan. 2, 1984, 75; Jeannye Thornton, "Family Violence Emerges from the Shadows," review of *Something about Amelia*, *U.S. News and World Report*, Jan. 23, 1984, 66; cf. John Corry, "TV View: Call It TV, but Don't Call It Reality," *NYT*, late city final ed., sec. 2, col. 1, 1.

67. Janet Maslin, "Film: *The Color Purple*, from Stephen Spielberg," review of *The Color Purple*, *NYT*, Dec. 18, 1985, http://movies.nytimes.com/movie/review?res=9F06E5DC153BF93BA25751C1A963948260 (accessed Apr. 5, 2008).

68. See Andrew Delbanco and Thomas Delbanco, "A.A. at the Crossroads," *New Yorker*, Mar. 20, 1995, 50; Janice Haaken, "From Al-Anon to ACOA: Codependence and the Reconstruction of Caregiving," *Signs* 18 (Winter 1993): 321; Robin Room, "'Healing Ourselves and Our Planet': The Emergence and Nature of a Generalized Twelve-Step Consciousness," *Contemporary Drug Problems* 19 (Winter 1997): 717.

69. Room, "Healing Ourselves," 737.

70. Bob McCullough, "The New Spin Is Spirituality," *Publishers Weekly*, May 16, 1994, 40; Maria Heidkamp, "Beyond Therapy," *Publishers Weekly*, Dec. 6, 1991, 31.

71. Vicki Bane, "I Am an Incest Survivor: A Star Cries Incest," *People Weekly*, Oct. 7, 1991, 84.

72. Heidkamp, "Beyond Therapy," 31.

73. Judith Rosen, "Women's Bookstores: 20 Years and Thriving," *Publishers Weekly*, May 11, 1992, 18; Ellen Bass and Laura Davis, *The Courage to Heal: A Guide for Women Survivors of Child Sexual Abuse* (New York: Harper and Row, 1988).

74. Ellen Bass and Laura Davis, *The Courage to Heal: A Guide for Women Survivors of Child Sexual Abuse*, 3rd ed. (New York: Harper Perennial, 1994).

75. Joe Manning, "Visit by Author on Incest Is Criticized," *Milwaukee Jour-*

nal Sentinel, Oct. 17, 1996, final ed., 3; "The Best of the Bunch," *USA Today*, Apr. 6, 1995, final ed., 5D; Maria Simson, "Banking on Brand Names; Perennial Favorites Dominate the Lists; The Red and the Black, Tallying the Books '93: Paperback Bestsellers," *Publishers Weekly*, Mar. 7, 1994, Special Report Supplement, S21; Heidkamp, "Beyond Therapy," 31.

76. Dorothy Allison, *Bastard out of Carolina* (New York: Dutton, 1992); Jane Smiley, *A Thousand Acres* (New York: Knopf, 1991). See also Alexis Jetter, "The Roseanne of Literature," *New York Times Magazine*, Dec. 17, 1995, 54 (profile of Allison). In 2007 Smiley said that while incest was a plot element, American agriculture was the theme of *A Thousand Acres*. Jane Smiley, "CEO President," *Huffington Post*, Aug. 20, 2007, www.huffingtonpost.com/jane-smiley/ceo-president_b_27658.html (accessed Aug. 20, 2007).

77. Katie Roiphe, "Making the Incest Scene: In Novel after Novel, Writers Grope for Dark Secrets," *Harper's*, Nov. 1995, 65; Laura Shapiro, "They're Daddy's Little Girls," *Newsweek*, Jan. 24, 1994, 66.

78. Susan Faludi, *Backlash: The Undeclared War against American Women* (New York: Crown Publishers, 1991).

79. E.g., *Forgotten Sins*, ABC-TV, Mar. 7, 1996, television film of Lawrence Wright, *Remembering Satan: A Case of Recovered Memory and the Shattering of an American Family* (New York: Knopf, 1994); Lawrence Wright, "Remembering Satan," *New Yorker*, May 17, 1993, 60, May 24, 1993, 54; Gary Indiana, "Believe the TV Movie," review of *Indictment: The McMartin Trial*, HBO-TV, *Village Voice*, May 23, 1995, 46; Mark Pendergrast, *Victims of Memory: Incest Accusations and Shattered Lives* (Hinesburg, VT: Upper Access, 1995); "Divided Memories," *Frontline*, PBS-TV (Journal Graphic Transcripts, #1313, Apr. 11 and 18, 1995); Richard Ofshe and Ethan Watters, *Making Monsters: False Memory, Psychotherapy, and Sexual Hysteria* (New York: Scribner, 1994); Elizabeth Loftus and Katherine Ketcham, *Myth of Repressed Memory* (New York: St. Martin's Press, 1994), 227–63; Debbie Nathan, "What McMartin Started: The Ritual Abuse Sex Hoax," *Village Voice*, June 12, 1990, 36.

80. Richard A. Gardner, *Sex Abuse Hysteria* (Cresskill, NJ: Creative Therapeutics, 1991), 121.

81. Carol Tavris, "Beware the Incest Survivor Machine," *New York Times Book Review*, Jan. 3, 1993, p. 1; cf. Judith Lewis Herman, letter to the editor, "Real Incest and Real Survivors: Readers Respond," and Ellen Bass and Laura Davis, letter to the editor, "Real Incest and Real Survivors," both in *New York Times Book Review*, Feb. 14, 1993, 3. Tavris published a similar critique in her book *Mismeasure of Woman* (New York: Simon and Schuster, 1992), 312–33.

Feminist analysis of the incest accusations has been divided. See Judith Butler, *Undoing Gender* (New York: Routledge, 2004), chap. 7; Walker, *Trauma Cin-*

ema, chap. 1; Judith Lewis Herman, "Afterword, 2000: Understanding Incest Twenty Years Later," in Judith Lewis Herman, *Father-Daughter Incest*, new ed. (Cambridge: Harvard University Press, 2000), 221–42; Diana E. H. Russell, introduction to *Secret Trauma: Incest in the Lives of Girls and Women*, rev. ed. (New York: Basic Books, 1999), xxiv–xl; Janice Haaken, *Pillar of Salt: Gender, Memory, and the Perils of Looking Back* (New Brunswick, NJ: Rutgers University Press, 1998); Elaine Showalter, *Hystories* (New York: Columbia University Press, 1997); and Louise Armstrong, *Rocking the Cradle of Sexual Politics* (New York: Addison-Wesley, 1994).

82. Tavris, "Beware," 1.

83. Jeanne Wright, "Commitments: People Who Read Too Much; Have All Those Psychology Books of the Last Decade Really Been of Much Help?" *LAT*, June 26, 1995, E2.

84. Martin Seligman, "Personal Responsibility and Individual Choice," Inaugural Symposium, University of Pennsylvania, *Almanac Supplement* 41, Nov. 1, 1994, *FMSF Newsletter*, Jan. 1, 1995, unpaginated, http://fmsfonline.org/fmsf95.101.html (accessed Aug. 16, 2007); Elayne Rapping, "Needed: A Radical Recovery," *Progressive*, Jan. 1994, 32; Wendy Kaminer, *I'm Dysfunctional, You're Dysfunctional: The Recovery Movement and Other Self-Help Fashions* (Reading, MA: Addison-Wesley, 1992); David Reiff, "Victims All? Recovery, Co-Dependency, and the Art of Blaming Somebody Else," *Harper's*, Oct. 1991, 49. See, e.g., Susan Forward, *Toxic Parents: Overcoming Their Hurtful Legacy and Reclaiming Your Life* (New York: Bantam, 1989).

85. Frederick Crews, "The Revenge of the Repressed," *New York Review of Books*, pt. 1, Nov. 17, 1994, 54; pt. 2, Dec. 1, 1994, 49. Crews expanded the articles into *Revenge of the Repressed* (New York: New York Review of Books, 1995).

86. Crews, "Revenge," pt. 2, 49.

87. Loftus and Ketcham, *Myth of Repressed Memory*, 207, 206–14, 21–22, 55–56, chap. 9; Elizabeth Loftus, "The Reality of Repressed Memories," *American Psychologist* 48 (1993): 518, 525–26; cf. Kenneth S. Pope, "Memory, Abuse, and Science: Questioning Claims about the False Memory Syndrome Epidemic," award address, *American Psychologist* 51 (Sept. 1996): 967–68.

88. See Mike Stanton, "U-Turn on Memory Lane," *Columbia Journalism Review*, July/Aug. 1997, unpaginated, http://backissues.cjrarchives.org/year/97/4/memory.org (accessed Apr. 4, 2008). E.g., Sandra E. Lamb, "Tragic Delusions: How 'Recovered' Memories Tear Families Apart," *Family Circle*, July 18, 1995, 55; Claire Safran, "Dangerous Obsession: The Truth about Repressed Memories," *McCall's*, June 1993, 98.

89. See Stanton, "U-Turn on Memory Lane"; Laura A. Carstensen et al., "Re-

pressed Objectivity," letter to the editor, *APS Observer* 6 (Mar. 1993): 23; and "A Mom," letter to the editor, *FMSF Newsletter*, Mar. 8, 1994, unpaginated, http://fmsfonline.org/fmsf94.308.html (accessed Aug. 16, 2007).

90. See False Memory Syndrome Foundation, "What Is False Memory Syndrome?" *FAQ*, www.fmsfonline.org/fmsffaq.html#WhatIsFMS (accessed July 12, 2007); Pope, "Memory, Abuse, and Science," 957, 957–59.

91. Stanton, "U-Turn on Memory Lane"; Jennifer J. Freyd, *Betrayal Trauma: The Logic of Forgetting Childhood Abuse* (Cambridge: Harvard University Press, 1996), 197–98; Jennifer J. Freyd, "Theoretical and Personal Perspectives on the Delayed Memory Debate" (paper presented at the Center for Mental Health at Foote Hospital's Continuing Education Conference: Controversies Around Recovered Memories of Incest and Ritualistic Abuse, Aug. 7, 1993, reprinted in *Family Violence & Sexual Assault Bulletin* 9 [Fall 1993]: 28); Pamela Freyd, "How Could This Have Happened?" in *Confabulations: Creating False Memories—Destroying Families*, ed. Eleanor Goldstein and Kevin Farmer (Boca Raton: SIRS Books, 1992), 27.

92. J. Freyd, "Theoretical and Personal Perspectives," 8.

93. "From Our Readers," *FMSF Newsletter*, Mar. 8, 1994, unpaginated, http://fmsfonline.org/fmsf94.308.html (accessed Aug. 16, 2007); see also *Front Page*, WNYW-TV, New York (Radio TV Reports Transcript, July 24, 1993), unpaginated.

94. E.g., Richard Ofshe and Ethan Watters, *Making Monsters: False Memory, Psychotherapy, and Sexual Hysteria* (New York: Scribner, 1994); Loftus and Ketcham, *Myth of Repressed Memory*, 150, 204, 258, 263; Wright, *Remembering Satan*; cf. Karen A. Olio and William F. Cornell, "The Facade of Scientific Documentation: A Case Study of Richard Ofshe's Analysis of the Paul Ingram Case," *Psychology, Public Policy and Law* 4 (Dec. 1998): 1182. See also Elizabeth Wilson, "Not in This House: Incest, Denial, and Doubt in the White Middle Class Family," *Yale Journal of Criticism* 8 (1995): 35.

95. Hollida Wakefield and Ralph Underwager, "Recovered Memories of Alleged Sexual Abuse: Lawsuits against Parents," *Behavioral Sciences and the Law* 19 (Autumn 1992): 483, 486.

96. See Herman, *Trauma and Recovery*, 96–114.

97. Judith L. Herman and Mary R. Harvey, "Adult Memories of Childhood Trauma: A Naturalistic Clinical Study," *Journal of Traumatic Stress* 10 (1997): 557, 559–60, 563.

98. After Jennifer Freyd and other academics published an article in the Policy Forum of *Science* calling for interdisciplinary research to better inform public policy on child sexual abuse, False Memory Syndrome Foundation board members responded by criticizing both their arguments about the incidence of child

sexual assault and psychological theories about repression. Jennifer J. Freyd et al., "The Science of Child Sexual Abuse," *Science*, Apr. 22, 2005, 501; "The Problem of Child Sexual Abuse," letters to the editor and response of the authors, *Science*, Aug. 19, 2005, 1182.

99. See "Divided Memories," *Frontline*, PBS, Apr. 4 and 11, 1995; "Search for Satan," *Frontline*, PBS, Oct. 24, 1995; Loftus and Ketcham, *Myth of Repressed Memory*, chaps. 8–10; and Loftus, "Reality of Repressed Memories," 526–28, 530, 534; cf. Stephen Porter and Kristine A. Peace, "The Scars of Memory: A Prospective, Longitudinal Investigation of the Consistency of Traumatic and Positive Emotional Memories in Adulthood," *Psychological Science* 18 (2007): 435, 435–36, 439–40; Lisa DeMarni and Jennifer J. Freyd, "What Influences Believing Child Sexual Abuse Disclosures? The Roles of Depicted Memory Persistence, Participant Gender, Trauma History, and Sexism," *Psychology of Women Quarterly* 31 (2007): 13, 19–20; Kristin Weede Alexander et al., "Traumatic Impact Predicts Long-Term Memory for Documented Child Sexual Abuse," *Psychological Science* 16 (2005): 33, 38; and Herman and Harvey, "Adult Memories," 558–61.

100. Richard Ofshe and Ethan Watters, "Reply to Letters to the Editor," *Society*, Nov.–Dec. 1993, 9, 12.

101. See Richard Ofshe, "Curriculum Vitae," Department of Sociology, University of California, Berkeley, http://sociology.berkeley.edu/faculty/ofshe/cv.htm (accessed July 10, 2007); Richard Jerome, "Suspected Confessions," *New York Times Magazine*, Aug. 13, 1995, 28; and John F. Kihlstrom, "Making Monsters: False Memories, Psychotherapy, and Sexual Hysteria," review of Ofshe and Watters, *Making Monsters*, *NEJM* 333 (July 13, 1995): 132.

102. See Don J. DeBenedictis, "McMartin Preschool's Lessons: Abuse Case Plagued by Botched Investigation, Too Many Counts," *ABA Journal* 76 (Apr. 1990): 28, and Steven Strasser and Elizabeth Bailey, "A Sordid Preschool 'Game,'" *Newsweek*, Apr. 9, 1984, 38.

103. See Ted Rohrlich, "McMartin Case's Legal, Social Legacies Linger," *LAT*, Dec. 18, 2000, Metro, B1; Patt Morrison, "The Victims Can't Be Counted," *LAT*, Dec. 18, 2000, Metro, B1; Debbie Nathan and Michael Snedeker, *Satan's Silence: Ritual Abuse and the Making of a Modern American Witch Hunt* (New York: Basic Books, 1995); Ruth Shalit, "Witch Hunt," *New Republic*, June 19, 1995, 14; and Robert Safian, "McMartin Madness: Ten Days in the Life of the Longest, Most Gruesomely Difficult Criminal Trial Ever Had," *American Lawyer* (Oct. 1989): 46. David Shaw received a Pulitzer for his four-part series "McMartin and the Media" in January 1990. As late as 2008 the *Times* was still affirming the lessons it had learned. In an editorial against capital punishment for child rapists, the *Times* adopted the defendant's position that "in cases based on a child's testimony

about sexual abuse, 'the risk of wrongful conviction is especially pronounced.'" "No Death Penalty for Rapists," *LAT*, Apr. 16, 2008, www.latimes.com/news/opinion/la-ed-death16apr16,0,5142846.story (accessed Apr. 16, 2008).

Between 1991 and 1997, the PBS series *Frontline* aired producer Ofra Bikel's three-part investigation into a similar prosecution of the staff of the Little Rascals Day Care Center in North Carolina. "Innocence Lost: The Plea," PBS, May 27, 1997; "Innocence Lost: The Verdict," PBS, July 20 and 21, 1993; "Innocence Lost," PBS, May 7, 1991.

104. Elizabeth Loftus, "Remembering Dangerously," *Skeptical Inquirer*, Mar.–Apr. 1995, 20; cf. Jennifer Hoult, "'Remembering Dangerously' and *Hoult v. Hoult*: The Myth of Repressed Memory That Elizabeth Loftus Created," 2005, www.rememberingdangerously.com (accessed Apr. 4, 2008).

When Loftus testified as an expert witness in 2006 on behalf of former White House aide Scooter Libby, she admitted under cross examination by U.S. Attorney Patrick Fitzgerald that the "methodology she had used at times in her long academic career was not that scientific" and that "her conclusions about memory were conflicting." Carol D. Leonnig, "In the Libby Case, a Grilling to Remember," *Washington Post*, Oct. 27, 1996, A21.

105. See Pope, "Memory, Abuse, and Science," 963–71.

106. Katy Butler, "Self-Help Authors Freed from Liability, Suit Involving Incest Claims Continues," *San Francisco Chronicle*, Sept. 6, 1994, final ed., A16; "Recent Cases: Briefly Noted: Torts/Publisher Misrepresentation Claim," *Entertainment Law Reporter* 16 (Apr. 1995): 411; "Judge Throws Out Another Lawsuit Brought against Authors of *The Courage to Heal*," *Business Wire*, Sept. 29, 1994; Barbara Steuart, "The Courage to Sue: New Chapter in Recovered Memory," *Recorder*, Aug. 23, 1994, 2; Alan Dershowitz, "The Newest Excuse: 'I Read a Book,'" *San Francisco Examiner*, July 23, 1994; Alan Dershowitz, "'Abuse Excuse' Extended into Bizarre Memory Suit," *Buffalo News*, July 2, 1994, final ed., Viewpoints, 3; Rosie Waterhouse and Phil Reeves, "Author of Child Sex Abuse Book Is Sued," *Independent* (London), May 15, 1994, Britain Page, 8; *Barden v. HarperCollins Publishers, Inc.*, 863 F. Supp. 41 (1994). See Judith M. Simon, "The Highly Misleading 'Truth and Responsibility in Mental Health Practices Act': The 'False Memory' Movement's Remedy for a Nonexistent Problem," *Moving Forward*, Fall 1995, 1, and Cynthia Grant Bowman and Elizabeth Mertz, "A Dangerous Direction: Legal Intervention in Sexual Abuse Survivor Therapy," *Harvard Law Review* 109 (Jan. 1996): 549.

107. See Jennifer J. Freyd and Kathryn Quina, "Feminist Ethics in the Practice of Science: The Contested Memory Controversy as an Example," in *Practicing Feminist Ethics in Psychology*, ed. M. Brabeck (Washington, DC: American Psychological Association, 2000), 101.

INDEX

Abbott, Grace, 143, 146
Abt, Isaac A., 95
Ackerman, Rhea C., 204
Acton, William, 71
Adams, John Quincy, 23
Addams, Jane: ASHA and, 142, 143; on Cook County Hospital, 90; National Conference on Charities and Corrections and, 128; nuclear family and, 155; publications of, 116, 139; social hygiene movement and, 122, 126
Additon, Henrietta, 200
age of consent: criminal prosecutions and, 38, 49–50; as irrelevant, 186; legislation for, 47–48. *See also* legal consent
alcohol, as metaphor for male corruption, 48–49
American Social Hygiene Association (ASHA): COS and, 165; description of, 120; B. W. Hamilton's instructions and, 152; influence of, 142–44; Interdepartmental Social Hygiene Board and, 200; school exclusion and, 176. See also *Journal of Social Hygiene*
Anthony, Susan B., 46
appellate courts: in *Hobday*, 188; incest cases and, 50; overturning of convictions by, 39–40; records of, 35
arrest statistics, 239 n. 45
ASHA. *See* American Social Hygiene Association
Atkinson, Isaac E., 81
Atler, Marilyn Van Derbur, 1, 2, 219

backlash in 1980s and 1990s, 220–25
Bailey, Abigail, 10, 42
Bailey, Asa, 42
Baker, Edith M., 148–49, 153, 199
Barr, Roseanne, 219
Barringer, Emily, 116
Bass, Ellen, 219, 221
Beck, Theodoric Romeyn, 67–68
Becker, Tracy C., 99, 100
Bederman, Gail, 107
Bellamann, Henry, 218
Bellevue-Yorkville Health Demonstration, 169–74
Benson, Reuel A., 179–80, 181
Binford, Jessie F., 146
Blackwell, Henry, 46
Bowen, Louise deKoven, 142
Branch, Geraldine, 211
Brown, Irene Quenzler, 10–11, 43
Brown, Richard D., 10–11, 43
Brownmiller, Susan, 221
Brunet, Walter, 141–42, 165, 170, 171, 173

Cabot, Hugh, 143, 144
Cabot, Richard, 134
Caldwell, Erskine, 333 n. 48
Casper, Johann Ludwig, 76, 77, 111
Chandler, Harry, 157
Chargin, Louis, 165, 166, 179
Charity Organization Society (COS), 165–69
child abuse: in day care, 223–24; as medical issue, 211. *See also* sexual abuse
Child Abuse Prevention and Treatment Act, 211
child guidance field, 190
child protection movement, 44–45
Chudacoff, Howard, 26
Clarke, C. Walter, 158–59, 176, 179, 198, 206

class: appellate courts and, 39–40; assumptions about sexual behavior based on, 82–83, 88, 101–2, 108–9, 124, 125; as cause of infection, 169; child protection movement and, 45; at end of nineteenth century, 40–41; False Memory Syndrome Foundation and, 222; Goldbergs and, 196–97; incest and, 113–14, 139–42, 186–87; newspaper and court reports and, 24, 27–28; toilets and, 179–80; treatment at venereal disease clinics and, 197–98
The Color Purple (Walker), 218
community, responses of, 33, 34, 194–95
Comstock, Anthony, 122
Comstock laws, 144
convictions, overturned, 35, 39–40
Cook, John C., 104
Cooper, Astley: evidence of miscarriages of justice and, 69, 70; Taylor and, 73; women and, 79, 101, 155
Cooper, Thomas, 60
COS (Charity Organization Society), 165–69
Cotton, Alfred Cleveland, 103–4, 126, 137
Cotton, John, 113
The Courage to Heal (Bass and Davis), 219–20, 221, 224
courts. *See* appellate courts; criminal prosecutions; trial court records
Cowan, Ruth Schwartz, 150
Crain, Gladys, 118, 142, 154, 180
credibility of girls and women: attacks on, at trials, 184–85, 186–88, 193–94; determination of, 220; diagnosis and, 99; doctors and, 10; Edgar, Johnston, and, 101; False Memory Syndrome Foundation and, 221–24; forensic examiners and, 190–91; gender bias and, 73; W. Healy and, 189–90; Herman and, 216; J. Walker and, 85–86; Wilde and, 72
Crews, Frederick, 221
crime of opportunity, incest as, 196–97
criminal prosecutions: age of consent and, 49–50; diagnosis and, 98–99; frequency of, 35; of R. McAfee, 182–85; medical views and, 56; proof and, 36; records of, 21–22; of Wheeler, 42–43

"dangerous things," 128, 207
Darrow, Clarence, 195–96
Davis, Katharine Bement, 142, 146
Davis, Laura, 219, 221
day care, child abuse in, 223–24
Dayton, Cornelia Hughes, 21
delinquency in girls, 139, 187
denial, consequences of, 213–15
Deutsch, Sarah, 47, 51
Devlin, Rachel, 11, 215
diagnosis of gonorrhea: Atkinson and, 81; bacteriologic, 87; Beck and, 67–68; complications in, 80; in eighteenth century, 58–59; errors in, 59–60; forensic value of, 131, 207–8; in genteel families, 65–66; medical jurists and, 58; J. Morris and, 81–83; in prebacterial era, 61–63; in Progressive Era, 6–7; M. Ryan and, 69–70; spread of disease and, 90–91; Taylor and, 66–67, 73–74; J. Walker and, 83–87; F. Wharton, Stillé, and, 76–78; Wilde and, 70–71, 72. See also *Neisseria gonorrhoeae;* nonsexual transmission
Diseases of Infancy and Childhood (Holt), 90, 207
Dock, Lavinia L., 129–30
doctors: authority of, as questioned, 80; duty of, to protect men, 53–58; mothers and, 78–80; physical evidence of assault and, 53, 61, 68, 72–73; social biases of, 119, 212–13; as witnesses in rape cases, 36–37. *See also* medical views; *specific doctors*
Dodge, Grace, 142
domestic employees, 149, 151
domestic violence: second-wave feminism and, 10; temperance reformers and, 46
Dummer, Ethel Sturges, 187, 189
Dunglison, Robley, 111
Durkee, Silas, 64–65, 80
Dworkin, Andrea, 221

economic alterity of male sexuality, beliefs about, 2
Edgar, James Clifton: career of, 99–100; forensic evidence of rape and, 99, 100–101; on gonococcus bacteria, 125; Krafft-Ebing and, 106; race and, 107–8; superstitious cure and, 109, 111–12
Eliot, Charles W., 143, 144
Ellis, Havelock, 118
Emerson, Haven, 142, 143, 146–47
ethnicity: assumptions about sexual behavior based on, 82–83, 101–2, 106, 109; child protection movement and, 45; incest and, 113–14; superstitious cure and, 109–10
eugenics, 157–58, 187
evidence of assault: doctors and, 53, 61, 68, 72–73; forensic, 72–73, 99, 100–101, 131, 207–8; gonorrhea as, 6, 14
experiential knowledge of women: gender, power, and, 225–27; ideology elevated over, 41, 44; paternal protection ideologies and, 224–25; social value of, 4–5; struggles over, 13; as trivialized by doctors, 78–80

Fairchild, Amy L., 107, 118
Falconer, Martha, 143, 165, 200–201
False Memory Syndrome Foundation, 1, 221–24
family: ASHA and, 145–46; home protection rhetoric of, 50–51; social hygiene movement and, 120–21, 123–24; views of, 33–34. *See also* fathers; mothers; siblings
fathers: attacks on character of daughters by, 184, 193–94; authority of, and coercion by, 37–38; Bellevue-Yorkville investigation and, 173; ideologies about, 224–25; J. Morris and, 82; in psychoanalytic theory, 215; Seippel and, 138; social privileges and role of, 30–32; treatment of daughters and, 178; "white father" symbol, 106–7, 108–9
feminism, second-wave, 10, 215–21
Ferriar, John, 55
films dramatizing incest, 217–19
Findlay, Palmer, 128
Finkelhor, David, 16
Fitzgerald, F. Scott, 218
fomites, definition of, 14
fomite transmission, methods of, 14, 15. *See also* objects, transmission by
forensic evidence, 72–73, 99, 100–101, 131, 207–8
Frances Juvenile Home, 136–37
Freud, Sigmund, 189, 217, 226
Freyd, Jennifer, 222
Freyd, Pamela, 221–22
Freyd, Peter, 221–22

Gagnon, John, 215
Gardner, Richard, 220–21
gender, power, and subjugated knowledge, 225–27
Gerry, Elbridge T., 45, 112
Gibb, W. Travis, 101–2, 113–14
Gibbons, James, 143
Gillis, Jonathan, 79
girls: blaming sexual contact on, 185–86, 215; coercion of, 37–38; dangers confronting, in nineteenth century, 47; delinquency in, 139, 187; feeble-minded, 135; as perpetrators, 135; as prey, 84; responses of, to incest, 31–32, 41–43; as sources of infection, 162, 164; susceptibility to casual contacts and, 8, 125; as threats to community, 8, 198–202; types likely to be attacked, 114; Vanderbilt Clinic and, 132. *See also* credibility of girls and women
Goldberg, Jacob A., 196–97
Goldberg, Rosamond W., 196–97
gonorrhea, vaginal (gonococcal vulvovaginitis): cure for, 210; "epidemic" of, 7–8, 89–95; in genteel families, 78–80; as physical evidence of assault, 6, 14, 207–8; prevalence of, 91, 181; symptoms of, 14, 61–62, 94; transmission of, 14–15; untreated, 14, 94; ward epidemics of, 96–98. *See also* diagnosis of gonorrhea; nonsexual transmission

Gonzalez, Thomas A., 208
Good Samaritan Dispensary, 102, 132
Gordon, Linda, 10
Gram, Hans Christian, 62–63
Gram stain, 62–63

Hale, Jasper I., 78
Hale, Matthew, 57
Hamilton, Alice, 127–29, 130, 202
Hamilton, B. Wallace, 131, 132–33, 152
Hampson, Jane, case: in Britain and America, 58–61; overview of, 53–58; references to, 68
Hardy, Harriet, 209
Harris, Louis I., 165
Harris, Norman MacL., 94, 96
Harvey, Mary, 223
Healy, Mary Tenney, 189
Healy, William, 116–17, 189–90, 200
Herman, Judith Lewis, 43–44, 214, 216–17, 222–23
Higham, John, 106
Hodder, Jessie Donaldson, 133, 134
Holt, L. Emmett, 90, 97–98, 202
home protection rhetoric, 50–52
Hopkins, Harry, 165
hospitals: fomite transmission and, 14–15; preadmission testing and, 98; ward epidemics at, 96–98, 105
housekeeping tasks, 149–50
Houston, John, 218
Hunner, Guy L., 94, 96
Hunter, John, 59, 65
hygiene, poor, as cause: displacing source of infection onto mothers, 147–55; speculation about, 95; Taussig and, 161; tenement living and, 100–102, 131; as unsubstantiated, 173
hysteria, breaking silence as, 220–21

immigration: home protection rhetoric and, 51–52; national identity and, 105–9. *See also* ethnicity
imprisonment: of fathers and daughters, 194–95; in nineteenth century, 38–39, 40, 41; in twentieth century, 186
incest: assumptions about, 2; as crime of opportunity, 196–97; laws prohibiting, 34. *See also* evidence of assault; responses to discovery of incest
infants, gonorrhea ophthalmia in, 94–95, 138, 237–38 n. 32
innocent infection, 124. *See also* nonsexual transmission; objects, transmission by
institutionalization: events contributing to, 265 n. 184; for isolation or quarantine, 198–203
Interdepartmental Social Hygiene Board, 200–203
investigations: by Bellevue-Yorkville Health Demonstration, 169–74; by Charity Organization Society, 165–69; by Herman, 216–17; by Interdepartmental Social Hygiene Board, 201–2; by New York Tuberculosis and Health Association, 179, 197–98; by New York Vaginitis Research Project, 206–7
isolation or quarantine of infected girls: in classrooms, 162; at Frances Juvenile Home, 136–37; if mothers failed or refused treatment guidance, 153; institutionalization, 198–203; recommendations for, 207; at Ruth Home, 203–5; suspension from school, 164, 174–81

Jacobi, Abraham, 97, 131
Jacobson, Matthew Frye, 117–18
Johnston, J. C.: forensic evidence of rape and, 99, 100–101; Krafft-Ebing and, 106; race and, 107–8; superstitious cure and, 109, 111–12
Jordan, David Starr, 142, 143, 157
Journal of Social Hygiene, 140, 144, 146, 157, 209
Judge Baker Clinic, 190, 200

Kempe, C. Henry, 211, 212
Ketcham, Katherine, 221
Keyes, Edward L., Jr., 116, 142, 206
Kimball, Reuel B., 94–95, 96
Kinsey, Alfred, 214
knowledge, subjugated, 225–27. *See also* experiential knowledge of women

Koplik, Henry, 96–97
Krafft-Ebing, Richard von, 106, 111, 115, 214

laws: prohibiting incest, 34; prohibiting sexual contact, 35–36
legal consent, 36–38, 186. *See also* age of consent
Leszynsky, William M., 92
Liston, Robert, 62
Loftus, Elizabeth, 221
Lovell, Bertha C., 148, 151, 162
lynch mobs, 3, 109

Male, George Edward, 58–59, 60, 67–68
male spectators at trials, 27, 49
manliness, ideologies about: alcohol and, 48–49; delay of exposure and, 31; as racialized, 188–89; shift from nineteenth-century to twentieth-century, 41; temporary loss of manly volition, as cause, 193
Massachusetts General Hospital, 133–34
Massachusetts Society for the Prevention of Cruelty to Children, 44, 133
Masson, Jeffrey Moussaief, 217
McAfee, Melba, 182–85
McAfee, Nancy, 185
McAfee, Robert, 182–85
media accounts: backlash in, 224; books and films, 217–20; in eighteenth century, 20–21; in *National Police Gazette,* 25, 26–29; in nineteenth century, 2–3, 22–26, 28–29, 31–33; of sex crimes, 214; themes of, 192; in twentieth century, 1, 3, 191–92
medical examiners: forensic value of diagnosis and, 207–8; nativism and, 117; in nineteenth century, 66, 99–101
medical jurisprudence, 66–68
Medical Jurisprudence (Taylor), 67, 72–73, 190–91
medical jurists: description of, 56; diagnosis of gonorrhea and, 58; Hampson case and, 60–61; miscarriages of justice and, 71–72
medical records, 155, 166, 179
medical views: criminal proceedings and, 56; of female knowledge, 5; history of, 12–13; in nineteenth century, 5–6; reversal in, 3–4; social biases and, 7–8, 9–10. *See also* doctors
Meigs, Charles D., 62
Menninger, Karl A., 190
Metalious, Grace, 218
Mohr, James C., 60, 67
morality: ASHA and, 142, 143–44, 145–46; scrofula and, 71–72; venereal disease and, 63–64, 65–66
moral reform movements, 46–52
Morris, Evangeline, 209–10
Morris, John, 62, 65, 81–83
Morrison, Toni, 218
Morrow, Prince A., 101, 122–25, 142, 202
Morse, John Lovett, 89–90, 117
mothers: attacks on motives of, 185; Bellevue-Yorkville clinic and, 171; doctors and, 78–80; presence of, as deterrence, 37; responses of, to incest, 32–33, 41–43; as sources of infection, 128, 147–55, 173, 174, 212; testing of, 172–73; treatment and, 132–33, 153

Nabakov, Vladimir, 218
nationality. *See* ethnicity
National Police Gazette, 25, 26–29
National Venereal Disease Control Act, 206
nativism, 106, 117–18
Neisser, Albert, 62
Neisseria gonorrhoeae, 14, 15, 62, 125
Nelson, Nels A., 118, 174–75, 177–78, 180–81
newspaper articles, accounts in: in eighteenth century, 20; in nineteenth century, 2–3, 23–26, 28–29; on responses of girls, 31–32; on responses of mothers, 32–33; themes of, 192; in twentieth century, 191–92
New York Society for the Prevention of Cruelty to Children, 44–45, 98–100
Nisbet, William, 71
Noble, Charles P., 89
nonsexual transmission: attempts to identify, 8–9, 125; confusion over, 80;

nonsexual transmission (*continued*)
with "dangerous things," 128, 207; Nelson and, 177, 178, 180; in nineteenth century, 8, 78–87; number of cases and, 126; questioning of, 211–12; B. Ryan and, 74–76; speculation about, in 1940, 207–8; Taussig and, 160–62; F. Wharton, Stillé, and, 77–78. *See also* school lavatories; seats of water-closets or toilets
novels dramatizing incest, 218, 220

objects, transmission by: "dangerous things," 128, 207; displacing source of infection onto, 93–94, 95; Morrow and, 124–25. *See also* school lavatories; seats of water-closets or toilets
O'Donovan, Charles, 89
Ofshe, Richard, 223
ophthalmia, gonorrhea, in infants, 94–95, 138, 237–38 n. 32

Pacific Protective Society, 203, 204
Painter, Nell Irvin, 41
pamphlets, 126–27, 140–41, 171, 174
Pancoast, Seth, 63
parents, complaints about, 137–38. *See also* fathers; mothers
Parker, Kathleen, 41
Parran, Thomas, 205–6
paternal protection ideologies, 30–32, 224–25
Paxton, Ruth, 211
Pelouze, Percy S., 143, 177–78, 206
Percival, Thomas, 54–55, 56–58, 60
peritonitis, gonorrheal, 94
physicians. *See* doctors; medical views; *specific physicians*
Pitré, Giuseppe, 110, 118
Pivar, David J., 121
Plant, Rebecca Jo, 149
Pleck, Elizabeth, 10
Pollack, Flora, 90, 113, 114–16, 138
Popenoe, Paul, 157–58
Pound, Roscoe, 143
poverty: of defendants, and appellate courts, 40; incest and, 114; scrofula, gonorrhea, and, 71–72; spread of disease and, 100–102, 103. *See also* hygiene, poor, as cause; tenement housing
power, gender, and subjugated knowledge, 225–27
Prettyman, J. S., Jr., 65–66
privies, urban, 159, 160, 161
progeny of incest, 195
Progressive Era: hallmarks of, 4; position on etiology of infection in, 104–5; professionals in, 120; social hygiene movement and, 122–26
protective workers, 146
psychiatry, role of, 214
psychoanalytic theory, 214, 215, 217, 226
psychological consequences, 17–18, 213–14
punishment: in nineteenth century, 38–39, 40, 43; in twentieth century, 186, 194–95
purity reformers, 121–22

race: assumptions about sexual behaviors based on, 82–83, 88, 107–8, 109; child protection movement and, 45; consequences of breaking social contract and, 29–30; masculinity and, 188–89; in nineteenth century, 24, 40–41
racial alterity of male sexuality, beliefs about, 2
Rankins, Mamie, 203
rape: doctors as witnesses in cases of, 36–37; forensic evidence of, 99, 100–101. *See also* statutory rape
recovery movement, 219–20, 221
reporting and data collection, 166, 174, 211. *See also* medical records; trial court records
repression of memory, 222–23
residential homes, 202–3
respectability, and behavior, 30, 31, 33
responses to discovery of incest: by community, 33, 34, 194–95; in eighteenth century, 19–20; by girls, 31–32, 41–43; by mothers, 32–33, 41–43. *See also* punishment
Rice, John L., 206

Riis, Jacob, 100, 102, 132
Robertson, Stephen, 47
Rockefeller, John D., Jr., 122, 142–43
Roiphe, Katie, 220
Rosenthal, Edwin, 104
Rubin, S. H., 175, 176
Ruth Home, 203–5
Ryan, Burke, 74–76, 80, 100
Ryan, Michael, 69–70, 71

Safer, Morley, 1
Sanford, Sarah, 49
Sanger, Margaret, 209
sanitation. *See* hygiene, poor, as cause
sanitation reformers: causes and, 126–30; effect of, on treatment, 130–38
school epidemics, 162–63, 164, 176
school exclusions, 164, 174–81
school lavatories: COS and, 165–66; epidemic and, 162–63, 164; Nelson and, 177; Spaulding and, 136; Stalnaker and, 176; Taussig and, 160, 161
scientific motherhood, 149
scrofula, 71–72
seats of water-closets or toilets: displacing transmission from person to, 159, 176; Dock and, 129–30; instruction on use of, 181; medical investigation of, 206–7; Nelson and, 180–81; "old wives' tale" of, 209–10; shape of, 165–66; warnings about, 179; Wehrbein and, 168–69. *See also* school lavatories
Seippel, Clara, 116, 126, 136–38, 162–63
sexual abuse: consequences of, 17; diagnosis of, 213; indicators for, 13–14; Pollack and, 115–16; prevalence of, 15–16, 213; psychological symptoms of, 17–18; rates of, across race, class, and ethnicity, 17; WCTU and, 48–49. *See also* child abuse
sexual behavior, assumptions about: class and, 82–83, 88, 101–2, 108–9, 124, 125; ethnicity and, 82–83, 101–2, 106, 109; race and, 82–83, 88, 107–8, 109
sexually transmitted disease. *See* gonorrhea, vaginal; venereal disease
sexually transmitted disease testing, indications for, 15
"sexual psychopath" label, 185
Sgroi, Suzanne M., 211–12
Sheffield, Herman B., 92–94
siblings: penalties for, 195; testing of, 15
silence: breaking of, as hysteria, 220–21; of professionals, 9
Smith, Richard M., 134–35, 199
Smith, Sidney, 76
Snow, William, 142, 143, 179, 200
social biases: appellate courts and, 39; of doctors, 119, 212–13; empirical data and, 141; of Goldbergs, 196–97; medical jurisprudence and, 67; medical views and, 7–8, 9–10; of J. Morris, 81–82; of Morrow, 124; in nineteenth century, 25–26; of J. Walker, 85–86
social contract, breaking of, 29–30
social hygiene movement: description of, 9, 120–21; family values and, 142–47; Morrow and, 122–26; as national policy, 205–8. *See also* American Social Hygiene Association; *Journal of Social Hygiene*
Something about Amelia (movie), 218
Sommerville, Diane Miller, 21–22
sources of infection: girls as, 162, 164; mothers as, 128, 147–55, 173, 174, 212; objects as, 93–94, 95
"source unknown," attribution of infections to, 131, 180, 201–2, 212
Spaulding, Edith, 126, 135–36, 139
Spencer, Anna Garlin, 143
Spielberg, Steven, 218
Stalnaker, Paul, 176
Stanton, Elizabeth Dacy, 46
statutory rape: consent and, 49, 186; genteel women and, 51; in nineteenth century, 35–36; punishment for, 38–39
STD testing, indications for, 15
Stearns, Peter N., 150
Steer, Arthur, 179–80, 181, 328 n. 125
Stillé, Moreton, 76–78, 80, 111
Stokes, John H., 143, 174
Stone, Lucy, 46

"stranger danger," focus on, 214
superstitious cure, 109–19, 138, 212
syphilis, 237 n. 25

Talbot, Fritz B., 134
Tardieu, Ambroise, 111
Taussig, Frederick, 117, 151, 160–62
Taves, Ann, 10
Tavris, Carol, 221
Taylor, Alfred Swaine, 66–67, 68, 72–74
temperance reformers, 46–47
tenement housing: privies and, 159, 160, 161; spread of disease and, 125; venereal disease and, 102, 166
Terrell, Mary Church, 145
Terry, Jennifer, 214
testing of family members, 15, 172–73, 212
Thomas, Norman, 170–71
Thomas, W. I., 140
Tobacco Road (Caldwell), 333 n. 48
toilets. *See* privies, urban; seats of water-closets or toilets
Tomes, Nancy, 147, 159
Towne, Arthur W., 140
treatment: clinic visits, weekly, 151–52; effect of sanitation theories on, 130–38; fathers and, 178; in nineteenth century, 64–65; rapid, 210; recommendations for, 152–53; at venereal disease clinics, 197–98. *See also* superstitious cure
Trenwith, W. D., 105
trial court records: advertisements for, 21; in nineteenth century, 2–3, 23
trials: attacks on credibility of girls and women at, 184–85, 186–88, 193–94; male spectators at, 27, 49
typhoid, 127–28
typhus, 54, 55

Valentine, Ferdinand C., 112–13
Vanderbilt Clinic, 131–32, 164
venereal disease: immorality and, 63–64, 65–66; as neglected condition, 90–91; Parran and, 205–6; purity reformers and, 121–22; rise in rates of, 210–11; social hygiene movement and, 122–26; social reform movement and, 9–10; superstitious cure for, 109–19, 138, 212; tenement housing and, 102; untreated, 14, 94. *See also* gonorrhea, vaginal; treatment
victims, attacks on character of, 184, 186–87, 189–95, 193–94
von Bulow, Dietrich, 22–23
Vonderlehr, R. A., 210
von KleinSmid, Rufus B., 157

Wagner, Edna Pearson, 154
Wald, Lillian, 170
Walker, Alice, 218
Walker, Jerome, 83–87, 105, 111
Ward, Michael, 54, 55–56, 57, 58, 60
ward epidemics, 96–98, 105
Watters, Ethan, 223
WCTU. *See* Woman's Christian Temperance Union
Webster, Ralph, 190
Wehrbein, Kathleen, 167–69
Weinberg, S. Kirson, 213
Welt-Kakels, Sara, 95, 104–5
West, John B., 118
Wharton, Francis, 76–78, 80, 111
Wharton, Lawrence, 198
Wheeler, Ephraim, 10–11, 20–21, 31, 42–43
"white father" symbol, 106–7, 108–9
White House Conference on Child Health and Protection, 198–99
Whitelock, Thomas Sidney, 183
Wigmore, John Henry, 190
Wilde, William R.: career and personal life of, 72; evidence of assaults and, 68, 72; evidence of miscarriages of justice and, 70–71; forensic value of diagnosis and, 131; superstitious cure and, 110, 111
Willard, Frances, 47, 49
Willard Parker Hospital, 170–71
Williams, Gurney, 117
Williams, Phyllis H., 118
Winslow, Charles-Edward Amory, 143, 298 n. 61
Witthaus, R. A., 99, 100
Wolbarst, Abraham L., 102, 103

Woman's Christian Temperance Union (WCTU): age-of-consent legislation and, 47; critique of male authority by, 51; influence of, 48–50; issues promoted by, 122
women: consequences of sexual abuse in, 17–18, 213–14; feminism and voices of, 215–20; home protection rhetoric and, 50–51; reports of sexual assaults in childhood by, 16–17; venereal disease in, 64. *See also* credibility of girls and women; experiential knowledge of women; girls; mothers
Wood, Kinder, 59
Wynne, Shirley, 170

Yarros, Rachel, 126
Yesko, Stephen, 152